Evolutionary Ecology across
Three Trophic Levels

MONOGRAPHS IN POPULATION BIOLOGY

EDITED BY SIMON A. LEVIN AND HENRY S. HORN

(list continues following the Index)

Evolutionary Ecology across Three Trophic Levels

Goldenrods, Gallmakers, and Natural Enemies

WARREN G. ABRAHAMSON
AND
ARTHUR E. WEIS

PRINCETON UNIVERSITY PRESS

PRINCETON, NEW JERSEY

1997

Copyright ©1997 by Princeton University Press
Published by Princeton University Press, 41 William Street, Princeton,
New Jersey 08540
In the United Kingdom: Princeton University Press, Chichester, West Sussex

Cataloging in Publication Data

Library of Congress Cataloging-in-Publication Data

Abrahamson, Warren G.
    Evolutionary ecology across three trophic levels : goldenrods,
gallmakers, and natural enemies / Warren G. Abrahamson and Arthur E.
Weis.
        p.   cm. — (Monographs in population biology ; 29)
    Includes bibliographical references (p.   ) and index.
    ISBN 0-691-03733-7 (cl : alk. paper). —ISBN 0-691-01208-3 (pb :
alk. paper)
    1. Eurosta solidaginis—Ecology.   2. Eurosta solidaginis—
Evolution.   3. Eurosta solidaginis—Host plants.   4. Eurosta
solidaginis—Parasites.   5. Goldenrods—Ecology.   6. Goldenrods—
Evolution.   7. Coevolution.   8. Gall insects—Host plants.
I. Weis, Arthur E. (Arthur Edward), 1951 –   . II. Title.
III. Series.
    QL537.T42A27   1997   96-42051

This book has been composed in Baskerville

Princeton University Press books are printed on acid-free paper and meet the
guidelines for permanence and durability of the Committee on Production
Guidelines for Book Longevity of the Council on Library Resources

Printed in the United States of America by Princeton Academic Press

10   9   8   7   6   5   4   3   2   1

10   9   8   7   6   5   4   3   2   1
(Pbk.)

Contents

Foreword

Although, as an ecologist with an interest in plant/insect interactions, I have at my disposal close to half a million species that I can legitimately claim as subjects for study, for the past two decades (effectively the entirety of my professional career to date), I have spent much of my time engaged in the study of only two species. One of these species is a small, physically unremarkable brown moth and the other is perhaps charitably described as a noxious weed. I spend my time so occupied not for want of imagination or lack of other opportunities but because I believe fervently that, just as the quality of a physical structure depends not only on its design but on the integrity of its constituent parts, the quality of conceptual knowledge depends on the integrity of its inputs. Not all ecologists or evolutionary biologists approach their disciplines in this precise manner; practitioners of the comparative approach often emphasize breadth over depth, highlighting the need to gather information on a wide range of systems in order to document robustly and unambiguously general patterns. Without question, both approaches are necessary, but this is not to say that everyone will embrace them with equal enthusiasm. I know that my dogged devotion to a tiny fraction of the world's biodiversity has baffled some of my colleagues. One anonymous reviewer of a grant proposal of mine, in fact, argued that, with 80% of what there is to know about these organisms already worked out, it was hardly worth the marginal return to be gained by investing the remaining 20%.

This book, *Evolutionary Ecology across Three Trophic Levels: Goldenrods, Gallmakers, and Natural Enemies,* by Warren Abrahamson and Arthur Weis, is the perfect rebuttal to such an argument. These two authors have personally invested a quarter-century of effort into the study of a wildflower, a fly, and the fly's small circle of enemies; taking into account the person-years invested by collaborators, students, postdoctoral fellows, and other associates, that quarter-century is multiplied manyfold. What these estimable scientists have for their effort is arguably the most

vii

thoroughly characterized three trophic level interaction on the planet, a paradigm system to be used with confidence in defining a conceptual framework for the study of ecological interactions in general. At first blush, *Eurosta,* the fly, and *Solidago,* the host plant, seem less likely candidates for intense scrutiny than my small brown moth and noxious weed, but I envy Abrahamson and Weis, and their fellow travelers, the appealing attributes of their system. The lifestyle of the gallmaking fly makes it exceptionally well suited for quantitative measurement and experimental manipulation; the genetics, physiology, and geographic distribution of the plant permit field, laboratory, and greenhouse study; the system's third trophic level is diverse but manageably so; and the history of the system itself—native to North America and remarkably uncontaminated by human interference—allow for reasonably unambiguous evolutionary interpretations of data.

Years of focused study have produced a wealth of information that is, ironically, staggering in its breadth, encompassing not only ecology and evolution but also systematics, physiology, genetics, molecular and developmental biology, behavior, and a range of other subdisciplines within biology. Synthesized in this volume is more than a century of literature, interpreted, and in most cases inspired by, the conceptual framework of natural selection. August Krogh, a great physiologist of an earlier era, wrote in 1929 that "for a large number of problems there will be some animal of choice, or a few such animals, on which it can be studied most conveniently." This principle is the guiding philosophy behind the use of model organisms that is so characteristic of contemporary cell and molecular biology and so generally eschewed by the majority of ecologists in favor of documenting the diversity generated by natural selection. *Solidago, Eurosta,* and *Eurosta*'s natural enemies may well come to be regarded as a model interaction, as it were, for the fields of ecology and evolutionary biology. Ideally, this book will serve not only as a useful source of information but also as a source of new ideas and insights, and as a shining example of the value of a thorough understanding of the organisms themselves to organismal biologists of any stripe.

May R. Berenbaum,
University of Illinois, Urbana

Preface and Acknowledgments

This book presents the results of two and one-half decades of our empirical work on plant-insect interactions. This book, like our work, centers on the ecology and evolution of the interactions among a host plant, goldenrod (*Solidago*), the parasitic insect that induces a gall on the plant stem, *Eurosta solidaginis* (Diptera: Tephritidae), and a suite of insects and birds that are natural enemies of the gallmakers. Although the specifics of this work are driven by the natural history of clonal plants, gallmakers in general, and this gallmaker in particular, our work addresses the theories and concepts that have guided research on plant-insect interactions over the past twenty-five years, and reaches beyond to address general problems in evolutionary biology. Because goldenrod and its insects are amenable to many types of experimental manipulations, we have employed it as a model system to study the ecology and evolution of specialized enemy-victim interactions. As we take readers through our empirical studies with the *Solidago-Eurosta*-natural enemy system, we hope to impress upon them the many steps involved in the interactive relationships among species. The reproductive success of a gallmaker depends on completion of many actions in sequence. At each step in the process of host finding, ovipositing, gall inducing, feeding, and maturing on goldenrod, host-plant characters can facilitate or thwart the gallmaker and thus impose a complex selective regime on the insect. From the plant's perspective, each gallmaker action is a potential selective pressure on one or more plant characters. It is by examining individual systems that the scope of details that impinges on the evolution of interactions becomes apparent.

Why should we study interactions centered on gall-inducing insects? Plant-herbivore interactions encompass a broad range of natural histories, grading from large migrating ungulates, to polyphagous and swift-moving orthopterans, to oligophagous and awkward Lepidoptera larvae to monophagous and sessile aphid nymphs. Although it is important to understand the ecology and evolution of plant relationships for herbivores at all levels of this scale, the greater number of taxa on the specialized, immobile end suggests that evolution here may be more rapid. Gallmaking insects occupy this end of the spectrum. Their natural histories make them apt subjects to explore issues in the evolutionary ecology of plant-insect interactions, and issues on the evolution of species interactions generally. They spend larval development encased in a tumorlike plant growth that they themselves induce. There they receive nutrients and protection from the elements. Conveniently, a gall flags the gallmaker's presence to the investigator and so allows rapid and accurate censusing. Female gallmakers leave a record of their egg-laying activity through oviposition scars they make in the host plant. Furthermore, since the interaction between plant and insect revolves around the induction, growth, and development of the gall, gall appearance can provide information on the success or failure of the inducing insect.

In the past decade an appreciation has grown for the importance of natural enemies as factors that can influence the evolution of plant-herbivore interactions, and the study of gallmakers has led the way in gaining these insights. Most gallmakers support an array of parasitoids and predators. The success or failure of attack by the third trophic level is often determined by features of the gall that control visibility of and access to the gallmaker.

This book addresses a number of major issues including (1) the physiological and demographic consequences of herbivory for host plants, (2) mechanisms of selection against plant susceptibility to attack, (3) the evolution of host-plant choice behavior, (4) the conditions of formation of genetically distinct host races and their potential for subsequent speciation, (5) the value of the extended phenotype concept in evolution of host manipulation, (6) the structure of selection on phenotypic plasticity, (7) the effects of natural-enemy community interactions on the selection regimes for plant-use traits of herbivores, and (8) the evolutionary responses to selection of native species in natural environments.

The scope of this book should interest not only researchers in plant-animal interactions, but also researchers in the more general areas of evolutionary ecology, ecological genetics, physiological ecology, botany, entomology, and parasitology. The depth of detail is sufficient to understand the nature of the questions at hand and the means of arriving at answers, but not so deep as to allow the reader to lose sight of the big picture.

We offer our deep appreciation to our friends, colleagues, and students too numerous to list, who encouraged us to undertake this task. WGA extends his appreciation to James Karr and Madhav Gadgil for posing some of the stimulating questions and ideas back in the early 1970s that initiated this research, and to Otto Solbrig, James Layne, and Chris Abrahamson for their belief in him. He also thanks the many undergraduate and graduate students who have taken his course entitled "Plant-Animal Interactions" at both Bucknell University and Northern Arizona University for their stimulating thoughts and questions. AEW is grateful to Peter Price, Carl Bouton, and Paul Gross for stimulating thoughts about interactions spanning three trophic levels, and to Stewart Berlocher and Michael Lynch for introducing the possibilities in a quantitative genetic approach to evolutionary ecology. We both owe a special debt of deep gratitude to our forerunners who have previously examined goldenrod and its gallmakers. Foremost among these is Lowell D. Uhler, whose doctoral work on this system was an indispensable guide. Several times along the way we made new "discoveries" only to find them concisely laid out in his monograph (Uhler 1951). Our synthesis draws heavily from the unpublished and/or ongoing studies of several of our coworkers. These collaborators include Jonathan Brown, Timothy Craig, John Horner, and Joanne Itami. We greatly appreciate their willingness to freely share their data and insights.

Many postdoctoral fellows, graduate and undergraduate students, and colleagues have contributed appreciably to the collection of the data and the development of ideas contained in this book. Most significant among these are Jonathan Brown, Tim Craig, Wendy Gorman, John Horner, Joanne Itami, Ken McCrea, and Rod Walton. We gratefully acknowledge the efforts of Chris Aadland, Chris Abrahamson, Jill Abrahamson, Warren Abrahamson, Sr., Mark Andersen, Stephen Anderson, Paulette

Armbruster, Andrew Aulisi, Karen Ball, David Barrington, Robert Bertin, Steve Bimson, Catherine Blair, Chris Bosio, David Bresticker, Carrie Bretz, Lori Bross, D. Gordon Brown, Jonathan Brown, Richard Brown, Alice Calaprice, Paul Carango, Judy Carter, Hal Caswell, Michelle Chernin, Mitch Chernin, Ann Chiama, Tim Craig, Cathy Crego, Jim Cronin, Jackie Cummins, Jennifer Dailey, M. Deller, Richard Ellis, Jeff Erdley, Bill Flather, David Fletcher, Matt Ford, Madhav Gadgil, Cathy Gammon, Stephanie Gebauer, Kerry Givens, Wendy Gorman, Valerie Gorski, Larry Greenwald, Steve Griffee, Michele Hardner, David Hartnett, Lori Hartzel, Christine Hawkes, Linda Hendrian, Matt Hess, Heidi Hollenbach, Linda Holler, John Horner, Heather Houseknecht, Susan Torgerson How, Daniel Howard, Joanne Itami, C. Stephen Jeffries, Ann Johnson, Audrey Kapelinski, Elaine Keithan, Brian Knapp, Irene Kralick, Gary Krupnick, Helen Lang, Bonnie Libby, John Lichter, Jim Luterman, Amy Mackay, Wayne McDiffett, G. David Maddox, Lynne Mecum, George Melika, Sharon Miller, Will Miller, Kathy Mockler, Ellen Montgomery, Patrice Morrow, James Nitao, Rebecca Packer, Jon Payne, Patricia Peroni, Jocelyn Perot, Mike Piro, Margaret Rader, Larry Ragard, Anantanarayanan Raman, Bill Raschi, Kevin Regan, Thomas Richardson, Sherry Roth, Michele Rymond, Mike Sackschewsky, Joan Sattler, Sam Scheiner, Lori Scozzafava, Cathleen Shantz, Karen Shrawder, Dina Snow, Ben Stinner, Douglas Sumerford, Eric Sundvall, Joan Surosky, William Tap, Debbie Tarentino, John Tonzetich, Ron Toth, Carl von Ende, Edward Voss, Doug Walker, Rod Walton, Gwen Waring, Patricia Way, Adam Weis, Patricia Werner, Amy Whitwell, Harry Wilson, Cheryl Wolfe, Wade Worthen, Laurie Vernieri, and Michael Zivitz.

We thank Joseph L. Miller for his excellent illustrations of host plants, galls, gallmakers, and natural enemies. Our deepest gratitude goes to Chris Abrahamson, Irene Kralick, and Karen Shrawder for their attention to detail throughout the book's production. Most of all we thank our wives Chris and Audrey for their love, encouragement, and unending support; and to our children Jill, Adam, and Alex for their patience and love.

Finally, we thank Bucknell University and its David Burpee Endowment for funds for a substantial portion of the research contained in this book. Bucknell University, Northern Illinois University, the University of California, Irvine, and the Cedar Creek Natural History Area of the University of Minnesota have provided facilities and resources. We gratefully acknowledge the extensive funds provided by the National Science Foundation through awards GB-27911, BMS-7412144, DEB-7606832, DEB-7804295, SER-7813910, DEB-8205856, RIT-8310370, BSR-8600429, BSR-8614768, BSR-8614769, BSR-8614895, BSR-8615896, BSR-9111433, and BSR-9107150; and the Council for International Exchange of Scholars through the Fulbright Program. We sincerely thank John Fitzpatrick and the Archbold Biological Station for their support and encouragement during the writing of this book. The Archbold Biological Station, Lake Placid, Florida, and the Bodega Marine Laboratory, Bodega Bay, California provided ideal opportunities for our discussions and writing.

Evolutionary Ecology across
Three Trophic Levels

CHAPTER ONE

Evolutionary Ecology and the Interactions of Plants with Insects

1.1 AN UNEXPECTED TWIST OF NATURAL HISTORY

For anyone likely to pick up this book, it is unnecessary to document the diversity in the natural histories of the several millions of species inhabiting earth. Yet, almost all students of biology have at some point been jarred to learn of some feature in some organism they did not anticipate from previous experience. However, when such surprises come, even beginning biologists can evaluate their discoveries in a framework that will often lead them to say, "I could have predicted that." That framework is of course evolutionary theory. Dobzhansky's oft-quoted aphorism about the necessity of the evolutionary perspective to make sense of nature (Dobzhansky 1973) may or may not be true. However, it is certain that biologists have successfully employed concepts in evolutionary theory to come up with powerful explanations for both global patterns in the general biota, and for the origin and maintenance of unique traits in single species.

Insects that induce plant galls are one of those unexpected twists of natural history. Gallmakers stimulate their host plant to develop unique, tumorlike growths, galls, that provide them with food and shelter. The relationship is parasitic (Price, Waring, and Fernandes 1987) since the plant receives no benefit in return for the service. Galls are not the masses of undisciplined cells seen in animal cancers or bacterial plant tumors. Rather, gallmakers induce highly specific structures. For instance, several members of the wasp genus *Andricus* (Cynipidae; Hymenoptera) induce galls on *Quercus robur,* the English oak, and on *Q. petraea,* the durmast oak. Both *A. fecundator* and *A. inflator* inject their eggs into buds, but induce rather different galls. The gall of the former resembles a larch cone while that of the latter a peaked globe, about the size of a pea, attended at its base by scales (fig. 1.1). The wasp *Cynipis quercus-folii* induces a globular gall on leaf veins that is referred to as the "cherry gall" because of its size and shape, while *Andricus testaceipes* induces inconspicuous spindle-shaped galls on the same structure. Each of these galls is the result of a specific interaction between the insect and host plant, yet commonalties are apparent. At the center of a gall is a central chamber where the immature stage of the insect develops. It is typically lined with specialized plant tissue, the nutritive tissue, whose cells are rich in cytoplasm (hence rich in protein and other nutrients) and enclosed in thin cell walls (Shorthouse and Rohfritsch 1992). Outer tissue layers are often much tougher, even woody, and protect the gallmaker from the elements and in some cases from its natural enemies. A baroque twist is added to the plant-gallmaker relationship by the animals that are natural enemies to the gallmaker (Askew 1961; Washburn and Cornell 1979; Weis 1982a; Hawkins and Gagne 1989; Cornell 1990; Plakidas and Weis 1994). Most gall inducers are host to a small complex of parasitoid wasps. Most often these wasps use their needlelike ovipositors to drill through the gall tissue, down into the central chamber, where they lay an egg on or in the gallmaker. When the egg hatches, the wasp larva consumes the gallmaker, pupates, and then emerges as an adult ready to repeat the life cycle. Penetrating the gall is a challenge to the wasps, and often successful attack is limited to a "window of vulnerability" (Washburn and Cornell 1979; Craig, Itami, and Price 1990) before the gall grows too thick or too tough. Besides parasitoids, insectivorous birds, usually woodpeckers or chickadees and tits, peck open some galls to extract the gallmakers when other foods are scarce (Tscharntke 1992; Burstein and Wool 1992). Finally, there are those insects, known as inquilines, that oviposit into existing galls and usurp the gallmaker. Their natural history ranges in complexity from beetles that merely eat the existing gall to some wasps (evolved from parasitoids) that induce secondary galls (Hawkins and Goeden 1982). Advancing from the baroque to the rococo, the galls induced by some insects secrete nectar, thereby attracting

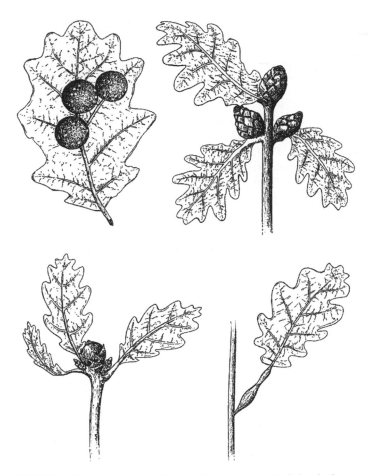

FIGURE 1.1. Four galls induced by cynipid wasps on the English oak, *Quercus robur* (Fagaceae) and the durmast oak, *Q. petraea*. *Top left:* The "cherry gall" induced by *Cynipis quercus-folii* (diameter 2 cm). *Top right:* The "larch-cone gall" induced by *Andricus fecundator* (length 2 cm). *Bottom left:* The globular gall induced by *Andricus inflator* (diameter 0.5 cm). *Bottom right:* The spindle-shaped gall of leaf petioles and midribs induced by *Andricus testceipes*. (Illustration by J. L. Miller)

ants that then deter parasitoids and inquilines (Washburn 1984; Seibert 1993).

Although galls are plant tissue, they can nonetheless be understood as insect adaptations. Natural selection acting on the insect species could favor individuals that upset plant development in ways that improve the nutritional quality or the protective properties of the feeding site. Some have argued that galls first arise as plant defensive structures that contain an otherwise pernicious threat to the vascular system in a nutritious enclosure and thereby obviate any need to bore into vital structures (Cockerell 1890; Hoffman 1985). Although a defensive growth response could provide the initial steps in gall evolution, selection on the gallmaker would easily subvert such a benign response to its own advantage.

This monograph summarizes the beginnings we have made in understanding key features in the evolutionary ecology of a plant-gallmaker-natural enemy interaction. For the past two and one-half decades we have investigated *Eurosta solidaginis* (Diptera: Tephritidae), a fly that induces a spheroid gall on the stem of tall goldenrod (*Solidago altissima;* Compositae). This insect belongs to the family of true fruit flies (i.e., not Drosophilidae), and so is related to the apple maggot, medfly, and other agricultural pests (Wasbauer 1972). The host plant is a common perennial herb of eastern North America. It can be found in old fields, road sides, floodplains, and other places of past disturbance. *Eurosta* is also found on related goldenrods near the edges of its range. A complex of carnivores attack *Eurosta* and its gall. Primary among these are two parasitoid wasps, an inquiline beetle, the downy woodpecker, and, on occasion, the black-capped chickadee. Throughout this book we will refer to the members of this assemblage as the "natural enemies" of *Eurosta*.

At various times we have examined this interaction from the perspectives of the plant, the gallmaker, or the natural enemies, and sometimes from all three. Few such plant-insect interactions have been as extensively investigated, save for some agricultural pests and their crop hosts. However, even after more than 20 years of research, we have made only a beginning. During our studies we have tried to use this system to explore general issues in evolutionary ecology. At the same time, this system has allowed us to investigate issues specific to the evolution of plant-insect interactions, especially those related to evolution of defense, host choice behavior, speciation, and the ecological structure of natural selection exerted by parasitoids and predators on plants and their herbivores.

1.2 PERSPECTIVE ON EVOLUTIONARY ECOLOGY

The evolutionary perspective has led to insights on nature from the level of the molecule to the level of the ecosystem. Foremost among the guiding concepts is that natural selection influences the direction and pace of evolutionary change. Even during times in the history of evolutionary biology when the importance of natural selection's innovative powers (relative to the other causes of evolution) were in doubt, it was still thought to define boundaries for evolutionary change (Provine 1971). The explanatory power of the evolutionary viewpoint, and in particular that of cumulative natural selection (Dawkins 1986), is so great that theory can often run far ahead of experimental confirmation. Theory sets out guidelines for understanding what is possible, or for predicting what is likely. Occasionally an empiricist may be able to decisively test a principle underlying theory. More often, the role of empiricists in evolutionary ecology is to generate the data needed to answer questions on "how strongly" and "how often" theoretical constructs apply to real systems. Chief among our goals has been to understand how selection has operated, or is operating, to shape key features of the goldenrod-gallmaker-natural enemy interaction.

The attempt to understand the ecology of organisms with natural selection dates, of course, to the time of Darwin and Wallace. However, it was with the field studies of David Lack (1947) and the experiments of Kettlewell (1973), and with others of their era (Cain and Provine 1992), that natural selection's power to shape natural history was fully appreciated. As John Endler's book *Natural Selection in the Wild* (1986) showed, selection pressures in natural populations are widespread, and often as strong as those imposed in artificial selection. Adaptive shifts in species following human-made disturbances to the environment (e.g., Antonovics, Bradshaw, and Turner 1971) and in experimental

7

arenas where natural selection was allowed to operate in the laboratory (e.g., Travisano et al. 1995) or in nature (e.g., Reznick and Endler 1982) have shown that selection can lead to very rapid evolutionary results. However, this leaves evolutionary ecology with a problem. If selection pressures are widespread, and potential for response is so great, why don't we see more species evolving on ecological timescales? In view of this problem, George Williams (1992) has warned that to understand natural selection, it may be as important to understand cases in which selection pressure does not cause evolution as to understand cases where it does. This makes the study of constraints an inherent feature in the study of evolutionary ecology.

Evolutionary ecologists working on a particular species can find themselves in the peculiar position of arguing that selection should be changing some character, but cannot due to countervening circumstances. Interactions between enemy and victim species do not result in endless "coevolutionary arms races," and so reasons for observed limits are sought (Thompson 1989, 1994). Finding the limits on selection can be a perilous enterprise since the list of possible constraints on selection is often long, the items on the list not mutually exclusive, and experimental verifications frequently beyond the limits of practicality. This leads to temptation for ad hoc storytelling that may be either adaptive or nonadaptive in flavor, depending on the proclivity of the teller.

Broad-scale, comparative studies may reveal patterns to suggest which constraints are important and which are not. The comparative method has its own problems (Harvey and Pagel 1991; Leroi, Rose, and Lauder 1994), but in the end its precision depends critically on the quality of data on the individual species being compared. In the long run, synthetic, big-picture studies will be what allow us to decide the answers to "how strongly" and "how often" questions on the evolution of enemy-victim interactions. But this can be achieved only through a dialogue between scientists knowledgeable on the natural history of specific systems who are informed by theory and those striving to outline the big picture.

During the time span of our studies, there have been several different phases in the study of plant-herbivore interactions

(Stamp 1992). In some circles the term "coevolution" has gone from buzz word to target of derision. Theories of plant defense have been refined several times (Feeny 1976; Coley, Bryant, and Chapin 1985; Herms and Mattson 1992). Much empirical work has revealed the complexity of insect-host-choice behavior (see Thompson 1988, 1994), and the role of population structure in fine-tuning insect performance (Edmunds and Alstad 1978; Karban 1989). Phylogenetic methods to understand patterns of radiation in herbivore and host (Futuyma and McCafferty 1990; Farrell, Mitter, and Futuyma 1992) were also introduced. Importantly, methods of quantitative genetics have been brought to bear on many questions in evolutionary ecology (see Stearns 1992), and have been particularly fruitful in understanding the structure of selection on plants and their herbivores (Berenbaum, Zangrel, and Nitao 1986; Marquis 1990; Simms 1990; see also Fritz and Simms 1992). In turn, insights into general problems concerning selection have emerged from work on plants and insects (Via 1984, 1991). Some of our work has contributed to these advances. The insights we have gained by studying a single system have given us a unique perspective for comment on others.

1.3 PLANT-INSECT INTERACTIONS AND BASIC QUESTIONS IN EVOLUTIONARY ECOLOGY

The ecological interaction between a plant and its herbivorous insects unfolds as a sequence of events in parallel ontogenies. A young plant is discovered and investigated by a female insect searching for oviposition sites. The female may reject the plant or lay an egg. The larva may eat heavily or may find its food distasteful. After feeding, the plant may regrow lost parts or not, may suffer reduced fecundity, or may complete its life cycle as if untouched. Meanwhile, the insect matures, pupates, mates, and, if female, initiates its own search for oviposition sites. At each event along this ontogenetic path, plant traits can thwart the insect, or at least reduce the impact on plant reproductive success. Conversely, at each step, insect characters determine whether it will be able to overcome barriers presented by the plant and successfully convert plant tissue into fertile eggs, placed in livable habitats. At the same time, the insect characters that determine success on the plant can affect its vulnerability to predator and parasitoid attack. The fundamental expectation for natural selection is that it will adjust the character states in a sequence that tends to increase fitness (see Frank and Slatkin 1992).

The sequential expression of the traits involved in the interaction imposes some important features on the structure of selection. The choice of host plant expressed by an ovipositing female will determine the quality and quantity of resources for her offspring, and so the evolution of the choice behavior depends in part on the subsequent performance of the offspring. In turn, the performance of the offspring evolves in an environment that is determined by parental choice. Similarly, plant infestation levels are determined in part by resistance mechanisms, but the fitness contributions made by a resistance mechanism will depend in part on the plant's ability to tolerate damage. Meanwhile, the selective value of tolerance will diminish as the degree of resistance increases. Because preference and performance, and resistance and tolerance are expressed in sequence, they have an epigenetic interaction concerning fitness (Atchley, Xu, and Vogl 1994). That is, the early expressed trait determines the environment in which the later is expressed, and the later expressed trait defines the selective environment for the earlier. A more formal treatment of the consequences for this type of interaction is presented at the end of the book, but it is a feature that will emerge throughout.

The next chapter presents background information on the *Solidago-Eurosta*-natural enemy interaction. We describe the basic ecology of each species, emphasizing the details that are the stuff of the natural history whose evolution we seek to understand. We acknowledge a debt of gratitude to our forerunners who have previously examined goldenrod and its gallmakers. Foremost among these is Lowell D. Uhler, whose doctoral work on this system was an indispensable guide along the way. Several times along the way we made new "discoveries" only to find them succinctly laid out in his monograph (Uhler 1951).

In the subsequent chapters we present the results of field, greenhouse, and laboratory experiments that have probed the potential for evolutionary change in those plant and insect characters most strongly involved in the interaction, including the potential for host race formation and gallmaker speciation. These chapters revolve around several related questions about the conditions for evolutionary change.

In chapter 3 we ask, What kinds of impacts do gallmakers have on the physiology and growth performance of this clonal host plant? Are the negative effects on growth and reproduction of the type that will result in natural selection on plant defense? On the way to answering these broad questions we describe what is known about *Eurosta*'s gall-inducing stimulus, the effect of gall induction on allocation of biomass, energy, and nutrients in goldenrod ramets, and the integration of these effects across the genet. We reach the tentative conclusion that gallmakers may act as agents of selection when their population levels are very high, or when they attack early in the first few years of the host plant's life.

The plant's perspective is continued in chapter 4, where we ask if individual variation in galling rates are due to genetic causes, and if the influence of genes can be modified by the availability of resources. Our work shows that host plant genotypes vary in their resistance to galling, and that resistance can be expressed at several points in the attack sequence. Some plants are less attractive to ovipositing females while others are less likely to initiate galls once oviposited upon. One of the more intriguing findings is the necrotic response, found to varying degrees in some plant genotypes: cells surrounding newly hatched larvae soon die, resulting in the death of the gallmaker. However, altering nutrient supply can change the relative genotypic differences in resistance. Thus, the potential for resistance evolution can depend not only on the abundance of gallmakers, but on the abiotic background environment as well. In summary, the realized resistance level of an individual host plant will depend on a hierarchy of traits that influence its acceptability and suitability to the gallmaker. Natural populations of goldenrod show ample genotypic variation for these traits to allow a selection response.

In chapter 5 we switch to the gallmaker's perspective. Here we ask to what degree variations in galling rates are the results of choices made by ovipositing females. Plant growth rate and secondary chemistry are used by the fly as cues when selecting among host individuals. However, we find the plant traits which influence *Eurosta*'s preferences do not reliably predict subsequent performance of the offspring. This chapter explores the reasons for weakness of correlation between female oviposition preference and offspring performance.

Gall size is an important factor that determines the strength of the sink that gall formation places on the plant. It is also important from the gallmaker's perspective because parasitoid and bird attack rates on *Eurosta* rise and fall with gall size. Thus both plant and gallmaker have a fitness stake in gall size. In chapter 6 we ask if plants, insects, or both make genetic contributions to variation in gall size. We conclude that although the gall is a piece of plant tissue, it can be considered a part of the gallmaker's extended phenotype. Thus selection acting on the insect can in principle result in evolution of gall size.

The specificity of host-choice behavior in herbivorous insects has long been thought to lend itself to the formation of host-associated races and to sympatric speciation. In chapter 7 we ask if *Eurosta* is divided into host-associated populations. Near the northern edge of *Solidago altissima*'s geographic range, the gallmaker is found on other goldenrods, particularly *S. gigantea*. Genetic differences in phenology and host-choice behavior are seen in flies from the two hosts. Molecular data further confirm some degree of reproductive isolation between both allopatric and sympatric host-associated populations. Our data indicate that some ecological factors, including escape from natural enemies, facilitate host shifting. The environment provided by the novel host plant can cause phenotypic changes in the flies and their galls that facilitate host shifting, such as altered phenology and escape from natural enemies.

The next two chapters examine the selective pressures that parasitoids and insectivorous birds place on the interaction between plant and gallmaker. First, chapter 8 shows that since none of *Eurosta*'s natural enemies are able to curtail the growth and development of the gall, their attack has no beneficial effect on goldenrod. However, gall size is a strong determinant of gallmaker survival. The parasitoid *Eurytoma gigantea* is unable to penetrate large galls, and so imposes a selective pressure on the gallmaker for increased gall size. An opposing selective force is exerted by birds, who preferentially attack large galls. The balance of selection is thus determined by the attack intensity of the two species. Some of the factors that influence attack intensity are examined in chapter 9. We examine the possibility of density- and frequency-dependent selection. This chapter also deals with the partitioning of selective effects between insect genetic variance in gall size and the variance due to the interaction between insect and plant genotype and concludes by asking if an evolutionary response to selection can be seen in *Eurosta*.

The final three chapters deal with evolutionary issues that emerge from our study of goldenrod, its gallmaker, and the natural enemies. A cautionary note is sounded in chapter 10, which shows two examples where environmental effects on growth and development can be mistaken for evolutionary response. Chapter 11 presents a conceptual framework that can bring together ecological and genetic approaches to the evolution of enemy-victim interactions. This framework builds on our view of the defense hierarchy, and the insects response to it. This developmental-genetic view shows how the epigenetic interaction between resistance and tolerance influences their evolution in plants, just as the interaction between insect preference and performance influences evolution of the herbivore feeding niche. We end with a chapter that comments on the relevance of our study of this single system to the evolutionary ecology of plant-insect interactions generally.

The Stem Gallmaker, Its Natural Enemies, and Goldenrod: A Model System of Tritrophic-Level Interaction

2.1 GALLMAKERS AS A SPECIALIZED HERBIVORE GUILD

Among the many arthropod feeding guilds, one of the most intriguing is that of gallmakers. Gallmakers are unique in that they alter the course of development of plant tissue to form a tumorlike growth from which the insect gains nutrition and protection from the environment (Abrahamson and Weis 1987; Weis and Berenbaum 1989). Gallmakers constitute a functional group broadly scattered across taxonomic lines and include representatives from four kingdoms, two phyla, two arthropod classes, and six orders of insects (Mani 1964; Abrahamson and Weis 1987). Although this feeding habit is found in mites, thrips, moths, weevils, flies, and wasps, it has had its broadest radiation in two families. The Cecidomyiidae (Diptera, gall midges) has more than 5000 gallmakers worldwide (Mani 1964) and the Cynipidae (Hymenoptera, gall wasps) has over 10,000 (Felt 1940). These two families alone account for over 70% of the nearly 1700 insect gallmakers in North America (Weis and Berenbaum 1989).

Many plant species become hosts to gallmakers, but most of the host plant species can be accounted for in relatively few plant families. For example, few gallmakers infest monocots, so that over 90% of the galls worldwide occur on dicots. Among dicots, most galls are found on species within the Rosaceae (rose family), Compositae (aster and sunflower family), and Fagaceae (chestnut and oak family) (Abrahamson and Weis 1987). Gen-

14

erally, gallmakers exhibit a degree of monophagy in host choice and restrict their attack to a single type of plant structure (e.g., bud, leaf midrib, petiole). This and the remarkable degree of specificity of the host plant's reaction to the gallmaker typically result in galls of distinctive morphology where several gallmakers utilize the same host species (Waring and Price 1989; Weis and Berenbaum 1989). Indeed, gallmakers are usually identified by the morphology of their plant gall (Felt 1940).

2.2 GALL PHENOTYPES: CONSEQUENCES OF TWO GENOTYPES

The development of the gall phenotype entails the interplay of two genotypes: that of the gallmaker, which codes for the gall-inducing stimulus, and that of the host plant, which codes for the growth response (Weis and Abrahamson 1986). Dawkins (1982) proposed that a gall structure be considered as an "extended phenotype" of the gallmaker's genome. As such, the ability to stimulate gall formation is an adaptive trait of the gallmaker. Gallmakers secrete substances that function as plant growth and differentiation regulators, overriding the host plant's normal growth systems to produce the gall (Carango et al. 1988; Hori 1992; Raman 1993). Several lines of evidence support this notion including (1) galls typically only develop from meristematic tissue; (2) gall formation involves a change in the pattern by which normally developed tissues are laid down; and (3) gall growth typically depends on the presence of a living gallmaker indicating that a constant supply of stimulant is required (Abrahamson and Weis 1987; Carango et al. 1988). In the subsequent sections, we will offer information about the host plant, gallmaker, and the gallmaker's natural enemies.

2.3 NATURAL HISTORY OF *SOLIDAGO*

Taxonomic Status

The taxonomic status and the phylogeny of the genus *Solidago* (the Latin name makes reference to its healing properties—*solidus* and *ago*—to make whole) are complicated and as yet unresolved (Croat 1972; Melville and Morton 1982; Semple and Ringius 1983; Gleason and Cronquist 1992). Although this native North American genus (except one species—the common Eurasian species *S. virgaurea;* Werner, Bradbury, and Gross 1980) contains well over one hundred species and is among the most ecologically studied wild plants in North America, no comprehensive taxonomic treatment of the genus exists. This is unfortunate given the ubiquitous nature of goldenrod distributions throughout much of North America and elsewhere (as an introduction), and its many interactions with pollinators, herbivores, and their natural enemies.

While goldenrod as a group is easy to recognize, especially during the autumn when their inflorescences dominate in old-field landscapes, species-level determinations have long been considered difficult because of the number of possible species within a region, the need to use technical characters to separate some taxa, and the occasional occurrence of interspecific hybrids (Semple and Ringius 1983).

Although the base chromosome number for the genus *Solidago* is 9 (*n*), many species are polyploids making multiples of nine common (Beaudry and Chabot 1957; Beaudry 1963; Melville and Morton 1982). For some species, ploidy level and infraspecific classification correlate, but classification of some *Solidago* species is complicated by the existence of more than one ploidy level (e.g., *S. gigantea* that occurs as diploid, tetraploid, and hexaploid taxa; Semple and Ringius 1983).

Solidago Flowering

Goldenrods are insect pollinated and typically self-incompatible (Mulligan and Findlay 1970; Melville and Morton 1982; Gross and Werner 1983). They possess heavy, sticky pollen that is carried by a wide array of invertebrates including honeybees, bumblebees, wasps, syrphid flies, soldier beetles (*Chauliognathus*), moths, and butterflies (Werner, Bradbury, and Gross 1980; Gross and Werner 1983; Semple and Ringius 1983). *Solidago* inflorescences frequently teem with these and other floral visitors during their blooming season even though their tiny florets contain tiny amounts (0.0001 mg sugar per floret; Heinrich 1976) of nectar. Floral visitors are likely encouraged by the high rate at which they can forage proximate florets within and among the numerous heads that compose an individual ramet's inflorescence. Bumblebees, for example, are estimated to handle up to 110 *Solidago* florets per minute (Heinrich 1976).

Goldenrods have frequently been maligned as the cause of late-summer hayfever (Wodehouse 1945). This is unfounded, however, because although the flowering of goldenrods is conspicuous during the late-summer hayfever season, *Solidago* pollen is too heavy to become airborne. Placing the blame for hayfever on goldenrod is a case of a correlation (of hayfever season with showy goldenrod flowers) that lacks cause and effect. Rather, it is ragweed (*Ambrosia* spp.) and other wind-pollinated plant species with their inconspicuous, green flowers, and not *Solidago,* that fill the air with light-weight pollen grains (Lewis and Elvin-Lewis 1977).

The goldenrod species that most often host the goldenrod ball gallmaker are members of the widespread *S. canadensis* sensu lato complex. An examination of the bloom phenologies of three common members of this complex showed that the flowering periods of these species markedly overlap but that they enter bloom in the distinct temporal sequence of *S. canadensis, S. gigantea,* and *S. altissima* (fig. 2.1; Givens 1982; Givens and Abrahamson, unpub. data). The hexaploid *S. altissima* blooms over a longer interval than the other two species, producing a greater niche breadth for *S. altissima* than for *S. gigantea* or *S. canadensis* (fig. 2.1; Table 2.1). Bloom-time overlap is greatest between *S. altissima* and *S. gigantea* in the Canadian data set but between *S. altissima* and *S. canadensis* in the Pennsylvania data set.

Host Plants for E. solidaginis

At least seven taxa of *Solidago* have been reported as hosts of *E. solidaginis,* including *S. altissima, S. gigantea, S. canadensis, S. rugosa, S. (Euthamia) graminifolia, S. serotina* (a synonym of *S. gigantea*), *S. rugosa,* and *S. ulmifolia* (Wasbauer 1972; Novak and Foote 1980; Ming 1989). While such reports provide some indication of the breadth of *Eurosta's* host preferences, our observations suggest that *E. solidaginis* is more narrowly oligophagous than the literature indicates. The two principal host plants of *E. solidaginis* are *S. altissima* and *S. gigantea;* both are putative

Solidago Phenology

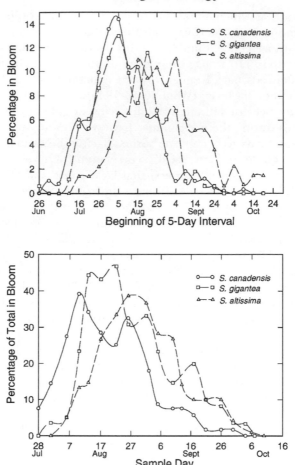

FIGURE 2.1. Bloom phenologies for *S. canadensis, S. gigantea,* and *S. altissima* across Canada (*top*) and in central Pennsylvania (*bottom*). Bloom dates for Canadian specimens were tallied from ≈1500 recently annotated specimens of the *S. canadensis* complex in the herbaria of the Ontario National Museum and the Ontario Department of Agriculture in Ottawa, Ontario, Canada. Only specimens in full bloom (or prior to it) were used. Phenology data for central Pennsylvania were recorded for each species along a 5-meter wide, 120-meter long belt transect at the Bucknell University Natural Area approximately every third day following the first appearance of flowers, from 28 July 1981 to 10 October 1981. A ramet was considered in bloom until approximately one-half of all its capitula were open. (Givens 1982; Givens and Abrahamson, unpub. data)

TABLE 2.1. Niche breadth (unweighted) of bloom phenology using Levins's (1968) formula and niche overlap using the formula of proportional similiarity (cf. Schoener 1970).

		Canada Data		Pennsylvania Data	
Niche breadths	*S. canadensis*	0.468		0.079	
	S. gigantea	0.478		0.061	
	S. altissima	0.556		0.093	
		S. can.	*S. gig.*	*S. can.*	*S. gig.*
Niche overlap	*S. gigantea*	0.339	—	0.142	—
	S. altissima	0.558	0.637	0.352	0.015

Source: Data from Canadian herbaria specimens and from ramets censused at the Bucknell University Natural Area (see legend, fig. 2.1; Givens 1982; Givens and Abrahamson, unpub. data).

members of the *S. canadensis* species complex (pers. obs.). In the following paragraphs we offer information about the taxa that have been most commonly reported to serve as hosts for *E. solidaginis*. This information has been summarized from Fernald (1950), Beaudry (1963), Werner, Bradbury, and Gross (1980), Melville and Morton (1982), Semple and Ringius (1983), and is supplemented by our personal observations.

Solidago altissima L. (syn. *S. altissima* var. *scabra),* late goldenrod, is the ancestral and primary host plant of *E. solidaginis* throughout the eastern portions of its range (Abrahamson, McCrea, and Anderson 1989; Waring, Abrahamson, and Howard 1990; Craig et al. 1993; Brown et al. 1995; Brown, Abrahamson, and Way 1996). A strongly clonal species, it frequently forms extensive, nearly pure clusters of ramets. Stems of ramets are covered with a short pubescence but hairs typically become deciduous near the stem's base (fig. 2.2). Leaves are strongly triple-nerved, possessing entire to serrate margins near the apex with a scabrous upper surface and a finely pubescent lower surface (the short leaf pubescence gives plants a gray-green tone). Individual capitula (reproductive heads) contain approximately 10–15 ray florets and 3–7 disk florets. This hexaploid ($2n = 54$) member of the *S. canadensis* complex is common in old fields, roadsides, floodplains, and disturbed sites. Morphologically, *S. altissima* is closest to *S. canadensis* var. *gilvocanescens.*

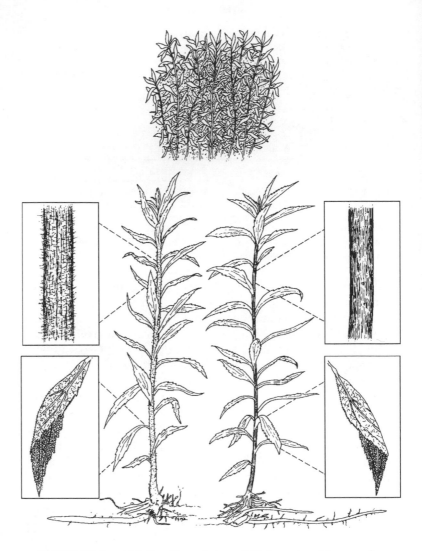

FIGURE 2.2. An old-field goldenrod clone (*top*) shown in midsummer before the onset of flowering. *Solidago altissima* (Compositae) ramet (*left*) with its relatively thick rhizomes. Insets illustrate the scabrous stem and the pubescent leaf underside of *S. altissima* ramets. *Solidago gigantea* ramet (*right*) with its relatively thin rhizomes. Insets illustrate the glaucous stem and leaf underside of *S. gigantea* ramets. (Illustration by J. L. Miller)

Solidago gigantea Ait. (syn. *S. serotina, S. shinnersii*), tall golden-rod, is a common but derived host plant of *E. solidaginis* across northern portions of the gallmaker's distribution (Waring, Abrahamson, and Howard 1990; Craig et al. 1993; Brown et al. 1995; Brown, Abrahamson, and Way 1996). Also a clonal species, its rhizomes are considerably thinner than those of *S. altissima* but like *S. altissima* can create extensive and nearly pure patches of ramets. Stems lack pubescence and are glabrous and glaucous up to the inflorescence. *Solidago gigantea* is the least pubescent member of the *S. canadensis* complex. Leaves are triple-nerved often possessing sharply serrate margins with hairs only on the veins of the leaf underside (fig. 2.2). Individual heads have approximately 7–15 ray florets and 6–12 disk florets. This taxon exists as a diploid ($2n = 18$), a tetraploid ($2n = 36$), and a hexaploid ($2n = 54$, N.B. often occurring as a broad-leaved form) that invades old fields, open woodlands, floodplains, and thickets across its range. The tetraploid taxon appears to be the most widespread. *Solidago gigantea* frequents somewhat more moist sites than *S. altissima* (pers. obs.).

Other species identified as hosts of *E. solidaginis* are less frequent hosts (pers. obs.). *Solidago canadensis* L. (syn. *S. scabra, S. pruinosa*), Canada goldenrod, is an occasional host plant of *E. solidaginis* in northern portions of its range (pers. obs.) and has supported gall formation by *E. solidaginis* in a greenhouse study (Abrahamson, McCrea, and Anderson 1989). Miller (1959) suggested that early reports of *E. solidaginis* galls on this goldenrod may have been due to confusion of *S. canadensis* and *S. altissima*. Another clonal species, *S. canadensis,* has shorter rhizomes than either *S. altissima* or *S. gigantea* and consequently grows with tightly clustered ramets in dense clumps. Stems are glabrous near their bottom but become pubescent from their midpoint to the apex. Leaves are triple-nerved and sharply serrate like those of *S. gigantea.* Leaves have glabrous undersides except for hairs on the major veins. *Solidago rugosa* Mill., rough-stemmed goldenrod; *S. ulmifolia* Muhl., elm-leaved goldenrod; and *S. (Euthamia) graminifolia* (L.) Nutt. (syn. *Chrysocoma graminifolia, S. lanceolata, S. nuttallii*), grass-leaved goldenrod have been reported to be hosts for *E. solidaginis.* We have never observed galls on the latter two taxa and only very infrequently seen them on *S. rugosa.*

21

Seed Dispersal, Germination, and Performance

The wind-dispersed fruit of *Solidago* is an achene with an attached pappus. Given the small size of these achenes, individual ramets of *Solidago* produce literally tens of thousands of fruits. *Solidago canadensis* and *S. altissima*, for example, have been estimated to produce between 5000 and 19,000 achenes (Bradbury 1973; Hartnett and Abrahamson 1979; Givens 1982). However, the number and quality of achenes can change as a consequence of gallmaker presence (see chapter 4).

Achene mass, pappus length, and fall velocity vary among species (Abrahamson, McCrea, Boomer, and Thum, unpub. data). Achenes of *S. gigantea* are significantly heavier (analysis of variance $F_{2,1227} = 1023, P < 0.001; 0.13 \pm 0.04$ mg, $\pm$ SD) than those of *S. altissima* (0.08 ± 0.02 mg) or *S. canadensis* (0.05 ± 0.01 mg), and the length of the pappus attached to achenes is longest in *S. gigantea*, intermediate in *S. altissima*, and shortest in *S. canadensis*. *Solidago altissima* and *S. canadensis* achenes fall significantly more slowly (analysis of variance $F_{2,1227} = 63.7, P < 0.001; 0.52 \pm 0.12$ m per sec and 0.52 ± 0.17 m per sec, $\pm$ SD, respectively) than the achenes of *S. gigantea* (0.61 ± 0.16 m per sec). Of the three species examined, the achenes of *S. altissima* and *S. canadensis* have the best chances of remaining airborne and being dispersed, while the heavier achenes of *S. gigantea* are likely less dispersible.

We examined the achene germination characteristics of these potential hosts of *E. solidaginis* by simulating April, May, and June night and day temperature regimes for central Pennsylvania in growth chambers (Abrahamson, McCrea, Boomer, and Thum, unpub. data). Temperature has a strong effect on germination rate in all species such that germination began earliest in the June treatment—within 5 or 6 days of placing achenes into the treatment. However, the final rates of germination in May and June temperature regime treatments did not differ. This suggests that the germination of these species most probably occurs during May under natural conditions.

Litter cover and the degree of canopy opening strongly affect seedling emergence, growth, and survival in *Solidago* (Goldberg and Werner 1983). This is not surprising given the colonizing strategies of the old-field *Solidago* species potentially attacked by

E. solidaginis. Unfortunately, relatively few details of *Solidago*'s seedling demography are known but several studies provide insights into the life history and adaptations of the genus.

Greenwald, McCrea, and Abrahamson (1985; unpub. data) explored the seedling performance of closely related *S. altissima* and *S. canadensis* (see Croat 1972) in a greenhouse competition experiment that varied seedling density, moisture level, and ratio of species. DeWit-style replacement diagrams (deWit 1960) suggested that *S. canadensis* had a competitive advantage over *S. altissima* under all experimental conditions, but that soil-moisture level had no significant effect on the outcome of seedling competition between these two species. Extrapolation of these greenhouse findings to natural field conditions where *S. canadensis* and *S. altissima* co-occur imply that *S. canadensis* should dominate and that *S. altissima* should be confined to those patches not exploited by *S. canadensis*. As is frequently the case with extrapolation of simple experimental conditions to the field, the reality throughout much of the range of these two potential *Eurosta* host species is quite different. *Solidago altissima* is often the more common and abundant species. Throughout much of the Mid-Atlantic region and New England, *S. altissima* dominates old fields and roadsides while *S. canadensis* and *S. gigantea* are confined to more mesic ditches or lower patches within old fields and floodplains (pers. obs.). Such field-distribution patterns suggest that there are variables beyond those examined in our greenhouse study that are important to how these species interact in nature. Some of these variables are likely related to climate since *S. canadensis* and *S. gigantea* generally range farther north than *S. altissima* and are normally more abundant in cooler climates (pers. obs.).

We examined the spatial distributions of *S. canadensis, S. gigantea, S. altissima,* and *S. juncea* at the Bucknell University Natural Area along a 5 m wide by 120 m long transect on a gently sloping hillside (Givens 1982; Givens and Abrahamson, unpub. data). The elevated end of the transect had well-drained soil, while soil of the lowest end of the transect was often waterlogged. Soil-moisture samples were collected at 15 m intervals at three dates during a single season.

The four *Solidago* taxa that occurred along the transect were not distributed uniformly. *S. altissima* clones were most dense at

23

the elevated end of the transect, while *S. gigantea* clones predominated toward the lower end. *S. canadensis* clones were restricted to a small region of the lower hillside and *S. juncea* (typically an indicator of dry soils; Werner and Platt 1976) dominated the area of the transect between *S. altissima* and *S. gigantea*. Soil-moisture levels averaged approximately 9% (with relatively high coefficients of variation over the upper half of the transect, 10–22%, because of drying and wetting cycles) at the elevated end of the transect and 17% (with relatively low coefficients of variation over the lower half, 3–10%) at the lowest end. These results suggest that these goldenrods segregate at least in part according to differing tolerances of variance in soil moisture. *S. gigantea* seems to occur primarily on damp to permanently wet ground, while *S. altissima* persists in soils that are (or become) quite dry. Werner and Platt (1976) demonstrated that *S. altissima* (called "*S. canadensis*" in their paper; P. A. Werner, pers. comm.) occupied the largest range of soil moistures of the five species of goldenrod they studied. *S. canadensis* seems to occupy soils with an intermediate range of soil moistures, never on the soggy muck often occupied exclusively by *S. gigantea,* nor on the well-drained sites sometimes occupied by *S. altissima* (Givens 1982).

In a second study, we compared the distributions of goldenrod species by determining the ramet density of each of five species (i.e., *S. altissima, S. gigantea, S. graminifolia, S. juncea, and S. rugosa*) in thirty old fields in central Pennsylvania. Five systematically located soil samples from each field were analyzed for organic matter, phosphorus, potassium, magnesium, calcium, hydrogen, soil pH, and cation exchange capacity (Abrahamson, Ball, and Houseknecht, unpub. data).

Of the five species, *S. altissima* was the most widespread (appearing in twenty-nine of the thirty fields) and the most abundant (densities of 0.2 to 52.9 ramets/m²). *Solidago gigantea* was the least common (appearing in only twenty of the thirty fields), and *S. juncea* had the lowest densities (ranging from 0.2 to 6.0 ramets/m²). A detrended correspondence analysis showed that *S. altissima* and *S. gigantea* had the most similar distributions (fig. 2.3). Furthermore, high abundances of these two goldenrods were positively correlated with calcium abundance and less acidic soil. *S. rugosa, S. juncea,* and *S. graminifolia* attained higher

24

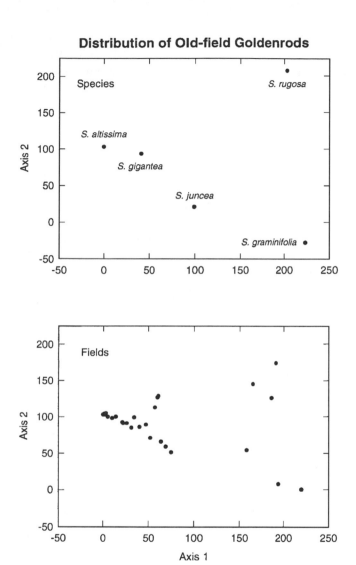

Distribution of Old-field Goldenrods

FIGURE 2.3. Species and field ordinations of central Pennsylvania old fields based on 1995 ramet densities of goldenrod species according to a detrended correspondence analysis using PC-Ord (McCune 1993). Each point represents the species position on the first two axes of the ordination. The x-axis is correlated with soil pH and calcium abundance such that fields with more calcium and high pH occur nearer the origin. (Abrahamson, Ball, Houseknecht, unpub. data)

densities in fields with more acidic soil. Our finding that *S. altissima* and *S. gigantea* are similar in distribution supports the suggestion of Schmid et al. (1988a,b) that these two goldenrods are ecologically similar.

Clonal Growth Form

The clonal growth of *Solidago* means that clones occur as a clumped resource for ovipositing *Eurosta*. Clonal expansion in the rhizomatous species of *Solidago* occurs by means of rhizome growth outward from the edges of the genotype. The consequence of this growth is a dense cluster of ramets that can grow to several meters in diameter. Clones of up to 10 m in diameter, for instance, have been reported for one *Solidago* taxon in virgin prairie (Werner, Bradbury, and Gross 1980). The clonal habits of the *Solidago* hosts of *E. solidaginis* give host-plant genotypes considerable longevity, albeit at a potential slowing of population-level responses to episodes of selection. Werner, Bradbury, and Gross (1980), for example, reported that clones of *S. canadensis* can persist for more than 100 years.

Schmid et al. (1988a) found that shoot growth and development, leaf size and physiology, and biomass allocation significantly differed among three potential hosts of *E. solidaginis* (i.e., *S. canadensis, S. altissima,* and *S. gigantea*) both in a common garden and in naturally occurring field populations. *Solidago canadensis* had many small and highly productive leaves that enabled rapid plant growth and early flowering followed by the highest resource allocation to seed maturation of the three species. *Solidago altissima* and *S. gigantea* maintained larger leaf areas but they had fewer, larger, less productive leaves and consequently grew slower with delayed flowering relative to *S. canadensis*. Unlike *S. canadensis,* both *S. altissima* and *S. gigantea* continued to strongly invest in vegetative activities (especially the production of long rhizomes) during seed maturation. Schmid et al. (1988a) interpreted these contrasts as differential investment in sexual reproduction, on one hand (*S. canadensis*), versus clonal growth and expansion, on the other (*S. altissima* and *S. gigantea*). As a consequence of these life-history differences, *S. canadensis* forms compact, more strongly integrated clumps in old fields, while *S. altissima* and *S. gigantea* are characterized by larger, more expansive, but less integrated clumps (Schmid et al. 1988a,b; pers. obs.).

The number of new ramets produced by the previous year's "mother" rhizome varies depending on species and resource availability. Data for potential *Eurosta* hosts *S. canadensis* and *S. altissima* indicate that the mean number of new ramets from a mother ramet ranges from two to over eleven (Werner, Bradbury, and Gross 1980). Smith and Palmer (1976) suggested that new rhizomes branch out from the previous year's ramet with somewhat regular patterns of geometry. Most new rhizomes were reported to extend either in the same direction as the mother rhizome or vary approximately 67° to the right or left of the mother's original direction (Smith and Palmer 1976). In contrast, Cain (1990a) found that branching angles, rhizome lengths, and numbers of daughter ramets varied widely among clonal fragments. Importantly, Cain showed that branching angles were independent of both previous branching angles and rhizome lengths and that the modal direction for *S. altissima* clonal growth was 0°. Cain concluded that clonal growth in this species is highly variable and more consistent with stochastic and random-walk models than with deterministic notions of clonal spread.

Seasonal Patterns of Clonal Growth

In a study of the seasonal growth activity and nutrient movement within clones of *S. altissima*, Abrahamson and McCrea (1985) showed that individual rhizomes were dormant over winter but quickly elongated and produced aerial stems the following spring. Nutrient concentrations in below-ground organs remained stable during winter months. However N, P, K, and Mg levels rapidly increased in new rhizomes approximately one month prior to the emergence of ramets from the ground. Although nutrient concentrations generally declined throughout the growing season due to dilution effects as ramets grew, total nutrient content increased due to rather constant uptake from soil reserves from May through September. *Solidago altissima* uses nutrients efficiently via recycling of many mineral elements among organs over time (Hirose 1975; Abrahamson and McCrea 1985). We suspect that many old-field members of this genus are similar to *S. altissima* in their generalized pattern of growth phenology and nutrient-use efficiency, particularly *S. gigantea* due to its ecological similarity to *S. altissima* (Schmid et al. 1988a,b).

Mother ramet size within *S. altissima* clones affected both the size and number of daughter ramets produced (Cain 1990b). Small rhizomes, for instance, tended to develop into small ramets, and smaller ramets had lower survival and fecundity than larger ramets. Given the relationship of ramet size and probability of survival, insect herbivores such as *E. solidaginis* that oviposit early in the growing season might be able to avoid ramets likely to die by ignoring smaller ramets (see chapter 5).

Herbivores of Solidago

Species of *Solidago* are attacked by a wide variety of insect herbivores. The insect fauna of *Solidago* in central New York is undoubtedly the best-studied example due to the many years of careful work by Richard Root and his students. *Solidago*'s herbivore fauna includes more than one hundred species distributed over at least five orders (Messina 1978; Messina and Root 1980). The subset that commonly attacks *Eurosta*'s principal host *S. altissima* (*S. gigantea* is more poorly studied) includes Hemiptera, Homoptera, Coleoptera, Lepidoptera, and Diptera (table 2.2).

2.4 NATURAL HISTORY OF *EUROSTA SOLIDAGINIS*

Taxonomic Treatment

Species in the genus *Eurosta* Loew are stem, rhizome, or crown gallmakers on various species of goldenrods (*Solidago,* Compositae). *Eurosta* may be most closely related to one of two genera, *Aciurina* Curran or *Valentibulla* Foote and Blanc, both gallmakers on Compositae of the western U.S. As many as twelve species of *Eurosta* have been recognized, all from North America. However, taxonomists have varied in the number of *Eurosta* species recognized (synonymies can be found in Foote, Blanc, and Norrbom 1993). Steyskal and Foote (1977) identified nine species based primarily on variations in the size and position of hyaline and gold areas on the dark background of the wing. More

28

TABLE 2.2.　Common insect herbivores of *Solidago altissima* by order, family, species, and functional feeding guild. Many other herbivores feed on *S. altissima*.

Order	Family	Species	Feeding Guild
Hemiptera	Miridae	*Lygus lineolaris*	flower feeder
		Slaterocoris spp.	foliage feeder
	Tingidae	*Corythuca marmorata*	mesophyll tapper
Homoptera	Cercopidae	*Philaenus spumarius*	xylem tapper
	Aphididae	*Uroleucon caligatum*	phloem tapper
		Uroleucon nigrotuberculatum	phloem tapper
Coleoptera	Chrysomelidae	*Exema canadensis*	Leaf chewer
		Microrhopala vittata	leaf miner (larvae), leaf chewer (adults)
		Ophraella conferta	leaf chewer
		Trirhabda virgata	leaf chewer
		Trirhabda borealis	leaf chewer
Lepidoptera	Gelichiidae	*Dichomeris* spp.	leaf chewer
		Gnorimoschema gallaesolidaginis	stem galler
	Tortricidae	*Epiblema scudderiana*	stem galler
		Epiblema spp.	stem borer
Diptera	Cecidomyiidae	*Asteromyia carbonifera*	leaf galler
		Rhopalomyia solidaginis	rosette galler
	Tephritidae	*Eurosta solidaginis*	stem galler
	Agromyzidae	*Ophiomyza* sp.	leaf miner
		Phytomyza sp.	leaf miner

Sources: Summarized from Messina 1978; Hartnett and Abrahamson 1979; Messina and Root 1980; Maddox and Root 1987, 1990; Meyer 1993; Meyer and Root 1993; Raman and Abrahamson 1995.

recently, Ming (1989) prepared a revision of the genus in which she recognized seven species based on morphological and life-history attributes. Ming's (1989) revision is also included in the *Handbook of the Fruit Flies (Diptera: Tephritidae) of America North of Mexico* (Foote, Blanc, and Norrbom 1993).

The goldenrod ball gallmaker was placed in the genus *Eurosta* Loew and designated *solidaginis* Fitch by Coquillett (1910). Later, Curran (1923, 1925) described two varieties: *fascipennis,* which he reported to occur on one or more prairie species of goldenrod; and *subfasciatus,* which occurred only on *S. canadensis* (more likely *S. altissima* by today's taxonomic treatments). Ming's (1989) revision of the genus *Eurosta* follows Curran to a degree in suggesting that *E. solidaginis* exists as two subspecies: *E. solidaginis* subsp. *solidaginis* (Fitch) throughout the eastern U.S. and *E. solidaginis* subsp. *fascipennis* Curran in the western U.S. The two subspecies can be distinguished as adults by differences in their wing patterns. The eastern subspecies has three distinctly separated, triangular hyaline areas that are broadly based on the wing margin. However, in the western subspecies, two of these hyaline areas are connected to form a band across the wing (Ming 1989).

Additional differentiation may be occurring within *E. solidaginis*. Both subspecies of *E. solidaginis* described by Ming (1989) are conspicuous gallmakers throughout their ranges; however, Stoltzfus (1989) reported that *E. solidaginis* can attack *S. canadensis* in Iowa without forming a gall. This non-gallmaking *E. solidaginis* occurred infrequently in the Iowa populations examined, with only sixteen adults emerging from some 2335 collected stems (<0.7%). Individuals of the non-gallmaker do not tunnel in the host plant's stem; instead they create a small cavity in the stem pith. Stoltzfus (1989) confirmed that the non-gallmaking and the gallmaking *E. solidaginis* can interbreed. Ovipositions by gallmaking females that were crossed to non-gallmaking males resulted in the initiation of two galls. However, no adults emerged from these galls. The non-gallmaking *E. solidaginis* may have moved into enemy-free space as Stoltzfus (1989) found no evidence of *Eurytoma* parasitoid or bird attack. However, it is possible that the lack of parasitoid attack is a consequence of Stoltzfus's (1989) small sample of non-galling *E. solidaginis*. This non-galling *E. solidaginis* may be rare in any given population, but it appears to be widespread. We have very infrequently encountered individuals of the non-galling *E. solidaginis* in Pennsylvania populations.

More recently our laboratory has shown that Ming's (1989) eastern subspecies of *E. solidaginis* is differentiated into two host races: an ancestral host race that infests *S. altissima,* and a derived host race that attacks *S. gigantea* (Waring, Abrahamson, and Howard 1990; Craig et al. 1993; Brown et al. 1995; Brown, Abrahamson, and Way 1996). It is likely that one of the factors facilitating this host shift is an increase in survivorship on the derived host plant relative to the ancestral host. This increase comes as a consequence of reduced mortality from natural-enemy attack in spite of higher early larval death on the novel host due to poor physiological adaptation (Brown et al. 1995; Brown, Abrahamson, and Way 1996). The evidence for and causes of this host shift are discussed in chapter 7.

The relationship of *E. solidaginis* to other species of *Eurosta* is critical to understanding the roles that host association and host shifts have played in the diversification of this genus. On the basis of morphological synaporphies, Ming's (1989) revision of *Eurosta* suggests that the rhizome-galling species (*E. comma, E. cribrata, E. fenestrata,* and *E. floridensis,* as well as the putative rhizome-galler *E. latifrons*) form a monophyletic group (although her conclusions were tentative because of missing data for some taxa). She considered the rhizome-galling habit derived because species of the presumed sister genus *Aciurina* are stem gallers like *E. solidaginis.* A preliminary phylogeny based on mtDNA variation conflicts with this hypothesis (Brown and Abrahamson, unpub. data). We conclude that *E. cribrata* is more closely related to *E. solidaginis* than to the rhizome-galling species *E. comma* and *E. floridensis* (fig. 2.4). Furthermore, our data suggest that the *E. floridensis* haplotype has emerged from within the *E. comma* clade. Because *E. floridensis* attacks a different *Solidago* species than *E. comma* (which itself varies in host association across its range; attacking at least three species of *Solidago*), we hypothesize that diversification in *Eurosta* has proceeded through host-plant shifts that follow major changes in life-history strategy (Brown and Abrahamson, unpub. data). Although much work remains to be done before we fully understand the roles of host association and host shifts in speciation within the genus *Eurosta,* we do understand the roles of host association and a host shift in host-race formation within *E. solidaginis* (see chapter 7).

Geographic Distribution

Both subspecies of *E. solidaginis* are common in North America. Ming (1989) reported that the eastern subspecies ranges from Maine west to North Dakota, and south only to Virginia and Kansas (or see Foote, Blanc, and Norrbom 1993). Uhler

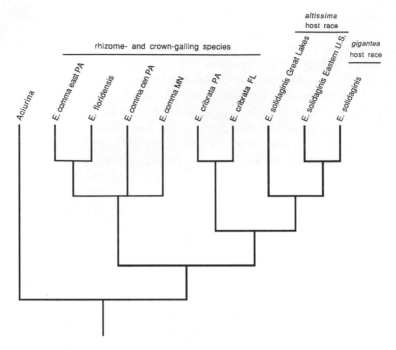

FIGURE 2.4. Preliminary phylogeny of *Eurosta* species based on 492 bp of sequence from mitochondrial cytochrome oxidase I and II genes (consensus of three most parsimonious trees is shown). We used direct sequencing of double-stranded fragments generated by the polymerase chain reaction, and sequences were aligned by eye. A most parsimonious network of relationships among alleles was determined using the EXHAUSTIVE search routine of PAUP3.1.1. Three species (*Eurosta latifrons, E. fenestrata,* and *E. lateralis*) are not included in this analysis. (Brown and Abrahamson, unpub. data)

(1951), however, cited specimens within this region but also specimens of what Ming would recognize as the eastern subspecies from as far south as Texas, Louisiana, and North Carolina. Miller (1959) added to Uhler's (1951) distribution range by confirming *E. solidaginis* populations from Georgia, Kentucky, Michigan, Mississippi, and West Virginia. Our research groups have located populations of the eastern subspecies in sites from New England to Minnesota, including southern Canada and south to northern Florida and Texas (pers. obs.).

Ming (1989) identified the distribution of the western subspecies from museum specimens collected from Washington east to Minnesota, and from Alberta south to Colorado. The two subspecies are sympatric across the upper Midwest, including North and South Dakota, Minnesota, and Iowa. Within the zone of sympatry, Ming (1989) did not find any evidence of intergrades between the two subspecies.

Sexual Dimorphism and Emergence

The sexes of adults of both subspecies are easily distinguished. The female's ovipositor gives her abdomen's apex a pointed appearance, whereas the abdomen of the male is bluntly rounded (fig. 2.5). Furthermore, the males are smaller in size on average than the females (Uhler 1951). In central Pennsylvania, adult *E. solidaginis* emerge from the gall in mid- to late May by breaking through the thin epidermal covering of the exit tunnel that was prepared as a larva (Anderson et al. 1989). In all populations we've examined (i.e., New England, Mid-Atlantic region, upper Midwest), males emerge significantly earlier than females (Weis and Abrahamson 1985; Craig et al. 1993; Abrahamson et al. 1994).

Mating Behavior and Oviposition

Unfortunately it is virtually impossible to follow individual flies in the field to observe their behavior. Individuals are quickly lost unless they are contained within a greenhouse or cage. Consequently our knowledge of *E. solidaginis* behavior comes from flies contained within cages or greenhouses.

Walton, Weis, and Lichter (1990) utilized scan sampling (recording the proportion of flies engaged in different behaviors at a given time) and intensive focal-animal sampling (all behaviors of an individual female were recorded for a 15-minute observation period) to characterize the behavior and oviposition of *E. solidaginis* on clonal fragments under greenhouse conditions. Behaviors were assigned to one of seven main categories, including courtship displaying, grooming, mating, flying, ovipuncturing, resting, or walking. The flies spent the largest proportion of their time resting (35%). However, flies spent a considerable proportion of their time moving among ramets or on ramets by

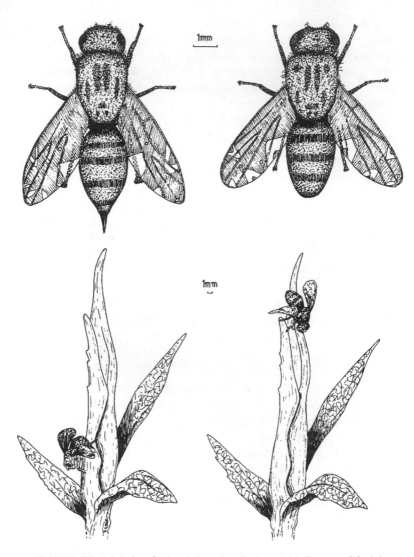

FIGURE 2.5. Adult female (*top left*) and male (*top right*) *Eurosta solidaginis solidaginis* (Tephritidae: Diptera) (*sensu* Ming 1989) showing the characteristic wing pattern of this eastern USA subspecies. Illustrations were drawn from central Pennsylvania specimens. Female in typical oviposition posture (*bottom left*) with her abdomen turned at right angles to the main axis of her body, and a male in typical waiting posture (*bottom right*), both on vigorously growing *S. altissima* buds. (Illustration by J. L. Miller)

either flying (21%) or walking (32%). Approximately 22% of their time was used in courtship displays (10%) and ovipuncturing (12%). The greatest period of activity occurred during the late morning and early afternoon and was probably dependent on temperature (Walton, Weis, and Lichter 1990; see also Uhler 1951).

Upon emergence, males position themselves on the apices of goldenrod. When approached by a female, males display by excitedly rocking their bodies from side to side. If a female is receptive, the male mounts her from the rear and clasps her with his front and middle pairs of legs. The male will curve the tip of his abdomen under the female's ovipositor in order to insert the aedeagus in her ovipositor (Uhler 1951). The length of time required for copulation varies by couple, from as little as 15 minutes to as much as an hour (pers. obs.). Uhler (1951) reported a mean of 41 minutes for twenty-four timed copulations. The intensive focal-animal study found that, on average, mating consumed only 0.4% of the fly's time (Walton, Weis, and Lichter 1990). Although we have observed females mate on the day of their emergence, they are more likely to mate on the day following emergence.

When females were active, they spent the majority of their time walking over the plant in a seemingly undirected search for oviposition sites (Walton, Weis, and Lichter 1990). Females often walk over a potential oviposition site many times before actually ovipuncturing, and such walks over buds are frequently punctuated with walks to nearby leaves and stem.

Female host-evaluation and oviposition behavior is strongly stereotyped (Abrahamson, McCrea, and Anderson 1989). Prior to oviposition, females climb to the apices of goldenrod buds and rapidly rub the apex with their forelegs, occasionally pulling the bud tip to their mouth parts. If not discouraged by the information these activities elicit, the females walk down to the base of the bud and walk around the base searching for a suitable oviposition site. If they find a suitable oviposition site, they orient head-downward on the bud, curl the tip of their abdomen under themselves, and force the sharp tip of their ovipositor into the leaves enveloped around the bud (fig. 2.5). Females may ovipuncture a single bud several times before either depositing an

FIGURE 2.6. Illustration of ovipuncture scars made by *Eurosta solidaginis* (Tephritidae: Diptera) on a bud of *Solidago altissima*. Individual females typically make their ovipunctures in vertical rows. (Illustration by J. L. Miller)

egg or terminating oviposition behavior (fig. 2.6; Uhler 1951; pers. obs.).

Although *E. solidaginis* are relatively strong fliers, ovipositing females typically fly only a meter or so to another goldenrod ramet and repeat their oviposition behavior (Uhler 1951). Adults usually make short, erratic flights in spite of being capable of sustained flight (pers. obs.). Such flight behavior may facilitate rapid oviposition since once a female has located a preferred host clone, short flights will allow her to oviposit on a number of ramets within that clone. However, variations in host clone size as well as heterogeneous distributions of potential host species may reduce a female's oviposition fidelity. Some fields, for example, contain few, very large clones of potential hosts. In other sites, clones of potential hosts frequently break up, leading to heterogeneous stands in which the ramets of clones are interdigitated. We have recently conducted a preliminary experiment to examine the oviposition fidelity of *E. solidaginis* by testing native, mated females in either a complex or a simple environment (Horner, Craig, Itami, and Abrahamson, unpub. data). Twenty ramets each of *S. altissima* and *S. gigantea* were placed in a cage with ramets in either two discrete blocks or mixed randomly. Trials of each environment were replicated with five females that had been reared from one of the two host plants. Although the frequency of oviposition mistakes (i.e., a female reared from *S. altissima* ovipositing on *S. gigantea*) was slightly higher in the complex environment, the difference was not statistically significant (Horner et al., unpub. data).

Females have considerable reproductive potential, as Uhler's (1951) dissections showed that females contain well over two hundred eggs. However, this potential may not be reached as we have rarely seen a single female ovipuncture more than 20 to 25 buds during years of controlled oviposition bouts in the greenhouse (pers. obs.).

Life Cycle and Phenology

Depending on temperature, eggs typically hatch in 5 to 8 days after oviposition. The larva bores several millimeters through the leaves that surround the bud to the base of the bud's meristematic region (Uhler 1951; Anderson et al. 1989). Once in the meristematic tissue, the larva creates a chamber as it feeds on the plant tissue. While many larvae die during or prior to this stage, often due to a hypersensitive, necrotic reaction by the host plant (Anderson et al. 1989; Hess, Abrahamson, and Brown, 1996), surviving larva induce the formation of a gall. The gall becomes apparent approximately three weeks after oviposition and reaches its maximum size about three weeks later (fig. 2.7; Weis and Abrahamson 1985). During early development a fraction of the larvae (5–40%) die from unknown causes (Cane and Kurczewski 1976). The surviving larvae continue to feed on a lining of nutritive cells around the gall's chamber (Abrahamson and Weis 1987; Bross, Weis, and Hanzley

37

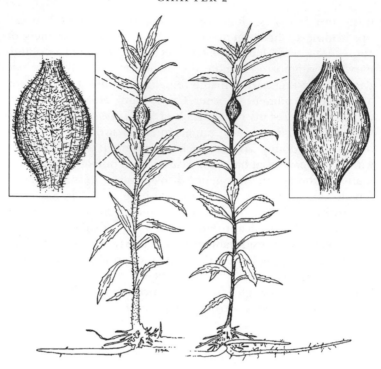

FIGURE 2.7. Ramet of *Solidago altissima* (*left*) with its scabrous stem. Inset illustrates the scabrous nature of *Eurosta solidaginis* stem galls on *S. altissima*. Ramet of *S. gigantea* (*right*) with its glaucous stem. Inset shows the glaucous character of *E. solidaginis* stem galls on *S. gigantea*. (Illustration by J. L. Miller)

1992) until September, when the larvae reach full size (Uhler 1951; Stinner and Abrahamson 1979). The larvae undergo two molts during their growth. The first to second instar molt occurs in approximately mid-July, and the second to third instar molt takes place in about mid-August (Uhler 1951). In approximately mid-September, the third instars begin to excavate an exit tunnel to but not through the gall's epidermis (fig. 2.8). Uhler (1951) examined 632 galls with completed exit tunnels. Of these, 57% were oriented upward (often at approximately 45°), just over 22% were downward, and approximately 20% were horizontal (see also Mecum 1994). After excavating the

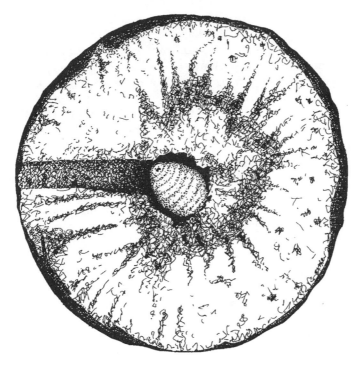

FIGURE 2.8. Cross-section of a ball gall formed by *Eurosta solidaginis* on *Solidago altissima* illustrating the third-instar larva, central chamber, and the exit tunnel that is excavated in early autumn prior to larval diapause. (Illustration by J. L. Miller)

exit tunnel, the larvae return to the gall's central chamber to overwinter.

Overwintering of Larvae

Although *Eurosta*'s gall is of some protection against abrupt changes of temperature, the gall has little insulative value (Uhler 1951; Layne 1993). Gall temperatures are highly variable, changing as much as $20°$ C on some days even during the winter (Layne 1993). While the internal temperature of a gall closely follows ambient temperature overnight, the larval chamber can be substantially warmer than ambient temperatures on

39

sunny winter and spring days (Layne 1993; Abrahamson et al. 1994). The protection the ball gall affords the overwintering larva against direct contact with rain, ice, snow, and sunlight is clearly much more important than its insulative value.

Overwintering *E. solidaginis* larvae are remarkably freezing tolerant due to temperature-dependent accumulation of the cryoprotectants glycerol and sorbitol and static, but elevated levels of trehalose. Northern and southern populations accumulate differential amounts of glycerol with Minnesota populations containing three to four times the levels found in Texas individuals (Baust and Lee 1982; Rojas et al. 1983). Even though the fat body cells of *E. solidaginis* may freeze and cause the coalescence of intracellular lipid droplets upon thawing, larvae frozen under conditions that caused extensive coalescence readily survived to complete their development and emerge as adults (Lee et al. 1993; Mugnano, Lee, and Taylor 1996). Storey and Storey (1986) showed for populations near Ottawa, Ontario, that glycerol production was stimulated by exposure to average daily temperatures of $10-15°C$ when minimum daily temperatures did not exceed $8°C$. Consequently, the highest glycerol synthesis rates in these Ottawa populations occurred during early October. Sorbitol production, on the other hand, required minimum daily temperatures of $3°C$ and was initiated in mid-November for these Canadian populations. Loss of sorbitol began in February, whereas glycerol levels did not begin to decline until mid-March (Storey and Storey 1986). Layne, Lee, and Huang (1990) suggested that freezing of body fluids in *E. solidaginis* is likely limited to the autumn when ice nucleation may occur across the cuticle of larvae. During autumn months, *Eurosta*'s galls have relatively high water contents, but as winter approaches gall moisture levels fall and thus the possibility of transepithelial ice inoculation is markedly reduced (Layne, Lee, and Huang 1990).

Gall Anatomy

Beck's (1947) study of the development of the ball gall detailed the anatomical changes in the various regions of the *Solidago* stem during gall formation. He showed that a cross section of normal stem typically reveals a central pith region, a circle of 18–30 vascular bundles, a pericycle (i.e., compact fibrous cells capping each bundle), a cortex of 6–9 cell layers, and an epidermis covered by a cuticle. However in the gall of *E. solidaginis,* the pith region (the region of cells eaten by the developing gallmaker) is several times larger than that of normal stem due to the numbers of cells, even though pith cells are smaller and richer in cytoplasm than those found in normal stem (fig. 2.8; Beck 1947; Bross et al. 1992). Over the course of gall formation, the numbers of lipidlike droplets and microbodies with crystalline inclusions increase within pith cells (Bross et al. 1992).

Eurosta galls have a peculiar meristematic zone that develops from the cambial region at the edge of each vascular bundle. These meristematic strands produce numbers of parenchyma cells that enlarge to approximately the size of pith cells. However, as the meristematic strands of aging galls diminish, they are typically replaced by resin ducts (Beck 1947). Hess (1993) has suggested that some mortality of second and third instar larvae may be due to resin accumulations within the larva's chamber.

The xylem, cambium, and phloem of galls each exhibit changes from their development in normal stems (Beck 1947). The number and size of xylem vessels, for instance, is increased in gall tissue and the cambium of the gall contains 4–6 cell layers compared to the two layers of normal stem. The gall's phloem cells are both more numerous and enlarged.

Although the pericycle of galls is virtually unchanged from that of normal stem, the cells that compose the cortex are more numerous, larger, and thicker-walled in *E. solidaginis* galls (Beck 1947; Bross et al. 1992). Subsequently, the cortex of these galls is much thicker than in normal stem (Beck 1947). Weis, Wolfe, and Gorman (1989) showed that the thickness of the cortex was the single greatest determinant of variation in gall diameter on *S. altissima* ramets. At the phenotypic level, the number of cortical cells and the amount of vascularization control the thickness of the cortex. However, among genotypes of *S. altissima,* the primary cause of variation in cortical thickness was differences in vascularization (Weis, Wolfe, and Gorman 1989).

There is little difference in the epidermal layer of *Eurosta* galls

41

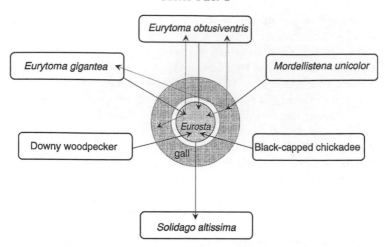

FIGURE 2.9. Food web associated with *Eurosta* galls: *Eurytoma obtusiventris* directly attacks the gallmaker; *E. gigantea* attacks the gallmaker, including those previously parasitized by *E. obtusiventris,* then eats gall tissue; *Mordellistena unicolor* is an inquiline that eats gall tissue, but will also kill any of the other insects it encounters in a gall; the two bird species peck open overwintering galls and extract the full-grown *Eurosta* larvae, but avoid parasite larvae.

compared to that on normal stem. The cells comprising the epidermis of galls are only slightly enlarged over those of normal stem (Beck 1947).

2.5 *EUROSTA SOLIDAGINIS'* PARASITES AND PREDATORS

Gallmaking insects support complex associations of parasitoids (Askew 1961; Washburn and Cornell 1979; Price 1980; Weis 1982a; Hawkins and Gagne 1989; Cornell 1990; Tscharntke 1992; Plakadis and Weis 1994), and *Eurosta* is no exception. Three insect and two avian carnivores (fig. 2.9) are the predominant source of mortality between oviposition in the spring and eclosion of the adult a year later (Uhler 1951; Cane and Kurczewski 1976). In central Pennsylvania populations, mortality rates of established larvae due to natural enemy attack have been as high as 96.1%, and have seldom dropped below 30% (Abrahamson et al. 1989; Weis, Abrahamson, and Andersen 1992). As we discuss the biology of *Eurosta*'s natural enemies in

THE SYSTEM

FIGURE 2.10. Adult female of *Eurytoma obtusiventris* (Eurytomidae: Hymenoptera) on the bud of *Solidago altissima*. *Eurytoma obtusiventris* is thelytokous (parthenogenic females, no males), and oviposits into the host egg or early larva while the host is still in the goldenrod bud (Uhler 1951; Weis and Abrahamson 1985). The females use host-plant cues during foraging, and preferentially search *S. altissima* buds (Brown et al. 1995). (Illustration by J. L. Miller)

the following sections, we acknowledge our debt to the entomologists and ecologists who have studied these species over the years (Harrington 1895; Ping 1915; Milne 1940; Uhler 1951, 1961; Judd 1953; Miller 1959; Cane and Kurczewski 1976; Schlichter 1978; Confer and Paicos 1985; Walton 1988; Cappuccino 1991, 1992; Sumerford 1991).

Insect Parasitoids and Inquilines

Two insect parasitoids, *Eurytoma obtusiventris* (figs. 2.10 and 2.11) and *E. gigantea* (figs. 2.11 and 2.12) (Hymenoptera; Eurytomidae), use *Eurosta* as their exclusive host species. Although

43

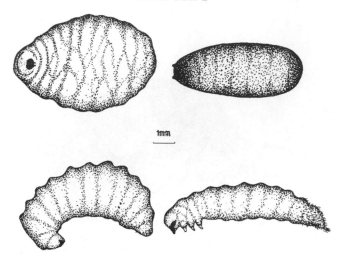

FIGURE 2.11. Inhabitants of autumn and overwintering galls of *Eurosta soli-daginis,* including a diapausing third-instar larva of *E. solidaginis* (*top left*), a larva of *E. solidaginis* parasitized by *Eurytoma obtusiventris* (*top right;* the larva of the parasite is inside the premature puparium of the host), larva of the external parasite *Eurytoma gigantea* (*bottom left;* this parasitoid consumes its host by the end of August), and the larva of the inquiline and predator, *Mordellistena unicolor* (*bottom right;* Mordellidae: Coleoptera). *Mordellistena unicolor* larvae burrow through the cortex and meristematic region of the gall during the growing season, leaving a fine, light-colored frass. In the fall and winter, the beetle larva moves from the gall's periphery into its central chamber. During the fall, winter, or spring the beetle encounters the *Eurosta* larvae and usually kills it. (Illustration by J. L. Miller)

these two are from the same genus, their natural histories are quite different. *Eurytoma obtusiventris* is thelytokous (partheno-genic females, no males) and oviposits into the host egg or early larva while the host is still in the goldenrod bud (Uhler 1951; Weis and Abrahamson 1985). The females use host plant cues during foraging, and preferentially search *S. altissima* buds (Brown et al.1995). The minute parasitoid larva that hatches from the single egg remains quiescent in the host until the early autumn, when it induces premature pupation in *Eurosta* and then begins its own development (fig. 2.11). Galls containing hosts attacked by *E. obtusiventris* often accumulate a resinous substance within the central chamber during autumn. The ma-

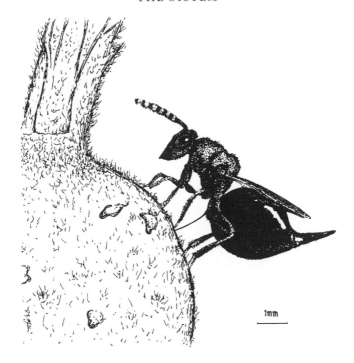

FIGURE 2.12. Adult female of *Eurytoma gigantea* (Eurytomidae: Hymenoptera) ovipositing (note the extended ovipositor) into a ball gall of *Eurosta solidaginis* on *Solidago altissima*. *Eurytoma gigantea* is aherynotokous (females from fertilized eggs, males from unfertilized eggs) and oviposits into the central chamber of fully grown galls. The success of an attempted attack depends on the thickness of the gall wall and on the parasitoid's ovipositor length (Weis and Abrahamson 1985). (Illustration by J. L. Miller)

ture parasitoid larva overwinters in the gall and then emerges the following spring before new galls start to form.

By contrast, *E. gigantea* is aherynotokous (females from fertilized eggs, males from unfertilized eggs) and oviposits into the central chamber of fully grown galls (figs. 2.11 and 2.12; Weis and Abrahamson 1985). The success of an attempted attack depends on the thickness of the gall wall and on the parasitoid's ovipositor length (Weis, Abrahamson, and McCrea 1985). Its attack season extends from the end of June into August. However, the hardening of the gall tissue at the end of July may curtail

45

further attack (Bross et al. 1992). Upon hatching, the parasitoid larva quickly consumes the host larva, and then continues to feed on the gall tissue, consuming both the nutritive and the meristematic regions (Uhler 1951; Weis, Wolfe, and Gorman 1989). It may also use host-plant cues in its search, but oviposition probes occur only when it walks across a gall surface. It will probe spindle galls induced on goldenrod stems by the moth *Gnorimoschema gallaesolidaginis,* but these galls are quickly rejected (Weis 1993a). There is a bimodal emergence of this species. Most individuals overwinter in the gall as larvae, then pupate and emerge the following spring after gall development begins. But, approximately one-tenth of the larvae complete development in the late summer (Uhler 1951; Weis, Abrahamson, and McCrea 1985) and quit the gall. The fate of these early emergers is unknown, as they do not attack the standing crop of gallmakers at that time. Furthermore, there are no records for alternate hosts and we have never reared them from other tephritid gallmakers. If they successfully overwinter as adults, they do not attack *Eurosta* any earlier the following season than those that overwinter as larvae.

The beetle *Mordellistena unicolor* (Coleoptera; Mordellidae) is an inquiline in the *Eurosta* gall (fig. 2.11). Females do not have an extended ovipositor. They embed the egg in shallow holes made in the gall epidermis (Ping 1915). Foraging by this beetle has not been studied, and it is unknown what search cues it uses. The larva emerges from the egg onto the gall surface where it excavates a tunnel into the gall cortex. Subsequent gall growth seals the entrance hole. The larva burrows through the cortex and meristematic region of the gall for the rest of the growing season, leaving a fine, light-colored frass. In the fall andwinter, the larva moves from the gall's periphery into its central chamber. During the fall, winter, or spring, it encounters the *Eurosta* larvae and usually kills it. *Mordellistena* pupates in the spring and emerges at about the same time as gall appearance. This beetle has it own natural enemies, *Schizoprymnus* sp., a brachonid parasitoid wasp, and *Zaleptopygus* sp. of the Ichneumonidae (Uhler 1951).

Although *Mordellistena* usually kills *Eurosta,* this is not always the case. During the summer months healthy inquiline larvae can sometimes be found in the periphery of galls in which *Eurosta* has succumbed to early larval death (J. P. Lichter, unpub. data), and so seems to prosper in galls where the gallmaker is already dead. We have occasionally found mature *Mordellistena* pupae in galls along with healthy *Eurosta* pupae in the late spring, and presumably both would have successfully emerged as adults. Hence we have concluded that unlike the parasitoids, *Mordellistena* does not depend on *Eurosta* for nourishment. However, to avoid torturous prose in subsequent chapters, we will refer to *Mordellistena* as a parasitoid.

These three enemy species enter the gall chamber in the sequence *E. obtusiventris, E. gigantea,* and finally *Mordellistena.* The latter two species are able to attack galls that have been previously attacked by earlier species (fig. 2.9). Consequently, when galls are opened for study, it is possible to determine only which species was the last to attack. Throughout our work we have calculated percent parasitism inflicted by these species as the percentage of galls in which they are found. Multiple attacks make this measure a biased estimate of the actual mortality inflicted. A *Eurosta* larva that made a gall containing *Mordellistena,* the last in the sequence, may have been already killed by one of the *Eurytoma* species. We calculated the extent of this bias (fig. 2.13) for a collection of over two thousand galls from twenty fields in 1984 (Abrahamson et al. 1989). Although 19% of the gall contained *Mordellistena,* our calculations suggest that only 14.3% of the gallmakers were actually killed by this beetle, while the remaining 4.7% were already dead from previous wasp attacks. In turn, attack *by E. obtusiventris,* first in the sequence, is underestimated in the percent parasitism data (fig. 2.13). Although these biases exist in the data, their magnitudes are small, and corrections unreliable due to the sampling error for percent parasitism. For this reason, we have proceeded on the assumption that percent parasitism calculated from gall contents is as good an estimate of mortality inflicted as can be had.

Several other insects are incidental enemies of *Eurosta,* and most of these are stem borers (Ping 1915; Uhler 1951). The most unusual case we have witnessed was that of a Megachild bee that excavated a brood chamber in a larger than average gall. After gallmaker eclosion, a variety of squatters take advantage of

Mortality Inflicted by Parasitoids and Inquilines: Correction for Multiparasitism

Observed parasitism rates

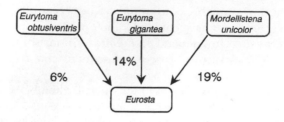

Actual parasitism rates

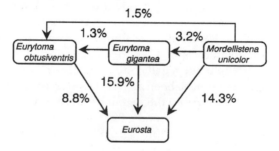

FIGURE 2.13. Calculation of the percent mortality inflicted by the three insect enemies of *Eurosta*. Observed parasitism rate (percentage of galls containing a parasite species) is a biased estimate of the mortality inflicted because of multiple attacks. Mortality caused by early-attacking species is underestimated while that caused by late-attacking species is overestimated.

the abandoned structure, from thrips to predaceous beetle larvae (Ping 1915; Judd 1953).

Bird Attack on Eurosta

Even if *Eurosta* manages to escape attack by parasitoids in the summer and early fall, it is still vulnerable to bird predation (fig. 2.14). Two common insectivorous bird species include *Eurosta* in their winter diets, the downy woodpecker (*Picoides pu-*

FIGURE 2.14. Black-capped chickadee (left, *Parus atricapillus*) makes a "messy, large, irregular hole by grabbing bits of the gall with [its] bill and tugging them free," and downy woodpecker (right, *Picoides pubescens*) drills a "tidy, narrow, conical hole" (Confer and Paicos 1985, p. 58). Downy woodpeckers tap on *Eurosta* galls to locate the epidermal cap of the gallmaking larva's emergence tunnel. *Eurosta solidaginis* larvae in large galls are more frequently eaten by these avian predators. (Illustration by J. L. Miller)

bescens) and the black-capped chickadee (*Parus atricapillus*). Woodpeckers were the predominant predator in the Pennsylvania fields that we studied most intensively, with predation rates sometimes exceeding 60% (Abrahamson et al. 1989; Weis, Abrahamson, and Andersen 1992; Mecum 1994). However, in other areas such as southern Ontario, chickadees can attack up to 50% of overwintering galls (Schlichter 1978). Both species open the gall with their beaks and extract the larva within, but they make distinctive holes (fig. 2.14). The woodpecker drills a "tidy, narrow, conical hole" while the chickadees make a "messy, large, irregular hole by grabbing bits of the gall with their bill and tugging them free" (Confer and Paicos 1985, p. 58).

A number of researchers have suggested that downy woodpeckers tap on *Eurosta* galls to locate the epidermal cap of the gallmaking larva's emergence tunnel (Moeller and Thogerson 1978; Confer and Paicos 1985; Confer, Hibbard, and Ebbets 1986; Troll 1990; Mecum and Abrahamson, unpub. data). Downy woodpeckers typically enlarge the emergence tunnel, which the gallmaker excavates just prior to entering diapause in the fall (Moeller and Thogerson 1978). Confer and Paicos (1985) found the larva had been extracted this way in 53% of the more than five hundred attacked galls they examined, and yet the entrance to the tunnel occupies about 1% of the gall's surface area. A woodpecker that selects galls with exit tunnels will not only spend less energy opening the gall, but it will be rewarded with a gallmaker larva. Gallmaker larvae have about ten times the energetic value of the enemy larvae (Stinner and Abrahamson 1979). Schlichter (1978) suggested that birds may be able to discriminate galls containing enemies from those with healthy gallmakers by the vibrations the insects make. However, Confer and Paicos (1985) examined not only the extraction holes, but the peck marks that woodpeckers make when testing a gall, and concluded that they are more likely to abandon a gall when there is no exit tunnel than when there is, and this was a more likely basis for discrimination. Our recent field experiments support this conclusion that downy woodpeckers use *Eurosta* emergence tunnels as a cue to the presence of a *Eurosta* larva (Mecum 1994; Mecum and Abrahamson, unpub. data). As a final note, Cane and Kurczewski (1976) found that in many cases where birds had attacked galls parasitized by *E. obtusiventris,* the parasite is not eaten even though it is in easy reach. They suggested that this insect enemy is distasteful due to the resin that accumulates in the central chamber of their galls.

Survivorship and Gall Size

The probability of gallmaker survival depends on gall size since larvae in small galls are more vulnerable to attack by the parasitoid *E. gigantea*. Furthermore, larvae in large galls are more frequently eaten by avian predators. The remaining common natural enemies, *Mordellistena* and *E. obtusiventris*, are not affected by gall size. The consequence is that *Eurosta* larvae in intermediate-sized galls have the highest probability of survival to adulthood (Weis and Abrahamson 1985; Abrahamson et al. 1989; Weis, Abrahamson, and Andersen 1992). We will show the significance of this differential attack on the gallmaker by its natural enemies in later chapters.

2.6 THE *SOLIDAGO-EUROSTA*-NATURAL ENEMY SYSTEM: A TOOL FOR UNDERSTANDING THE EVOLUTION OF SPECIES INTERACTIONS

The *Solidago, Eurosta,* and natural-enemy system has been amenable to a multitude of the studies aimed at understanding the ecology and evolution of a tri-trophic-level interaction. The results of the comprehensive natural history work of Lowell D. Uhler (1951) and others as well as the work in our laboratories conducted over more than two decades offers a strong empirical base for understanding the interactions among the host plant, gallmaker, and natural enemies. This system truly deserves the designation of a "model system." In the subsequent chapters, we will take you through experimental and observational studies and we hope we'll leave you with an appreciation of the scope of factors that govern the evolution of this plant-animal interaction.

CHAPTER THREE

Eurosta's Impact on Goldenrod

Gallmakers are unique among herbivores in that they alter the development of their host's tissue to form a tumorlike growth from which the gallmaker gains both nutrition and shelter (Abrahamson and Weis 1987). Consequently, the impacts of gallmakers can be considerably more severe than those of free-feeding herbivores when measured on a per herbivore basis (Stinner and Abrahamson 1979). Gallmakers not only "rob" the host plant of the consumed tissue while procuring their nutritional requirements but they also cause their host to alter tissues that would otherwise serve more productive functions in plant growth and reproduction (Abrahamson and McCrea 1986a). As parasites, gallmakers can create an appreciable sink for host-plant resources as they stimulate the development of their gall and acquire plant nutrients needed for their own growth and maturation. Thus, gallmaker herbivory has the potential to cause a number of changes in and impacts to their host plant as gallmakers redirect plant-tissue development. As a result, gallmakers often negatively impact on characters that influence host-plant fitness.

Several well-known examples illustrate how parasitization of agricultural crops by gallmakers can have severe economic consequences. The grape phylloxera, for example, destroyed nearly one-third of the French vineyards during the later half of the nineteenth century. This gallmaker destroyed as many as 88% of the vines in some localities (Riley and Howard 1891). On the Great Plains, the Hessian fly severely reduced wheat production until the introduction of resistant host-plant varieties (McColloch 1923). However, the frequency and degree to which gallmakers affect their hosts can vary markedly depending on herbivore infestation levels, host-plant organs attacked, and host responses (Abrahamson and Weis 1987). For example, fitness losses caused by galls developing on vegetative host organs may be slight under endemic infestation levels (Abrahamson and Weis 1987); however, when reproductive organs are affected or when gallmaker densities become extreme, the potential is high for appreciable losses of host fitness (Harris 1980; Collins, Crawley, and McGavin 1983; Birch, Brewer, and Rohfritsch 1992).

A gallmaker's impacts on its host occur at several levels. At the cellular and tissue level, a gallmaker redirects the differentiation of the host's tissues (Dreger-Jauffret and Shorthouse 1992). Indeed, the specificity of a gallmaker's stimulus and host's reaction allows positive identification of many gallmakers by the morphology of their gall (Felt 1940). At the ramet level, a gallmaker can alter physiological processes, ramet growth rate, and the allocation of resources within the ramet to various vegetative and reproductive activities (Abrahamson and McCrea 1986a,b). At the genet level, gallmaking insects can modify the architecture of their host (Craig, Price, and Itami 1986) and can potentially reduce their host genotype's fecundity and fitness. The following paragraphs are intended to summarize what is known about *Eurosta's* impacts at each of these levels.

3.1 CELL AND TISSUE-LEVEL IMPACTS

Küster (1930; see also Meyer and Maresquelle 1983; Ananthakrishnan 1984; Meyer 1987; and Dreger-Jauffret and Shorthouse 1992) described two principal types of galls: kataplasmas and prosoplasmas. The tissues of relatively simple kataplasmic galls are scarcely differentiated (calluslike galls), and as a result they lack a consistent external morphology. Kataplasms of the same type exhibit considerable variation in both shape and size (Brues 1946). Among arthropods, only a few species of Hemiptera and Homoptera induce such galls; the cause is typically an agent other than an insect (Dreger-Jauffret and Shorthouse 1992). The more complex prosoplasmic galls, such as Eurosta's, have characteristic external morphologies (depending on host-plant infested, species, sex, or stage of gall insect, and the organ attacked) with gall tissues differentiated into well-defined zones of nutritive and sclerenchyma cells (Dreger-Jauffret and Shorthouse 1992).

Stages of Gall Development

Prosoplasmic galls go through four basic stages of development, including initiation, growth and differentiation, maturation, and dehiscence (Rohfritsch 1992). The initiation stage is called *hypoplasy* (the inhibition of normal cellular differentiation), which involves the isolation of a group of cells (Rohfritsch and Shorthouse 1982; Elzen 1983). The physiologically modified isolated cells are the source of nearly all the cells that eventually form the gall (Rohfritsch 1992). Typically, initiation is associated with the activities of the first-instar larva. However, there is at least one case (e.g., sawflies) in which the ovipositional fluid of the adult female provides the initial stimuli (Hovanitz 1959). Consequently, initiation is a subtle process usually composed of specific larval behavior that physically stimulates host tissues and larval secretions that induce highly individual tissue-recognition events.

Rohfritsch (1992) described gall growth and development as both hyperplasy (or cell division) and hypertrophy (the overgrowth or enlargement of the cells stimulated by the gallmaker). She further pointed out that gall development and growth rates are frequently dependent on gallmaker feeding. As gallmakers either feed on solids or suck liquids from plant cells, they pour salivary fluids on the damaged host cells. Host-plant cells characteristically react to this stimulation with cell proliferation and differentiation. Because of this reaction, the initiating gall forms a nutritive zone on the inner surface of the larval chamber (Rohfritsch 1992).

The maturation stage involves the formation of a sheath of sclerenchyma about the nutritive layers, effectively creating two regions, the so-called inner nutritive zone and the outer cortex zone. Rohfritsch (1992) suggested that the gallmaker is in primary control of the inner region while the outer cortex region is principally influenced by the host plant; however, evidence described in chapter 6 suggests that the thickness of the outer layers of *Eurosta* gall are influenced by insect genotype. Gall maturation typically occurs during the gallmaker's last larval instar. Consequently, it takes place when the gallmaker consumes the most food per unit time and potentially is having its greatest impact on its host.

The fourth stage involves gall dehiscence (opening) and, in many galls, occurs during the latter portion of the maturation stage. Rohfritsch (1992) noted that dehiscence often produces major changes in the gall's physiology and its chemical composition. Sap and water no longer move into the gall and the gall no longer functions as a nutrient sink. While gallmakers vary in whether or not they pupate in the gall, most of the dehiscence-stage changes facilitate gallmaker emergence. *Eurosta* galls do not have a dehiscent stage per se, but at season's end their cells die with the rest of the plant stem. As autumn ends, the larva chews an exit tunnel to, but not through, the epidermal layer just before diapause (i.e., during September and October; Uhler 1951). Thus the adult, after a brief spring pupation, is able to emerge from the gall by popping the remaining epidermal cap even though it lacks chewing mouthparts.

Nature of the Stimuli

The causes of the developmental changes that occur during gall development are not fully understood (Raman 1993). However, clearly the stimuli involve more than the simple mechanical irritation of feeding. In most cases, the stimuli appear to be related to secretions from the gallmaker since gall development slows or ceases and a rapid regression of gall characteristics occurs if the gallmaker dies (Beck 1947; Bronner 1977; Rohfritsch 1977, 1992). Carol Mapes (pers. comm.) examined the role of the living larva in gall growth for *E. solidaginis* by either killing the larva with the systematic insecticide Oxamyl or physically removing it from rapidly growing galls. In both cases, gall growth did not immediately stop, but instead, gall growth continued at a reduced rate even without a larva. However, in both treatments final gall diameters were appreciably less than for galls containing healthy *Eurosta*.

It seems likely that both the source of the stimuli as well as the specific chemicals involved would vary among gallmakers. Salivary secretions and excreta have been implicated as possible stimuli sources for some gallmakers. However, it seems that amino acids and plant-growth regulators such as auxins, auxin precursors, or cytokinins are instrumental in promoting gall growth and development (McCalla, Genthe, and Hovanitz 1962; Miles 1968; Heady, Lambert, and Covell 1982; Elzen 1983; Hori 1992; Raman 1993).

Hori (1992) reported that the galls of plant-sucking insects and mites result from the injection of phytotoxic saliva during feeding. The injected saliva or its specific components quickly diffuse from the feeding site to the surrounding tissues and create hormonal and metabolic imbalances. In some cases, the diffusible components of the saliva have come from the host plant by way of the insect's hemolymph (Miles 1968). Indeed, the salivary glands of hemipterans and homopterans are known to have excretory functions. Saliva produced by these insects contains high levels of amino acids (particularly free amino acids that are nonessential to insect metabolism) as well as other substances (Miles 1968; Hori 1992). While amino acids apparently do not perform a principal role in gall initiation, they do affect host plant growth through their effect of increasing the permeability of host cells, rates of respiration, transpiration rates, and amount of protoplasmic streaming (Hori 1992). The consequence is that the injected amino acids could "condition" host tissues to make them more sensitive to the gallmaker (Hori 1992). Beck (1947) and Uhler, Crispen, and McCormick (1971) suggested that the secretions of *Eurosta* create a tissue state within the affected regions that is similar to that found in the ramet's apical meristem.

While free amino acids may be involved in *Eurosta* gall development (Uhler, Crispen, and McCormick 1971), research using explants of *E. solidaginis* galls on nutrient medium has suggested that the larvae act as auxin sources (indole acetic acid, IAA) during gall development (Mapes and Davies 1992; C. Mapes, pers. comm.). Carol Mapes and her associates have found that the presence of *Eurosta* larvae enhanced the growth rates of central gall explants on zero-hormone media to the same extent as did the addition of auxin/cytokinins to cultures lacking *Eurosta*. Central gall explants grown on zero-hormone media in the presence of *Eurosta* larvae exhibited enhanced root initiation, further suggesting that the gallmaker was providing auxin (Mapes and Davies 1992). The levels of auxin present in *E. solidaginis* galls are sufficient to affect other parts of the ramets. For example, the auxins from the *Eurosta* gall inhibit the release of lateral buds (C. Mapes, pers. comm.), unlike many other insect stem galls that cause release lateral buds from apical dominance (e.g., elliptical goldenrod gall of the moth *Gnorimoschema gallaesolidaginis;* Hartnett and Abrahamson 1979). Yet the auxin level in *Eurosta* larvae and their galls changes during gall development. Levels are high in growing larvae and developing galls but auxin levels fall over time as gall growth slows. During winter months, the diapausing *Eurosta* larvae contain low amounts of auxin. Mapes's data also suggest that the larvae can synthesize plant cytokinins; however, the levels of these hormones in *Eurosta* larvae are low compared to the amounts measured in galls.

A number of insects besides *Eurosta* secrete auxins such as IAA or other plant growth-promoting substances. Hori (1992) noted that auxins occur in a number of insects, including aphids and lygus bugs. However, the source of auxins varies with the herbivore. In some insects, auxins originating from the host plant are believed to be accumulated and redirected back to the host while in other insects IAA is produced along with the saliva. Regardless of the source of auxins, Miles (1968) and others (e.g., Hori 1992) have suggested that insect-related IAA functions as an "inducer" and is thus directly involved in the formation of galls. Hori (1992) pointed out, however, that some observations are inconsistent with an IAA-inducer theory, including the observation that certain homopteran galls contain much higher concentrations of IAA than does the saliva of the gallmaker.

Other researchers (reviewed by Hori 1992) have suggested that gallmakers may increase IAA activity in the affected host plant tissues with IAA synergists that block the breakdown of auxins by IAA oxidase and/or that activates the host plant's own auxins. Synergists, unlike IAA or auxin inducers, would operate independently of the concentration of IAA in affected host tissues since they would promote whatever IAA activity was taking place, functioning as "maturators" of galls (Hori 1992).

Polyphenols, including cytokinins and gibberellin, may act as synergists to IAA-induced growth of plants, and both groups of compounds, including phenolics and phenolases (polyphenol oxidases and peroxidases), are frequent constituents of gallmakers' saliva (van Staden, Davey, and Noel 1977; van Staden and Davey 1978; Elzen 1983; Hori 1992). However, the actions of individual phenolic compounds vary considerably. Monophenolic compounds, for example, are known to activate IAA oxidase and thus inhibit IAA activity, whereas diphenols often constrain IAA oxidase and consequently contribute to IAA activity.

The phenol-polyphenol oxidase pathway also can produce quinones that may be involved in host-plant defense against gallmakers through plant-tissue necrosis (the death of plant cells injured or near the gallmaker larva). Necrotic reactions typically kill the gallmaker very early in the gall initiation process and consequently prevent gall formation. The occurrence of hypersensitive, necrotic reactions of host plants to gallmakers is widespread and may represent an efficient form of plant defense against gallmakers (Anderson et al. 1989; Fernandes 1990). However, some gallmakers may redirect these quinones. Miles (1968) suggested that polyphenol oxidase in gallmaker saliva could alter harmful quinones into nontoxic compounds that facilitate gall induction. Whatever the case, it is likely that a delicate balance exists in the interaction between the phenol-polyphenol oxidase systems of gallmakers and their host plants (Hori 1992). The balance itself may govern whether the attack produces necrosis of plant tissues or a gall.

We first observed hypersensitive, necrotic responses in *S. altissima* during a field study that examined resistance and susceptibility to *E. solidaginis* (Anderson et al. 1989). When oviposited by *E. solidaginis,* some resistant genotypes of *S. altissima* frequently exhibited necrosis in host-plant bud tissues. We have also observed a similar reaction in susceptible genotypes that were especially heavily ovipunctured during separate greenhouse studies using caged *Eurosta* females (How, Abrahamson, and Craig 1993; Hess, Abrahamson, and Brown, 1996). In *S. altissima,* the necrotic region is a darkly pigmented cell mass that forms around the first-instar *Eurosta* larvae. In all cases, necrosis was associated with larval mortality. Necrotic reactions to herbivory are known in a variety of host plants (e.g., Fernandes 1990; Westphal et al. 1992; Ecale and Backus 1995; Bentur and Kaslode

1996). Because phenolics have been implicated in such necrotic reactions in other host-plant species (Kosuge 1969; Rohfritsch 1981; Westphal, Bronner, and LeRet 1981; Bazzalo et al. 1985), we hypothesized that phenolic defenses were affiliated with the necrotic reaction of *S. altissima* to *E. solidaginis*. Phenolic compounds, well known as plant defenses, can function as toxic agents, growth inhibitors, or digestibility reducers (Rhoades and Cates 1976; Zucker 1982; Bryant, Chapin, and Klein 1983).

We explored the relationships of host-tissue necrosis and phenolic levels first by measuring the total phenolics in galled and ungalled ramets of four host genotypes that varied in degree of resistance, and second by determining the distribution of the increases of phenolics within rapidly growing galls and their adjacent stems in three susceptible host genotypes (Abrahamson et al. 1991). Weekly tissue samples collected from two resistant and two susceptible host genotypes were analyzed for total phenolics by the Folin-Ciocalteau procedure. These samples indicated that phenolic content increased over time in all host genotypes. Unexpectedly though, we found that phenol levels in unattacked ramets were significantly higher in susceptible genotypes than in resistant host genotypes. Furthermore, we were unable to detect any increase in phenolics that was associated with necrosis (fig. 3.1). Instead, we found significantly higher phenolic concentrations in ramets that had initiated normal gall development.

Our examination of rapidly growing galls in three additional susceptible host genotypes clearly demonstrated that the increase in phenolics was localized in the gall tissue. No increases could be detected in stem tissues even within 1 cm of the gall (Abrahamson et al. 1991). In striking contrast to our hypothesis, there was a strong positive relationship between total phenolic levels and actively developing *Eurosta* gall (fig. 3.2). Newly initiated galls had about twice the concentration of phenolics as ungalled tissue and rapidly growing galls had phenolic levels of three to five times that of normal stem tissues of similar developmental stage.

Such elevated levels of phenolics in gall tissue have been reported in several other galls (Purohit, Ramawat, and Arya 1979). Hartley's (1992) survey of phenolic levels in galled and ungalled

Phenols and Necrosis

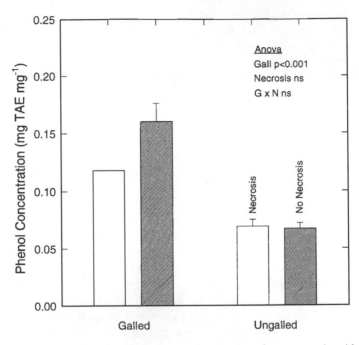

FIGURE 3.1. Total phenolic concentrations expressed as mg tannic acid equivalents per mg dry mass (post-extraction, $\pm$ standard error) in the water fraction of meristem extracts from ramets with and without visible signs of gall initiation (galled vs. ungalled) and visible necrosis surrounding the *Eurosta* larvae (necrosis [open bars] vs. no necrosis [cross-hatched bars]). Data are for 3-week post-oviposition only. Two-way analysis of variance results are on natural-log transformed data. (Redrawn from Abrahamson et al. 1991, "The role of phenolics in goldenrod ball gall resistance and formation," *Biochemical Systematics and Ecology* 19:615–622, with kind permission from Elsevier Science Ltd., The Boulevard, Langford Lane, Kidlington OX5 1GB, UK)

tissues for a wide range of gallmakers found that gall tissues contained higher levels of phenolics, or were not significantly different from ungalled tissues, in an overwhelming majority of cases. Furthermore, Hartley (1992) showed that these elevated levels of phenolics occur within the nutritive-tissue layer fed upon by the developing gallmaker.

Phenols in Rapidly Growing Galls

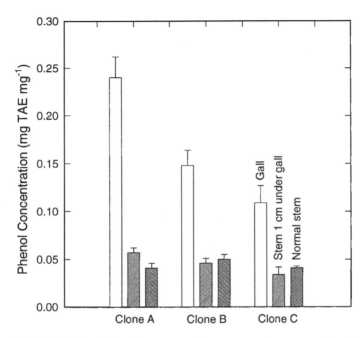

FIGURE 3.2. Total phenolic concentrations expressed as mg tannic acid equivalents per mg dry mass (post-extraction, ± standard error) in the water fractions from rapidly growing gall tissue (open bars), normal stem tissue 1 cm below the gall (cross-hatched rising right bars), and normal stem tissue (cross-hatched rising left bars) on an equivalent position on unpunctured, ungalled ramets. Data are from three susceptible clones different from those in figure 3.1. (Redrawn from Abrahamson et al. 1991, "The role of phenolics in goldenrod ball gall resistance and formation," *Biochemical Systematics and Ecology* 19:615–622, with kind permission from Elsevier Science Ltd., The Boulevard, Langford Lane, Kidlington OX5 1GB, UK)

We have offered several possible explanations of the changes in phenolics associated with gall development (Abrahamson et al. 1991). One explanation is that phenolic compounds are involved in the protection of the gallmaker from its parasitoid natural enemies or from fungi (Cornell 1983). Gallmakers might manipulate the development of their host plant to their own benefit by elevating phenolics in the outer regions of their

61

galls. However, this runs contrary to our findings in *Eurosta* galls since phenolic levels were elevated in nutritive tissues (Abrahamson et al. 1991). More likely, phenolic compounds may be associated with the normal initiation, growth and differentiation, and maturation stages of the *Eurosta* gall. It is possible that the gallmaker's stimulation increases phenolic levels while altering the delicate phenol-polyphenol oxidase balance in favor of gall development rather than necrosis. In our study, the host plants with necrotic reactions that killed gallmaker larvae did not have elevated levels of phenolics. The higher endogenous levels of phenolics that we found in susceptible *S. altissima* genotypes suggest a positive role for phenolics in *Eurosta*'s gall formation. It may be that the gallmaker's ability to initiate gall formation is at least partially based on the endogenous levels of phenolics present in its host, especially if some of the measured phenolics function as synergists for gall development. A host with higher endogenous levels of phenolics may require less stimulation to produce a gall than would a resistant genotype with much lower phenolic levels.

Unfortunately, the precise mechanisms of gall formation are not well understood; however, we do know that gall development entails the gallmaker stimulating cell division and elongation (Abrahamson and Weis 1987). Young galls in particular have high levels of enzymatic activity and protein synthesis making their metabolism exceptionally rapid (Bronner 1977; Rohfritsch and Shorthouse 1982; Hartley 1992). Phenolic compounds may play a role in this enhanced metabolism in that phenols have been implicated in the enhancement of plant-tissue response to hormones (Ray, Guruprasad, and Laloraya 1980; Ray 1986; Sharma, Sharma, and Rai 1986). Their net effect seems to be to increase the rate of host growth. This effect suggests the possibility that gallmakers, by stimulating an increase in phenolic compounds, may make host-plant tissues more sensitive to their elaboration of growth-stimulating, auxinlike agents. As noted above, Mapes and Davies (1992) have suggested that *Eurosta* larvae act as a source of auxin to stimulate tissue development.

Most of the work on the mechanisms of gall initiation, growth and differentiation, and maturation has focused on the effects of larval secretions on the phenotypic expression of the host tissue's genetic code. Cornell (1983), however, proposed that a virus or viroid could transfer DNA to the host plant that was encoded with the information needed to produce a gall. He suggested that the transposed DNA might act like regulatory genes that supersede the host plant's genetic code and consequently promote the biochemical sequences necessary for gall formation. Such transposed DNA elements would be dependent on alterations of host tissues by the gallmaker's secretions since gall development typically ceases when the gallmaker dies. Such dependency could involve insect secretions that render the host-plant tissues more reactive or more susceptible to viral attack. Although no one has demonstrated such a mutualistic relationship between a virus and a eukaryotic organism in a gallmaker/host-plant system, the concept is known in other organisms (Edson et al. 1981).

To explore the possibility of virus or viroid involvement in the *Eurosta-Solidago* interaction, we examined the mRNA of both galled and ungalled tissues for differences in protein production (Carango et al. 1988). Although we found no differences in proteins at one week after oviposition, we did record a hyperinduction of a 58,000 dalton protein in galled tissue during the period of most active gall growth and differentiation. The appearance of this hyperinduced native plant protein in actively growing and differentiating gall tissue suggests that some components of the larval secretions may serve as a transacting gene regulator that overrides the host plant's normal growth sequences to produce a gall (Carango et al. 1988). We concluded that *Eurosta*'s gall on *S. altissima* results from the actions of larval secretions to alter existing plant growth mechanisms, rather than from the insertion of new genetic information to the host plant as hypothesized by Cornell (1983). This conclusion is based on several bits of circumstantial evidence. First, the *Eurosta* ball gall, like other galls, must be formed in rapidly dividing, undifferentiated tissues (Beck 1947; Dieleman 1969). Moreover, gall growth and differentiation do not involve the formation of new tissue types. Instead they primarily produce a change in the organization of tissues (Beck 1947; Rohfritsch and Shorthouse 1982). Finally, continued long-term gall growth depends on the secretions of a living *Eurosta* larva. On the basis of what is known about viruses as infective agents, it would be unlikely that continuous larval secretions would be required for a viral agent, since a single viral infection is normally sufficient to transfer the genetic information to the host plant's cells (Carango et al. 1988).

3.2 RAMET-LEVEL IMPACTS OF GALLMAKERS

Allocation of Limited Resources

The concept of resource allocation assumes that each organism has a finite amount of resources available to devote to various activities, including growth, maintenance, and reproduction (Abrahamson 1989). As a consequence, resources spent on growth, for example, are unavailable for allocation to maintenance or reproduction, and vice versa. The losses of resources to a gallmaker and/or its gall diminish the resources that normally would be devoted to host-plant growth and development. Such losses can have severe ecological and evolutionary consequences through the reduction of host-plant fitness (Abrahamson and Gadgil 1973; Abrahamson and Caswell 1982; Whitham et al. 1991; Weis 1993b).

Researchers interested in resource-allocation questions in plants have debated which resource (e.g., mineral elements, energy, biomass) is the most critical currency to measure alterations of allocation patterns. Harper (1977), for example, suggested that the green plant is "a pathological overproducer of carbohydrates and the resource that needs critical allocation may often be something other than time or energy." Hickman and Pitelka (1975) and Hartnett and Abrahamson (1979) directly compared energy and biomass allocation patterns and found that they were so similar in plants that one currency could be substituted for the other. However, Thompson and Stewart (1981) offered that mineral-element allocation may be the most critical currency to allocate because activities such as reproduction require mineral elements, but reproductive structures make no contribution to the pool of elements. To directly compare these different currencies, we tested the assumption that biomass allocation, energy allocation, and nutrient-element allocation were equivalent for ecological-strategy studies in plants using five species of *Solidago* and *Verbascum thapsus* (Abrahamson and Caswell 1982). While our analyses of *Solidago* allocation revealed the expected strong correlation between biomass allocation and energy allocation, our results with both *Solidago* and *V. thapsus* showed that the mineral elements examined were allocated differently than biomass. Furthermore, we found significant nutrient × population interactions. This interaction indicates that between-population trends in biomass allocation do not identify a given mineral element, biomass, or energy as the single best currency of resource allocation patterns. As a consequence, we have used a variety of resource currencies to examine *Eurosta*'s impacts during our studies.

Growth Patterns

The presence of *Eurosta* and its gall did not affect the height of *S. altissima* ramets. There were no significant differences in mean ramet height for *Eurosta* galled ($n = 195$) and ungalled ramets ($n = 127$) at any of eight sample dates between 20 June and 15 July (the period of most active gall growth). However, the stem growth rate of *Eurosta*-galled ramets was significantly slower than the growth rate of ungalled ramets (McCrea, Abrahamson, and Weis 1985). Furthermore, the difference between stem growth rates for *Eurosta*-galled and ungalled ramets was greatest during the period of most rapid gall volume growth (the time when galls most rapidly accumulate carbon; fig. 3.3). The reduction in stem-growth rate in galled ramets is most likely due to competition for carbon and other resources between the developing gall and the apical bud.

Yet the effects of different gallmakers on the same host plant and host organ can vary. For example, the spindle gall of *Gnorimoschema gallaesolidaginis* (Lepidoptera: Gelechiidae) significantly decreased height in galled ramets compared to ungalled ramets (133 cm in galled ramets versus 161 cm in ungalled ramets) because of the loss of apical dominance and the release of lateral shoots (Hartnett and Abrahamson 1979).

Gall and Ramet Growth

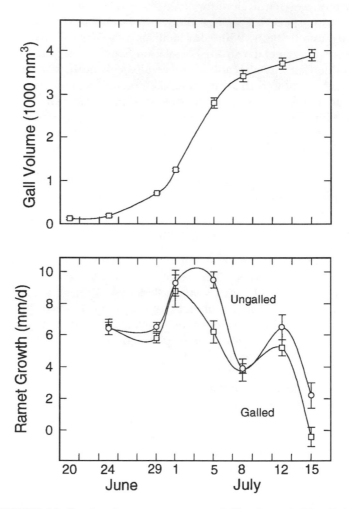

FIGURE 3.3. Results of measurements on a field cohort of 195 galled (squares) and 127 ungalled (circles) ramets. Data are means and standard errors. *Top:* Gall volume as calculated from gall diameter. *Bottom:* Growth in stem height per day, plotted at the end of each sample interval. (Redrawn from McCrea, Abrahamson, and Weis 1985, "Variation in herbivore infestation: Historical vs. genetic factors," *Ecology* 68:822–827, with kind permission from The Ecological Society of America)

Biomass Patterns

In our earliest work on *Solidago* allocation, we showed that the reproductive allocation within ramets of goldenrod populations generally declined with the increasing successional maturity of its vegetative community (Abrahamson and Gadgil 1973). The relative allocation to reproductive functions was significantly higher in old-field populations and species than in woodland populations and species. Subsequent studies showed that the probability of ramet flowering significantly increases with greater ramet biomass and height (Bresticker and Abrahamson, unpub. data). Furthermore, Abrahamson and Gadgil (1973) documented that allocation among vegetative organs varied according to the nature of the limiting resource. In presumed light-limited *Solidago* populations, the ratio of stem biomass/total biomass was low when goldenrods were growing under plants of much greater stature (i.e., trees). Yet the relative leaf allocation was high when competitors had the same stature (goldenrods with goldenrods). These patterns are adaptive given that biomass allocated to stem in old-field goldenrods allows them to compete for light with other perennials. Likewise, woodland goldenrod populations benefit from allocation patterns that provide relatively more photosynthetic ability than tall ramets. Woodland goldenrods obviously have no chance of outcompeting shrubs or trees for light.

This early work on biomass allocation patterns in *Solidago* ignored the potential for allocation variation among ramets created by differential herbivore attack. Some genets of *S. altissima*, for example, are commonly attacked not only by free-feeding herbivores such as *Trirhabda* beetles but also by several gallmakers. We examined the amount of variation in reproductive allocation attributable to the effects of three gallmakers: *E. solidaginis*, the moth *Gnorimoschema gallaesolidaginus* (elliptical gall), and the Cecidomyiid midge *Rhopalomyia solidaginis* on *S. altissima* (referred to as *S. canadensis* in our early papers; Hartnett and Abrahamson 1979).

In this study, we learned that infestation of *S. altissima* by gallmakers was very common, with up to 38.5% of the ramets attacked in old fields in central Pennsylvania (Hartnett and Abrahamson 1979). Furthermore, the presence of any of the three

Galls Affect Allocation

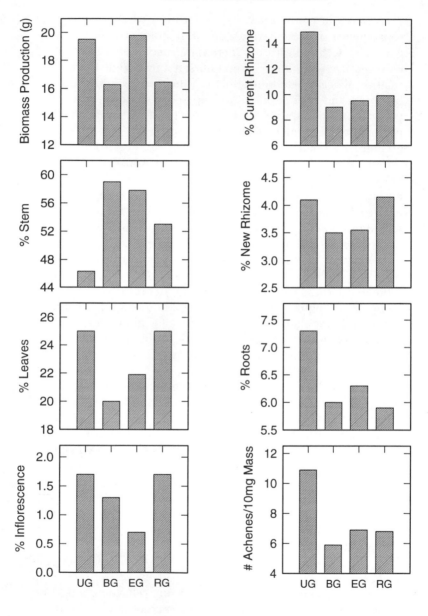

gallmakers reduced ramet production as well as biomass allocation to leaves and rhizomes, but increased allocation to stem. Most importantly, the presence of a gall of *E. solidaginis, G. gallaesolidaginis,* or *R. solidaginis* resulted in a significant reduction of reproductive allocation as measured by inflorescence production and propagule production in ramets of *S. altissima* (fig. 3.4). Reproductive biomass allocation, for example, was decreased by 32% and 43%, respectively, for *Eurosta*-attacked and *Gnorimoschema*-attacked ramets. Ramets attacked by any of the three gallmakers examined produced significantly fewer achenes per flower head, fewer heads per ramet, and fewer achenes per ramet (table 3.1). However, most of the achene loss per ramet was due to a decrease in the numbers of flowering heads produced within inflorescence of infested ramets. The net result was that the number of achenes per gram ramet biomass decreased by 45% in *Eurosta*-infested ramets and 28% in *Gnorimoschema*-attacked stems. An unexpected consequence of gallmaker presence was the gallmaker-specific alteration of achene mass for propagules on attacked ramets. The presence of *Eurosta*'s gall was correlated with a decline in achene mass, while the presence of *Gnorimoschema*'s gall was associated with heavier achenes. The presence of an elliptical gallmaker significantly increased lateral branch formation because of their propensity to cause

---

FIGURE 3.4. Total ramet biomass production and biomass allocation among organs in ungalled and galled *Solidago altissima*. UG = ungalled ramets, BG = ball gall-bearing ramets (*Eurosta solidaginis*), EG = elliptical gall-bearing ramets (*Gnorimoschema gallaesolidaginis*), and RG = rosette gall-bearing ramets (*Rhopalomyia solidaginis*). *Top left:* Total ramet biomass production (g dry mass). *Upper middle left:* Percentage of total ramet biomass production in stem. *Lower middle left:* Percentage of total ramet biomass production in leaves. *Lower left:* Percentage of total ramet biomass production in inflorescence. *Top right:* Percentage of total ramet biomass production in current rhizome. *Upper middle right:* Percentage of total ramet biomass production in new (daughter) rhizomes. *Lower middle right:* Percentage of total ramet biomass production in roots. *Lower right:* Number of propagules produced per gram of ramet dry mass. (Redrawn from Hartnett and Abrahamson 1979, "The effect of stem gall insects on life history patterns in *Solidago canadensis,*" *Ecology* 60:910–917, with kind permission from The Ecological Society of America)

TABLE 3.1. *Solidago altissima* propagule characteristics.

Ramet Type	Propagule per Head	Heads per Ramet	Propagule per Ramet	Mass (μg)	Fall Velocity (m/sec)	Pappus Length (mm)	Viability %	Germination
Ungalled	13.2	1439	19,034	69.3 ± 0.9	0.30	2.69 ± 0.04	70.0	30.7
Eurosta Galled	12.1 $P < 0.0001$	874 $P = 0.01$	10,343 $P = 0.001$	54.2 ± 2.0 $P < 0.01$	0.25 $P < 0.01$	2.62 ± 0.06 n.s.	65.0 n.s.	10.2 $P = 0.05$
Gnorimoschema Galled	12.6 $P = 0.03$	820 $P = 0.002$	10,832 $P = 0.005$	72.8 ± 0.8 $P = 0.01$	0.34 $P < 0.01$	2.60 ± 0.06 n.s.	68.0 n.s.	37.1 n.s.
Rhopalomyia Galled	12.4 $P = 0.001$	721 $P < 0.001$	9,372 $P < 0.001$	66.9 ± 1.2 n.s.	0.28 n.s.	2.66 ± 0.06 n.s.	65.0 n.s.	33.3 n.s.

Source: From Hartnett and Abrahamson 1979.

Notes: Levels of significance (P) are given for comparisons between particular gall-bearing ramets with non-gall-bearing ramets, as determined by Anova. n.s. = not significant ($P > 0.05$).

the apical meristem to lose dominance over lateral buds. As a consequence, elliptical gall-bearing ramets were significantly shorter than non-gall-bearing ramets.

The results of this study clearly showed that each gallmaker species induced different consequences to its host plant in spite of its attack of the same host-plant organ. Furthermore, gallmaker attack could have evolutionary as well as ecological consequences since gallmakers cause an appreciable decrease in seed reproductive allocation and the numbers of achenes produced per ramet. However, host-plant fitness will only be affected if these reproductive losses are appreciable at the genet level. *Solidago altissima* is a highly clonal plant and consequently any reproductive losses of infested ramets are integrated over all ramets, infested as well as unattacked, of a clone. As a consequence, the evolutionary impacts of gallmakers on host-plant life-history traits will depend on the level of attack both among and within genets as well as on the degree of reproductive allocation losses. The implication for evolutionary impacts of gallmakers on host plants is discussed in section 3.3.

Energy Patterns

We developed an energy budget for *E. solidaginis* and its natural-enemy guild to understand the energetic costs of gallmaker infestation to host plant ramets (Stinner and Abrahamson 1979). To do so, we examined the growth and respiration of larval *E. solidaginis* over the course of the growing season. Larval growth in *E. solidaginis* followed the characteristic "insect-type" growth curve (Bertalanffy 1957). Larval dry mass increased exponentially over time until mid-September or October, when the third-instar larva entered diapause (Stinner and Abrahamson 1979). Consequently the larva's most rapid growth and consumption come well after the gall is fully grown (fig. 3.5).

As expected, larval metabolism increased rapidly during gall maturation largely because the rate of respiration in *Eurosta* larvae is a function of larval mass (as measured by oxygen consumption). Oxygen consumption (μl/hr/larva, Y) by *Eurosta* larvae and hence *Eurosta* metabolism is proportional to larval dry mass (mg, X). Since the slope of this regression (0.95) approaches unity, we concluded that metabolic rate for this

71

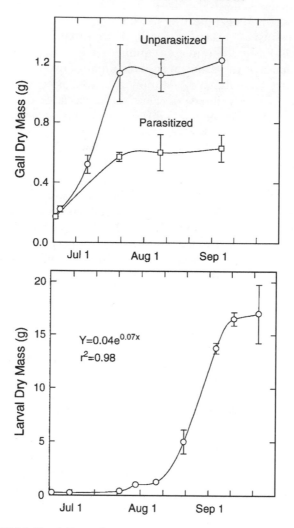

Gall and Larval Growth

FIGURE 3.5. *Top:* Gall growth curves for unparasitized (circles) and parasitized (squares) *Eurosta* galls. *Bottom:* Larval growth curve for *Eurosta solidaginis.* Each point is a mean ($\pm$ standard error) of dry masses of thirty galls or larvae. (Redrawn from Stinner and Abrahamson 1979, "Energetics of the *S. canadensis*-stem gall insect-parasitoid guild interaction," *Ecology* 60:918–926, with kind permission from The Ecological Society of America)

gallmaker is proportional to its mass rather than its surface area (two-thirds power rule). The oxygen consumption (Y) and larval dry mass (X) relationship is best described as

$$\log Y = 0.95 \log X + 0.64$$

$$r^2 = 0.94.$$

Egestion rates for *Eurosta* larvae closely follow the exponential functions obtained for larval growth rates. Fecal dry mass $(Y$ in mg) is related to time (x) as

$$Y = 0.30 \ e^{0.02x}$$

$$r^2 = 0.83.$$

Taken together, these data can be used to create energy budgets that indicate that over 25 kJ or $\approx 7\%$ of mean ramet production were allocated to development of the gall and the support of the gallmaker (fig. 3.6). On average, a host plant expended $\approx 26,100$ kJ in producing a ball gall and its inhabitant and $\approx 25,600$ kJ for an elliptical gall and its larva. The cost of $\approx 7\%$ of plant production to a single herbivore is appreciable given typical values of $5-10\%$ for the total of all grazing in old fields or grasslands (Wiegert and Evans 1967).

Both gallmakers metabolized a high percentage of plant carbohydrates based on mean respiration quotients of 0.89 for *Eurosta* and 0.88 for *Gnorimoschema*. Furthermore, the growth efficiencies (production/ingestion) that we measured for *Eurosta* (0.40) and *Gnorimoschema* (0.35) are near the upper end of the range of reported values for herbivores (Stinner and Abrahamson 1979). These high efficiencies may result from the relatively sedentary lifestyle of gallmakers as they expend little energy on movement compared to free-feeding herbivores. An important outcome of these high efficiencies of host-plant utilization is a lessened impact on the host. Any reduction of the impact should reduce the intensity of any selective pressures on the host to evolve defenses against the gallmaker. Yet any selection pressures for high efficiency of host utilization among herbivores must be balanced against selection on the herbivore for consumption. Slansky and Feeny (1977), for example, have argued that selection should maximize consumption rather than effi-

Energy Budget of *Eurosta* Galls

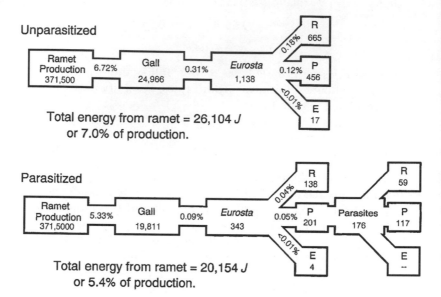

FIGURE 3.6. Energy flow diagram (in joules) for *E. solidaginis* galls expressed on a per ramet basis. The percentages indicate the ratio of each component to ramet production. P = production, R = respiration, E = egestion, and J = joules. (Redrawn from Stinner and Abrahamson 1979, "Energetics of the *S. canadensis*-stem gall insect-parasitoid guild interaction," *Ecology* 60:918–926, with kind permission from The Ecological Society of America)

ciency in herbivores so that herbivores are able to secure sufficient quantities of limited mineral elements.

Our study of gallmaker impacts using biomass currencies
found an approximately 15% decrease in total net production
for ramets attacked by gallmakers (Hartnett and Abrahamson
1979) but our energetic study (Stinner and Abrahamson 1979)
could account for only ≈7% in costs. Consequently, some 8% of
total ramet net production remained unaccounted for. While
some of this reduction in ramet production may be attributable
to the redirection of growth hormones into the gall and away
from the ramet's active growth regions (Mills 1969), additional

decreases may be caused by the influence of the gallmaker to concentrate ramet nutrients into the gall's nutritive region and thus away from the active regions of the host plant. However, we suspected that an appreciable portion of the reduction in production could be directly ascribed to the alteration of host-plant leaf characters.

We explored changes in host-plant leaf characters by comparing seven *Eurosta*-attacked and ungalled *S. altissima* ramets paired on the basis of similar developmental stage for leaf number, leaf chlorophyll content per leaf area, mean and total leaf areas, and total leaf mass (table 3.2; McCrea and Abrahamson, unpub. data). While leaf numbers and chlorophyll content/leaf area were not significantly different in galled and ungalled ramets, mean and total leaf area as well as total dry leaf mass were significantly reduced in galled ramets. Leaf areas of *Eurosta*-infested ramets averaged only 75–80% of the areas of unattacked ramets, and leaves of attacked ramets accumulated only about 80% of the dry mass of ungalled ramets. These alterations to the photosynthetic machinery are more than sufficient to markedly reduce production in gall-infested ramets since *Eurosta* presence resulted in a 19.4 ± 3.8% reduction in ramet leaf area.

Whereas the energetic cost to the host ramet of *Eurosta*-containing galls was ≈25 kJ, the cost of galls that contained parasitoids was only 15–20 kJ. The lower cost associated with natural-enemy-attacked galls may be a function of early gallmaker death. *Eurosta* parasitized by *Eurytoma obtusiventris,* for example, stop feeding earlier in the growing season than unparasitized *Eurosta* and enter a premature pupation stage. Consequently, the duration of gallmaker ingestion is reduced, which in turn lessens ramet costs. However, much of the reduction of parasitized gall cost can be attributed to *E. gigantea*'s attack of smaller and consequently less costly *Eurosta* galls (Weis, Abrahamson, and McCrea 1985; Abrahamson et al. 1989).

Not surprisingly, the parasitoid/inquiline guild had high growth efficiencies (0.66 for *Eurosta* natural enemies and 0.72 for *Gnorimoschema* natural enemies) compared to insects in general (Stinner and Abrahamson 1979). Two factors contributed to these high efficiencies. First, the natural-enemy guild expended

TABLE 3.2. Comparison of S. *altissima* ramets paired based on similarity of development stage.

	Leaf Number	Chlorophyll Content (absorbance @ 660 nm/leaf area)	Average Leaf Area (cm^2)	Total Leaf Area (cm^2)	Total Leaf Mass (g)
Normal ramets	67 ± 23.2	0.14 ± 0.03	4.5 ± 0.68	289 ± 82.4	1.69 ± 0.49
Galled ramets	71 ± 22.6	0.14 ± 0.01	3.4 ± 0.90	236 ± 76.9	1.39 ± 0.56
Significance of t-test	n.s.	n.s.	$P = 0.016$	$P = 0.0001$	$P = 0.007$

Source: From McCrea and Abrahamson, unpub. data.

Notes: Seven pairs of ramets were harvested and analyzed in early August 1984 (mean ± standard deviation). n.s. = not significant.

little energy on maintenance because of its members' sedentary lifestyle. Second, the natural enemies feed on a diet with relatively high nutrient and energy content (gallmakers are rich in proteins and fats; Abrahamson and McCrea 1986b). The mean respiration quotient for the parasitoid/inquiline guild was 0.82, which reflects the high proportion of protein in the natural-enemy guild diet. Although we did not determine respiration quotients for the two parasitoids and inquiline separately, we can predict that parasitoids would have lower respiration quotients than *Eurosta* or the mordellid inquiline since the parasitoids feed on food sources with a relatively high nutrient content per unit mass compared to the plant tissues used by the *Eurosta* or *Mordellistena unicolor*. The mordellid feeds almost exclusively on stem or gall tissue during the summer months but frequently consumes the gallmaker at or near the end of the growing season (Stinner and Abrahamson 1979; Abrahamson et al. 1989).

Gall Impacts on Photoassimilate Movements

We analyzed the effects of *Eurosta* galls on photoassimilate translocation and growth in ramets of *S. altissima* by introducing $^{14}CO_2$ ($\approx$370 kBq per ramet) into individual leaves of ungalled and galled ramets $\approx$10 cm above and below both small (<17 mm) and large (>19.5 mm) galls (McCrea, Abrahamson, and Weis 1985). By examining ramets with either small or large galls, we could assess the impacts of potentially different strength resource sinks on photoassimilate movements.

We hypothesized that there were two principal ways for the gallmaker and its gall to affect carbon flow within a ramet (McCrea, Abrahamson, and Weis 1985). First, a growing stem gall might act to partially block photoassimilate translocation through it by accumulating photoassimilate compounds as the gall grows. In this way, the gall would intercept carbon-based compounds as they are translocated through it, but the gall would not redirect compounds from other parts of the plant. A gall with this effect on source-sink relations is appropriately termed a nonmobilizing sink. Alternatively, the gall might act as a mobilizing sink by actively redirecting carbon-based compounds to itself from other plant organs. By doing so, host-plant resources are actively directed to the gall and are lost for use in other activities. Galls having the latter effect would be likely to have a greater impact on their host plant than galls with the former effect.

Gall size and labeling position had no significant effect on the amount of ^{14}C translocated out of the labeled leaf (McCrea, Abrahamson, and Weis 1985). However, the allocation of ^{14}C-containing photoassimilates among host-plant organs was affected significantly. Both gall size and labeling position influenced the percentage of label translocated to underground organs. If the label was introduced above the gall, less ^{14}C made its way to underground organs but the amount of label translocated to the apical meristem was not different from the quantities found in ungalled ramets. Furthermore, ramets with large galls allocated less label to below-ground organs than did ramets with small galls. Similarly, the amount of label translocated to the apical bud was significantly affected when the label was introduced below large galls but not small galls.

We also found a significant negative relationship between gall mass and percentage of label translocated to the apical bud when label was introduced below ($r^2 = 0.71, P < 0.001$), but not above, the gall. Similarly, this negative relationship was significant for below-ground organs when labeling was above the gall ($r^2 = 0.51, P < 0.01$), but not when labeling was below the gall (fig. 3.7; McCrea and Abrahamson 1985).

These results show that *Eurosta*'s gall behaves as a nonmobilizing sink, at least for photoassimilates produced ≈ 10 cm from the gall (McCrea, Abrahamson, and Weis 1985). These findings suggest that the gall position and host organ attacked may determine whether the nutritional needs of the gallmaker can be met simply by intercepting the normal translocation past the gall or if additional resources must be mobilized to the gall site from elsewhere. Gallmakers infesting locations with insufficient flow of nutrients to supply the developmental needs of the gall and gallmaker must have a means to actively direct resources away from other host activities and to the gall. Whereas gallmakers located on stems and leaf midribs are in near ideal positions to intercept host resources, leaf gallmakers located away from major leaf veins would likely need to actively mobilize resources to secure enough of them. For example, Billett and Burnett (1978) showed that maize leaves galled by the maize smut fungus import

^{14}C Translocation and Galls

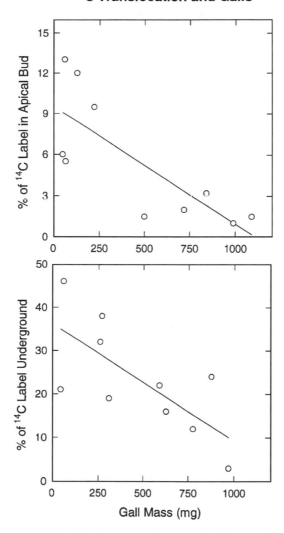

FIGURE 3.7. Relationships between gall mass and arcsine-transformed percentage of label in apical bud and underground organs. *Top:* Label ending up in apical bud when labeled leaf was below gall (arcsine $Y = -0.00015X + 0.18$, $r^2 = 0.71$, $P < 0.001$). *Bottom:* Label in underground organs when labeled leaf was above gall arcsine $Y = -0.00019X + 0.37$, $r^2 = 0.51$, $P < 0.01$. (Redrawn from McCrea, Abrahamson, and Weis 1985, "Goldenrod ball gall effects on *Solidago altissima:* ^{14}C translocation and growth," *Ecology* 66:1902–1907, with kind permission from The Ecological Society of America)

more photoassimilates than did ungalled leaves. Likewise, beech leaves infested with cecidomyiid galls of *Mikiola fagi* are mobilizing sinks for photoassimilates from neighboring leaves (Kirst and Rapp 1974).

Mineral-Element Patterns

The effects of *Eurosta*'s gall on ramet biomass production, biomass allocation, photoassimilate production, and photoassimilate translocation are negative and potentially costly to the host ramet (Hartnett and Abrahamson 1979; Stinner and Abrahamson 1979; McCrea, Abrahamson, and Weis 1985). Yet if *Solidago* ramets are "pathological overproducers of carbohydrates" (Harper 1977), the loss of carbon compounds to a gall and gallmaker may have only limited ecological consequences. However, the loss of even small amounts of a mineral element that is in very limited supply may have major ecological consequences. We examined the movements of mineral elements within host plants and the redirection of mineral elements by ball galls in a series of studies (Abrahamson and McCrea 1985, 1986b) because some galls had been shown to be strong accumulators of nitrogen and phosphorus (Abrahamson and Weis 1987).

Several studies have suggested that while biomass and energy have similar allocation patterns (Hickman and Pitelka 1975; Abrahamson and Caswell 1982), allocation of mineral elements may be independent of these currencies (van Andel and Vera 1977; Lovett Doust 1980; Lovett Doust and Harper 1980; Abrahamson and Caswell 1982). Furthermore, a consideration of mineral element allocation is even more crucial in studies involving multitrophic levels since herbivores are typically nutrient (particularly nitrogen) limited rather than energy or carbohydrate limited (Slansky and Feeny 1977). Since host-plant growth can be limited by mineral-element availability, the supplies of mineral elements to the host plant can become especially important to gallmaker feeding if their growth occurs at a time of low nutrient availability. Gallmakers that attack host-plant tissues at times of low nutrient content may increase their consumption to accumulate sufficient quantities of nutrients. The outcome could be greater impacts on the host plant.

To understand the patterns of mineral-element allocation within *S. altissima,* we monitored macronutrient movement among plant organs throughout the course of a year (Abrahamson and McCrea 1985). While mineral-element concentrations (nitrogen, phosphorus, potassium, calcium, and magnesium as % of dry mass) of above-ground organs varied as much as 6-fold during the course of a year, similar variations in below-ground organs were only 2–3-fold. Among organs, we found the highest mineral-element concentrations in the youngest, actively growing organs and in inflorescences. The seasonal pattern of mineral-element concentrations in *S. altissima* was similar to that found in most other herbaceous plants (Chapin, Johnson, and McKendrick 1980; Gay, Grubb, and Hudson 1982; Whigham 1984) in that peak concentrations occurred in rapidly growing organs but concentrations in these organs waned as growth rates fell. Nutrient concentrations in overwintering below-ground organs remained constant. However, nitrogen, phosphorus, potassium, and magnesium levels in new rhizomes increased rapidly during March and early April, over a month before ramet emergence. In spite of uptake of mineral elements from soil reserves (Hirose and Monsi 1975; Abrahamson and McCrea 1985), nutrient levels in ramets declined throughout the growing season due to the dilution of the element concentrations as ramets added biomass.

The primary sinks of nitrogen and phosphorus during the growing season were the metabolically active leaves (50–70% of the ramet's mineral resources), and later in the growing season ramet inflorescences became major mineral-element sinks (Abrahamson and McCrea 1985). However, whether an organ served as a sink or a source varied depending on the season. Leaves, for example, were sinks during the early portions of the growing season but became sources as inflorescences were initiated in late summer. Developing inflorescences gained nitrogen, phosphorus, potassium, and magnesium from leaves and calcium from stem reserves. This exchange of mineral elements from leaves and stems to inflorescences is possible in part because of *S. altissima*'s late-season flowering. This ability to internally translocate mineral elements appreciably reduced the need for nutrients from soil reserves and may facilitate *S. altis-*

81

sima's broad tolerance of a variety of edaphic conditions (Abrahamson, Armbruster, and Maddox 1983). However, the losses of elements from leaves were insufficient to account for all the gains in inflorescences. The remainder of inflorescence mineral-element needs came through uptake from soil reserves. The very limited autumn translocation of mineral elements from above-ground to below-ground organs was surprising given the amount of translocation among ramet organs. It is possible that soil stores are typically more than adequate to provide for the pronounced September and October uptake of nitrogen by below-ground organs.

We examined the allocations of biomass, nitrogen, phosphorus, and potassium in a three-factor experiment that compared (1) ungalled ramets and those attacked by *Eurosta* larvae, (2) unfertilized and fertilized ramets, and (3) rhizome-connected ramets and ramets whose rhizome connection to their clone was severed (Abrahamson and McCrea 1986b). Our findings were similar to the results of earlier studies (Hartnett and Abrahamson 1979) in that the presence of *Eurosta* galls increased the ramet's biomass allocation to stem but decreased its allocation to leaves and seed reproduction, and reduced its number of new rhizomes (9.1 and 6.9 for ungalled and ball-gall infested, respectively) and matured new rhizomes (4.2 and 2.7 for ungalled and ball-gall attacked, respectively). Although ball-gall presence did not alter total biomass production, it did reduce the absolute amount of biomass allocated to inflorescences (1.77 to 0.90 g), leaves (5.21 to 3.86 g), and new rhizomes (1.29 to 0.88 g).

Some of the mineral elements added to the soil by fertilizer additions (using 10-10-10, % total nitrogen, % available P_2O_5, and % soluble K_2O, added 29 June and 16 August at the rate of 10.4 kg/100 m² per application) were taken up by fertilized ramets. Fertilized ramets had higher levels of nitrogen, phosphorus, and potassium in all organs (except potassium in current rhizome) than unfertilized ramets. Furthermore, ramets attacked by *Eurosta* had significantly higher levels of nitrogen in below-ground organs but lower amounts in leaves than ungalled ramets. The *Eurosta* galls on fertilized ramets contained higher levels of nitrogen, phosphorus, and potassium than galls from unfertilized ramets. However, gall mass did not increase with fertilization, suggesting that either the production of *Eurosta* galls is not nutrient limited or the effects of fertilization occurred after most gall growth had ceased. We suspect that both control and fertilized clones were above any minimum nutrient level critical for gall development since *Solidago* readily takes up available mineral elements at levels well beyond concentrations necessary for maximum growth (Abrahamson and McCrea 1986b). However, we have shown elsewhere (Weis and Abrahamson 1985) that maximum gall growth rates occur during June and that galls reach peak diameter by approximately 10 July. It is possible that the lack of biomass increase of fertilized galls (Abrahamson and McCrea 1986b) was due to the lateness of the first fertilizer treatment (29 June). *Eurosta* growth, on the other hand, does not become exponential until well after peak gall diameter has been reached and larval growth remains exponential until early September (Stinner and Abrahamson 1979; Weis and Abrahamson 1985). Consequently, both fertilizer treatments could have affected larval development.

Since *S. altissima* is highly clonal, physically connected ramets may be physiologically integrated. To examine this possibility we isolated ramets from their parental clone in early summer by removing a short segment of the rhizome connection between the ramet and the remainder of the clone (Abrahamson and McCrea 1986b). Severed ramets compensated for their lack of connection to clonal resources by allocating a significantly higher proportion of their total biomass to roots, current rhizome, and new rhizomes and a lower fraction to leaves and inflorescences. Thus, below-ground allocation was strongly emphasized over above-ground appropriation in severed ramets. Ramets were relatively independent of their parental clone by early summer as severing at this time did not cause appreciable ramet mortality nor alter the total biomass production of the ramet nor the mass of the mature *Eurosta* gall. However, the degree of integration among ramets of a physically connected clone varies with time over the growing season. Hartnett and Bazzaz (1983a) isolated ramets at several times during the growing season and found that the negative effects of isolation (lower growth, survivorship, and reproduction than connected ramets) were reduced as the growing season progressed. Their study

83

TABLE 3.3. Nitrogen, phosphorus, and potassium concentrations for various plant or animal components of ball and elliptical gall-bearing goldenrods.

	Eurosta solidaginis			*Gnorimoschema gallaesolidaginis*		
	%N	%P	%K	%N	%P	%K
Stem	0.77	0.14	1.60	0.67	0.21	2.60
Gall outside	0.67 (0.9×)	0.15 (1.1×)	1.96 (1.2×)	0.77 (1.1×)	0.23 (1.1×)	4.00 (1.5×)
Gall interior	1.02 (1.3×)	0.23 (1.6×)	2.72 (1.7×)	0.81 (1.2×)	0.21 (1.0×)	2.20 (0.8×)
Gall larva	4.43 (5.8×)	1.02 (7.3×)	0.78 (0.5×)	7.63 (11.4×)	0.55 (2.6×)	2.00 (0.8×)
Frass	1.76 (2.3×)	0.21 (1.5×)	1.06 (0.7×)	1.20 (1.8×)	0.27 (1.3×)	0.30 (0.1×)
Parasitoids	6.05 (7.9×)	0.55 (3.9×)	1.28 (0.8×)	8.09 (12.1×)	0.55 (2.6×)	1.04 (0.4×)

Source: From Abrahamson and McCrea 1985.

Notes: Values for times of concentrations over stem are given in parentheses. Data were determined from galls collected during winter diapause for *Eurosta* and just following the peak of larval growth for *Gnorimoschema*.

confirms suspicions that new ramets are physiologically dependent on their parental clone for resources, but this dependency rapidly declines as the ramet ages. Even though maximum *Eurosta* larval growth occurs during mid- to late season when ramets are more independent, gall growth, the major cost of gallmaker presence (Stinner and Abrahamson 1979), occurs early in the season when host-plant ramets are most dependent on clonal resources.

Our study of mineral-element costs to galled ramets showed a pronounced concentration of nutrient concentrations going up the gallmaker food chain (Abrahamson and McCrea 1986b). Apparently, gallmakers were redirecting considerable proportions of the ramet's mineral elements. Nitrogen and phosphorus, but not potassium, were strongly concentrated by both *Eurosta* and *Gnorimoschema* larvae (table 3.3). While *Eurosta* larvae contained 5.8 times more nitrogen and 7.3 times more phosphorus than ungalled stem tissue, *Gnorimoschema* larvae accumulated 11.4 times more nitrogen and 2.6 times more phosphorus than contained in stem. In addition, the inner nutritive zone of *Eurosta* galls was especially rich in mineral elements. This zone accrued 1.3 times more nitrogen, 1.6 times more phosphorus, and 1.7 times more potassium than comparably aged stems of unattacked ramets. The nutritive region of *Gnorimoschema* galls amassed only nitrogen (1.2 times more than stem).

By coupling these nutrient concentration data with biomass data, we can represent the movement of mineral elements from the host plant to the gallmaker to the guild of natural enemies as a flow model (fig. 3.8). This flow model illustrates the extent to which galls and gallmakers are sinks for nitrogen and phosphorus. Ramets attacked by *Eurosta* lost approximately 8 mg of nitrogen (3.4% of the total available) and 2.6 mg of phosphorus (5% of total) to the development of the gall and its gallmaker (Stinner and Abrahamson 1979; Abrahamson and McCrea 1986b). *Gnorimoschema*-attacked ramets forfeited about 11.4 mg of nitrogen (4.6% of total) and 3.2 mg of phosphorus (5.5% of total) to the gallmaker and its gall. The frass of both gallmakers contained low amounts (<0.01% of total) of mineral elements suggesting that the gallmakers conserve these important nutrients. The outcome is that the nitrogen and phosphorus costs of these goldenrod gallmakers to their host ramet under conditions

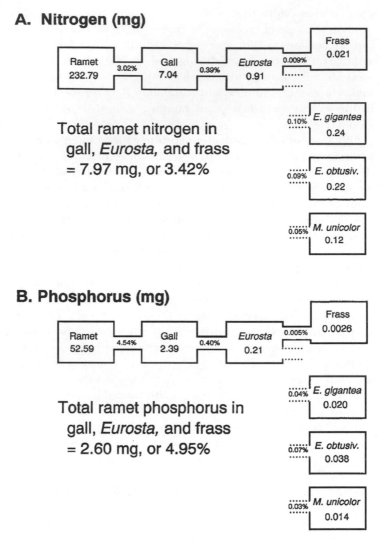

A. Nitrogen (mg)

B. Phosphorus (mg)

FIGURE 3.8. Nutrient flow diagram (in mg) for *Eurosta solidaginis,* its galls, frass, parasitoids, and inquiline expressed on a per ramet basis. The percentages indicate the ratio of each component to ramet mineral-element content. (Redrawn from Abrahamson and McCrea 1986b, "Nutrient and biomass allocation in *Solidago altissima:* Effects of two stem gallmakers, fertilization, and ramet isolation," *Oecologia* 68:174–180, with kind permission from Springer-Verlag)

of nonlimiting nutrient availability are similar to the energy and biomass costs determined for these gallmakers (Abrahamson and McCrea 1986b).

3.3 GENET-LEVEL CONSEQUENCES

Clone Architecture and Integration

Genotypes of *S. altissima* are highly clonal as they are composed of varying numbers of ramets that may or may not be physically and/or physiologically connected to one another (Abrahamson et al. 1991). There are probably many reasons that plants such as goldenrod maintain clonal connections. For example, vascular connection between sister ramets within a genet could enable the sharing of photoassimilates, water, and/ or mineral elements (Alpert and Mooney 1986; Schmid and Bazzaz 1987; Schmid et al. 1988a,b). Resource sharing among connected ramets of a genet could enhance genet growth. Ramets in sites with greater resource availability could export resources to ramets in sites with lower resource availability (Pitelka and Ashmun 1985; Hutchings 1988; Alpert 1991; Caraco and Kelly 1991). Furthermore, such clonal integration could allow for the support of new ramets during their early development (Lovett Doust 1981; Bloom, Chapin, and Mooney 1985) and for resource storage later in the growing season to facilitate the subsequent season's growth. Several authors have argued that physiological integration late in the growing season could be as beneficial as early integration since it enables control of spatial relationships within the clone (Bradbury and Hofstra 1975; Hartnett and Bazzaz 1985; Cook 1985). This control may optimize resource-use efficiency and reduce competition for light, water, and mineral elements among ramets.

However, several authors have argued that clonal integration can have costs (Salzman and Parker 1985; Caraco and Kelly 1991). The support of ramets stressed by shading, herbivory, or disease may reduce genet fitness (Pitelka and Ashmun 1985) unless the translocation of resources to stressed ramets is eliminated by physical or physiological isolation (McCrea and Abrahamson 1985; How, Abrahamson, and Zivitz 1994). As a result, we might predict that the extent and duration of clonal integration will vary among species in response to its relative benefits and costs (Abrahamson 1980).

We examined the extent and duration of clonal integration in sister ramets of *S. altissima* by monitoring mineral element sharing through ramet growth responses (Abrahamson et al. 1991). Clonal integration has been examined by most workers by using ^{14}C tracers (e.g., Colvill and Marshall 1981), severing ramet connections (e.g., Hartnett and Bazzaz 1983a; Abrahamson and McCrea 1986b), or analyzing mother-daughter sharing (e.g., Alpert 1991). However, the movement among sister ramets of mineral elements that are critical to genet growth has been less well studied (e.g., Giddens, Perkins, and Walker 1962; Slade and Hutchings 1987a–c; Friedman and Alpert 1991). The sharing or nonsharing of mineral elements may provide a more critical measure of clonal integration than the sharing of photoassimilates, particularly when the production of carbohydrates represents a small cost to the genet.

We used three genotypes of *S. altissima* to create two-ramet clonal fragments, each composed of a one-year-old mother rhizome and two new daughter rhizomes at $\approx 60°$ from one another (Abrahamson et al. 1991). Each two-ramet fragment was placed in a triad of pots so that the middle pot contained the mother rhizome and its associated roots and the left and right pots each contained a daughter ramet. Nutrient treatments of weekly application of 50 ml of ¼-strength Hoagland's solution (Hoagland and Arnon 1950) were randomly assigned to triads.

Mother rhizomes and their associated roots freely shared nutrients with their developing daughter ramets. The mother rhizome and associated roots' pot and the daughter ramets' pots were functionally a common environment. Ramet growth was similar regardless of whether nutrient additions were provided directly to a daughter ramet or indirectly through its mother rhizome. This was expected given the results of earlier studies (e.g., Hartnett and Bazzaz 1983a; Abrahamson and McCrea 1986b) and that the mother rhizome of *Solidago* ramets can only store resources for use by a daughter ramet (the mother rhizome itself cannot use materials stored in one year in a subsequent year).

Integration of Sister Ramets

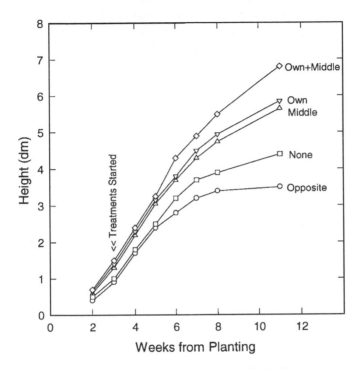

FIGURE 3.9. Mean ramet heights (dm) over time by fertilizer treatment. Treatments indicate which pot(s), if any, received nutrient addition. No height measurements were taken in the ninth and tenth weeks. (Redrawn from Abrahamson, Anderson, and McCrea 1991, "Clonal integration: Nutrient sharing between sister ramets of *S. altissima* [Compositae]," *American Journal of Botany* 78:1508–1514, with kind permission from the Botanical Society of America)

More important was our finding that sister ramets did not share mineral elements with one another (fig. 3.9). Instead, our data suggested that sister ramets compete for nutrients from their common mother rhizome. Both the height and biomass of a sister ramet had negative effects on ramet height, growth, and biomass. Furthermore, the finding that the biomass of two-ramet clonal fragments was much less than twice that of one-ramet clonal fragments suggested the presence of strong competition between sister ramets in multiple-ramet clones. Competition between sister ramets may occur as the taller ramets, having more leaves, transpire more and create a directional osmotic gradient for nutrient movement (Abrahamson et al. 1991).

The advantage of physically connected sister ramets not sharing nutrients might be in facilitating the genet's control over spatial relationships within the clone (Abrahamson et al. 1991). Ramets within a genet in nutrient-poor microsites are not supported and hence will produce fewer new rhizomes than sister ramets in nutrient-rich microsites. As a consequence, the genet would expand in more fertile microsites but would not expand in less fertile microsites. Similarly, a genet's fitness may be higher if it produces fewer but larger ramets instead of more but smaller ones. We could imagine that microsite variations within a field site would provide an array of architectures among *S. altissima* clones. Such variation may be responsible for the considerable variation in the numbers of new rhizomes and in rhizome lengths and angles measured by Cain (1990a,b). Genets able to produce more new rhizomes and/or shorten their rhizome lengths upon meeting a resource-rich microsite would remain in resource-rich patches and should gain fitness.

Influence of Eurosta on Clonal Architecture and Connections

There are potential advantages and disadvantages for a genet that maintains the clonal connections among its ramets. In the case of goldenrod clones, interconnected ramets may share photoassimilates and thereby ameliorate stresses such as the shading or defoliation of a single ramet (Hartnett and Bazzaz 1983a). Alternatively, adverse conditions affecting a single ramet have the potential to drain sufficient resources away from an integrated, sharing genet so that the genet's reproductive output is reduced (McCrea and Abrahamson 1985). In this case, the sharing of resources may result in a greater loss of fitness for the genet than if the reproductive output of the affected ramet was lost entirely. A genet that could isolate such an affected ramet could potentially have an advantage over genets that do not possess the means to eliminate resource sharing to affected ramets. Some goldenrod genotypes may have the means to abandon stressed ramets (McCrea and Abrahamson 1985; How, Abrahamson, and Zivitz, 1994). Such abandonment could be advantageous given that *Eurosta* gall development occurs while host ramets are still strongly dependent on their parental clone (McCrea and Abrahamson 1985) and that *Eurosta* gall development is costly to the affected ramet (Hartnett and Abrahamson 1979; Stinner and Abrahamson 1979; Abrahamson and McCrea 1986a,b).

We became aware of this possibility in an experiment designed to determine the effects of varying levels of attack by *E. solidaginis* on biomass allocation, leaf senescence rate, and rhizome connections in small clones of *S. altissima* (McCrea and Abrahamson 1985). While the results related to the effects of varying attack levels were not particularly informative, we made the unexpected finding that galled ramets were more than twice as likely to become detached from the remainder of their clone than were ungalled ramets (galled 36% connected, ungalled 76% connected, $\chi^2 = 23.7$, $P < 0.001$). This outcome suggested that some *Solidago* genotypes may possess a mechanism to isolate stressed ramets and consequently reduce gallmaker impacts to the genet (McCrea and Abrahamson 1985). This study also showed that presence of a ball gall lowered the leaf senescence rate at the ramet and clone levels. Isolation of individual ramets decreased their leaf senescence rate and clones with more rhizome connections (i.e., less fragmentation) had higher leaf senescence rates. This outcome may be a consequence of the gall blocking the downward flow of plant hormones such as auxins (auxins are known to inhibit the formation of abscission layers in leaves; Addicott 1970). *Eurosta*'s gall has been shown to at least partially block the translocation of photoassimilates (McCrea, Abrahamson, and Weis 1985) and may function similarly with auxins. Auxins or other plant growth regulators may play a role in maintaining rhizome connections within clonal plants such as *S. altissima*.

We further explored this premature deterioration or sloughing of rhizome connections in a second study using four genotypes in a greenhouse experiment and five additional field genotypes excavated at the end of the growing season (How, Abrahamson, and Zivitz 1994). Both the greenhouse and field experiments revealed that sloughing of rhizome connections was significantly more frequent among galled than among ungalled ramets. However, individual genotypes showed great variability in this trait, with some clones even being more likely to isolate ungalled than galled ramets. One of the greenhouse clones responded to *Eurosta* oviposition with a hypersensitive, necrotic reaction to the *Eurosta* larva and subsequently sloughed each ramet that exhibited hypersensitivity. This hypersensitive response at the ovipuncture site caused the loss of apical dominance and subsequently caused stem branching. Sloughing of galled ramets was significant in only one of the field clones examined. However, the pooled sample of galled and ungalled ramets from all five field clones showed that sloughing occurred more often ($\chi^2 = 10.9, P < 0.001$) than expected in galled ramets.

The greenhouse results indicated that sloughed ramets were shorter, slower growing, and, as we had found in the original greenhouse study (McCrea and Abrahamson 1985), isolated ramets were slower to senesce their leaves. Among field clones, ramet height was a significant covariate of sloughing in the presence of a gall such that shorter ramets were more frequently isolated (How, Abrahamson, and Zivitz 1994). Unfortunately, we cannot discriminate if this reduced leaf senescence rate is simply an artifact of the gallmaker's presence or a consequence of an adaptive response to manipulate its host by the gallmaker. In either case, this host plant reaction could provide more resources for gallmaker development and/or minimize the gallmaker's impact on its host.

When the ramets of all studies (McCrea and Abrahamson 1985; How, Abrahamson, and Zivitz 1994) are combined, the rate of isolation of galled ramets (44% sloughed) is significantly higher than among ungalled ramets (34% sloughed; table 3.4). Yet there is considerable variability in clonal fragmentation among *Solidago* clones (Maddox et al. 1989; Cain 1990a; Farrell, McCrea, and Abrahamson, unpub. data). Sloughing seems to be initiated in clones under a variety of conditions, of which galling is only one factor. The mechanism causing the premature disintegration of rhizome connections in some genotypes is un-

TABLE 3.4.   Effects of the goldenrod ball gall on clonal connections among combined ramets in a greenhouse and a field study, and in a second greenhouse study.

| | Number of Ramets | | |
	Connected	Sloughed	
Ungalled	309	158	66% connected
Galled	138	110	56% connected
		$\chi^2 = 7.7$, df $= 1$,	
		$P < 0.01$	

Source: How, Abrahamson, and Zivitz 1994, and McCrea and Abrahamson 1985, respectively.

known. Yet because sloughing is more likely to occur in shorter, slower-growing, and slower-to-senescence ramets (i.e., more immature ramets), we suspect that plant growth hormones trigger rhizome disintegration (How, Abrahamson, and Zivitz 1994).

The fate of sloughed ramets within a clone has yet to be studied; but given their smaller size and relative immaturity and Cain's (1990b) findings, we would predict that they have lower survivorship and produce fewer daughter ramets. It is conceivable that sloughing of galled ramets is an adaptation or preadaptation that increases the overall fitness of a clone by eliminating support of smaller ramets (How, Abrahamson, and Zivitz 1994). Yet as with any character, we cannot necessarily assume that ramet sloughing is adaptive. Sloughing may be simply an artifact from the reactions of certain host genotypes to *Eurosta*'s effects.

The variation in the occurrence of sloughing from clone to clone is not surprising given the degree of variation among *S. altissima* clones for the degree of resistance and susceptibility to *Eurosta* (McCrea and Abrahamson 1987; Anderson et al. 1989; Horner and Abrahamson 1992), growth and phenology (How, Abrahamson, and Craig 1993), and biomass allocation (Bresticker and Abrahamson 1984).

Lifetime Fitness Costs

The loss of reproductive output due to herbivory within a genet is perhaps the most critical question about herbivore impacts on host plants. We might expect the evolution of host plant resistance mechanisms if reproductive losses over the lifetime of a genet are appreciable and if impacts are variable among genets due to differences in characters that impart some resistance in hosts. On the other hand, if herbivore impacts are variable over time or if host plants lack heritable variation for resistance to an herbivore, we might expect little if any response by the host plant. Unfortunately, answers to questions about lifetime fitness losses at the genet level have been evasive in most herbivore/host-plant systems (Fritz and Simms 1992) and our efforts with *E. solidaginis*-attacked *S. altissima* genets are no exception. Nevertheless, we have gathered limited evidence to suggest that *Eurosta*'s impact to its host plant is both localized within the affected ramet and minimized within the genet due to the host's clonal growth habit.

To clarify *Eurosta*'s impact on the physiological integration of clones, we followed photoassimilate translocation between sister ramets in young, actively growing, two-ramet clones using ^{14}C labeled CO_2 introduced into one leaf on one of the ramets (McCrea and Abrahamson, unpub. data). Unlike the ramet-level $^{14}CO_2$ experiment described in section 3.2, in this experiment we introduced label to a leaf within 1 cm of a *Eurosta* gall across four treatments that varied the site of label introduction: galled clones in which the label was introduced 1 cm (1) above or (2) below the gall, (3) clones in which we introduced the label to a leaf on the ramet opposite the galled ramet, and (4) defoliated clones that had all the leaves on one ramet removed except those around the apical bud. Label was introduced to a single leaf of the foliated ramet in defoliated clones. A fifth treatment group consisted of ungalled control clones in which we introduced label on a leaf of similar developmental age as those labeled on galled clones.

Gallmaker presence could cause several outcomes affecting the translocation of ^{14}C-labeled photoassimilates among ramets within a goldenrod clone: (1) galled ramets could act as a sink by actively importing photoassimilates or other resources from adjacent ungalled ramets, (2) galled ramets could be physiologically isolated from the remainder of the clone and thereby minimize the gall's impact to the clone, or (3) gall presence might have no effect on resource movements within the clone.

94

¹⁴C Translocation within Clones

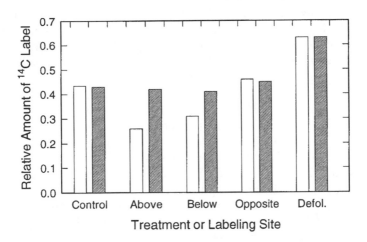

FIGURE 3.10. Relative amounts of ¹⁴C translocated to a sister ramet. Label was introduced 1 cm above or below the gall. Clones in which the label was introduced to a leaf on the ramet opposite the galled ramet, and defoliated clones that had all leaves on one ramet removed except those around the apical bud. Label was introduced to a single leaf of the foliated ramet in defoliated clones. Ungalled control clones in which label was introduced on a leaf of similar developmental age as those labeled on galled clones. Data shown by open bars are corrected for the amount of label remaining in the labeled leaf, and data illustrated by the cross-hatched bars are corrected for the amount of label remaining in the labeled leaf and in the gall. (From McCrea and Abrahamson, unpub. data)

We gained three important insights from this ¹⁴C clone-level experiment (fig. 3.10). First, galls readily take up photoassimilates produced by leaves in their immediate proximity (within 1 cm). This result contrasts with our finding that *Eurosta* galls functioned as nonmobilizing sinks in ramets where label was introduced ≈10 cm above or below galls (McCrea and Abrahamson 1985). Taken together, however, these results indicate that *Eurosta* galls have a very localized effect in mobilizing only nearby photoassimilates and intercepting more distant carbon compounds. Second, *Eurosta* galls do not sufficiently disrupt the normal translocation physiology of an affected ramet to cause

95

translocation of photoassimilates between ramets. In spite of us-ing clones with actively growing ramets, the pattern of photoas-similate movement suggested that ramets within both control and galled clones were independent of one another. The lack of a gallmaker effect on carbon movement within clones suggests that the gallmaker's impact is localized, possibly as a result of either the host's reaction and/or the gallmaker acting as a "good" parasite (Abrahamson and McCrea 1986a). While suc-cessful parasites need not evolve to be harmless, May and Ander-son (1983) have argued that parasite genotypes of intermediate virulence should predominate over those of extreme virulence. Finally, we learned that some stresses are sufficient to cause a physiological reintegration of ramets within a clone and initiate resource sharing. Defoliation of a ramet significantly increased the import of labeled photoassimilates from other parts of the clone (McCrea and Abrahamson, unpub. data). This outcome suggests that severely stressed ramets can be supported by adja-cent unstressed ramets.

We also examined carbon translocation between sister ramets of *S. altissima* in naturally occurring field clones late in the grow-ing season (Farrell, McCrea, and Abrahamson, unpub. data). We introduced $^{14}CO_2$ into a single leaf on one ramet of connected ramet pairs during early autumn. Rhizome connections were de-termined before labeling by excavating the uppermost portions of rhizomes with a jet of water, and once connections were docu-mented, rhizomes were reburied. The unlabeled ramet was de-foliated immediately before labeling in one half of the two-ramet clones. After 5 days, the ^{14}C content of all plant organs was measured by liquid scintillation.

We found that photoassimilates are quickly translocated out of aging goldenrod leaves and translocated to below-ground or-gans in autumn. Over 90% of the ^{14}C was translocated out of the labeled leaf within 5 days, and 95% of the translocated label was found in the below-ground organs of both ramets. Given the lateness of the growing season, it came as no surprise that the highest concentrations of label were located in the labeled ra-met's new rhizome (34%) and its connected current rhizome and associated roots (19%). Unlike the results found in younger clones with actively growing ramets, defoliation of older, senescing ramets had no significant effect on the ^{14}C translocation patterns between ramets. In spite of this result, a small but appreciable amount of label (7%) was translocated from the labeled ramets to the underground portions of the unlabeled ramets. These findings indicate that ramets within a clone share in the storage of late-season photoassimilate gains and that new rhizomes are the principal storage organ. Additional but lesser amount of storage was provided by the current rhizome and its associated roots.

We have also investigated gallmaker impacts on reproductive allocation across clonal fragments but found no significant effects. In one study, we created three-ramet clones and induced zero to three galls per clone. After growing these clones to reproductive maturity, we harvested clones to determine biomass allocation to ramet organs, leaf senescence rates, and rhizome connections (McCrea and Abrahamson 1985). We found no significant gallmaker effects on biomass allocation to any organs at the clone level. This finding, using only one host genotype, may be a consequence of the isolation of galled ramets since *Eurosta*-attacked ramets in this genotype lost their physical connection to their mother rhizome (McCrea and Abrahamson 1985).

In a more extensive field study, we determined biomass allocation, reproductive output (sexual and vegetative ramet production), fruit characteristics, and flowering phenology for established clones and their component ramets of *S. altissima* (Bresticker and Abrahamson, unpub. data). Even though we identified significant variation among clones in allocation to stem, leaf, inflorescence, current rhizome, new rhizome, and root, as well as for the number of new rhizomes, ramet height, number of achenes/inflorescence, and dates of flowering, we found no significant clone-level gallmaker effects on any of these measures. *Eurosta* presence affected allocation patterns only within the galled ramets. Ramets that were adjacent and physically connected to the galled ramet at the time of harvest were unaffected.

Because the current year's new rhizomes develop to become the subsequent growing season's ramets, the findings that *Eurosta*'s presence reduced both the numbers of new rhizomes and the amount of biomass allocated to new rhizomes suggests that

97

Eurosta's impact may carry over to the subsequent growing season (Hartnett and Abrahamson 1979; Abrahamson and McCrea 1986a,b). Fewer new rhizomes in the current season may mean that there will be fewer new ramets in the following season. Consequently, clones of goldenrod in young fallow fields that are highly attacked by *Eurosta* may experience reduced clonal expansion and hence lose access to resources that are usurped by competing, but less attacked, plant genotypes. Clone expansion losses in the first few years after field invasion may never be recovered.

To examine the possibility of such a carryover of a gallmaker's impact into a subsequent season, we induced galls in forty-five small, two-ramet clones growing in $50 \times 50 \times 30$ cm flats (McCrea and Abrahamson, unpub. data). After overwintering out-of-doors, these clones were grown without gallmakers under greenhouse conditions. Clones with one or two galls in year 1 had significantly fewer ramets in year 2 than ungalled control clones (table 3.5). The potential repercussion of this finding is that genets that lose rhizome production because of *Eurosta* attack during their first year or two of establishment could potentially suffer lifetime fitness losses. Assuming genets are the same in all other respects, over a number of years genets that are consistently attacked would expand less and consequently would be composed of fewer ramets than if no *Eurosta* attack had occurred. Although the reduction of clone expansion would vary from genet to genet, the clones with fewer ramets should have, on average, fewer achenes to disperse. The potential loss of up to 20% of the annual gain of ramets in rapidly expanded genets could appreciably reduce the genet's future and lifetime achene production since the average unattacked *S. altissima* ramet is capable of producing over 19,000 achenes (Hartnett and Abrahamson 1979).

We can make a crude estimate of these losses by assuming two extreme genet types and extrapolating data available from our previous studies. Both genets begin as seedlings (only one ramet) that do not flower or set seed in the first growing season. One genet is never attacked by the gallmaker and consequently is able to produce 4.1 new ramets per parental ramets in subsequent years. The second genet is attacked by *Eurosta* in its first

TABLE 3.5. Number of ramets produced in year 2 from two-ramet clones galled by *E. solidaginis* in year 1.

Number of Galls in Year 1	Number of Ramets in Year 2 (number of clones)
0	8.14 ($N = 14$)
1 or 2	6.58 ($N = 31$)

Source: From McCrea and Abrahamson, unpub. data.
Note: Anova F = 5.65, df = 1,44, $P = 0.022$

year, and in subsequent years it has approximately one-half of its ramets infested by *Eurosta*. It produces 3.3 new ramets per parental ramet per year because of the effects of *Eurosta* presence. For both genets, each unattacked ramet is able to produce 19,000 achenes, while gallmaker-attacked ramets produce 9500 achenes. Over a genet lifetime of only 5 years, the unattacked genet should produce ≈7,073,700 achenes, while the attacked genet should generate ≈2,408,250 achenes (table 3.6). Even over this short lifetime, the unattacked genet has produced nearly three times more achenes than the unattacked genet. Although these genets represent extreme examples, they do illustrate the potential effect of gallmaker presence on genet fitness.

Despite this potentially large impact of *Eurosta* on *Solidago* clonal expansion, our reanalysis of earlier data indicates little or no negative impact of galling on clonal expansion. Between 1982 and 1984, one hundred clones occurring in a roadside habitat were monitored for infestation and stem number. Thus, we had the opportunity to examine clonal expansion from the beginning of the first year to the beginning of the third. The only observed mortality over the period was from herbicide treatment that is often used in roadside maintenance. The rate of clonal expansion ($\log_n$[ramets 1982/ramets 1994]) was regressed over the yearly gall loads (galls per stem). This is equivalent to the selection-gradient analysis of Lande and Arnold (1983), where clonal expansion is used as the surrogate measure for fitness. The significantly negative regression coefficients for the gall loads could be interpreted as evidence for selection that favored resistance to *Eurosta*. However, the standardized regression coefficients for gall load were nonsignificant for both years

99

TABLE 3.6. Hypothetical achene production in two genets established from seed that differ in reproductive output due to the presence of *Eurosta solidaginis* galls.

	Year 1	Year 2	Year 3	Year 4	Year 5
Unattacked Genet					
Ramet number	1	4.1	16.8	68.9	282.5
Achene increment	0	77,900	319,200	1,309,100	5,367,500
Lifetime achenes	0	77,900	397,100	1,706,200	7,073,700
Attacked Genet					
Ramet number	1	3.3	10.9	36.0	118.8
Achene increment	0	47,025	155,325	513,000	1,692,900
Lifetime achenes	0	47,025	202,350	715,350	2,408,250
% more lifetime Achenes for unattacked Genet	—	166%	196%	239%	294%

Sources: Numbers from Hartnett and Abrahamson 1979; Bresticker and Abrahamson, unpub. data.

Notes: One genet is never attacked by the gallmaker and consequently produces 4.1 new ramets per parental ramet in each growing season (one-half the number produced under experimental conditions for two-ramet clones; McCrea and Abrahamson, unpub. data). The single first-year ramet of the second genet is attacked by *Eurosta*, and in subsequent years this genet has approximately one-half of its ramets infested by *Eurosta*. It produces 3.3 new ramets per parental ramet (one-half the number of *Eurosta*-attacked two-ramet genets under experimental conditions). For both genets, each unattacked ramet is able to produce 19,000 achenes while gallmaker-attacked ramets produce 9500 achenes. *Solidago altissima* ramets established from seed do not flower in their first year.

(1982, $b = 3.69 \pm 6.64$; 1983, $b = 0.80 \pm 6.45$). These suggest that any negative effects that *Eurosta* has on clonal expansion is nullified by other factors.

Why the disparity between our crude estimate for the potential effect of *Eurosta* and the lack of effect detectable in the field? The estimate did not account for important differences among genets. These differences include resistance to herbivory by *E. solidaginis* and its associated costs, ramet growth rate, or ramet fecundity. Yet differences in such variables are common among *S. altissima* genets (Maddox and Root 1987; McCrea and Abrahamson 1987; Anderson et al. 1989). For example, resistance to *E. solidaginis* may come with a cost of reduced ramet growth rate and reproductive output. Anderson et al. (1989) measured ramet growth in replicates of thirty genets of known resistance under common-garden conditions. We found that susceptible genets grew significantly faster than resistant genets. Ramets that were ovipunctured by *Eurosta* were growing significantly faster and were taller at the time of oviposition than unpunctured ramets (Anderson et al. 1989). This suggests that faster-growing ramets will be more reactive to the gallmaker's stimuli and that genotypes that grow faster initially may be more susceptible to *Eurosta* attack. There is, however, the caveat that gallmaker densities can fluctuate widely from year to year and site to site (Abrahamson et al. 1989; Weis, Abrahamson, and Andersen 1992; Sumerford and Abrahamson 1995; Sumerford, Abrahamson, and Weis, unpub. data). Consequently, variations in gallmaker abundance may alter the relative fitness advantage of susceptible and resistant genotypes. Over time, such variations in relative fitness among genotypes could promote an array of diverse genotypes ranging from susceptible to resistant.

Taken together, our results of ramet-level and genet-level studies suggest that the impact of *Eurosta* galls on *S. altissima* are localized within the infested ramet. As a consequence, we can argue that there is little impact at the genet or evolutionary individual level among established clones. If true, this would virtually exclude any direct evolutionary responses by the host plant to the impacts created by the gallmaker. However, because the impact of gall presence can come very early in the development of a host-plant clone, *Eurosta*'s presence in recently established, attacked genets could reduce the number of ramets composing that genet in later growing seasons and diminish its lifetime reproductive output. Under conditions of high *Eurosta* abundance, the reduction of new rhizome production coupled with the localized decrease of achene production in galled ramets could cause attacked genets to suffer a relatively lower rate of lifetime fecundity than unattacked genets. The outcome could be that attacked clones occupy less area and are composed of fewer ramets than unattacked genets. While there are several poorly tested assumptions in this simple scenario, it does serve to illustrate the suitability of this host-plant and gallmaker system for such an evolutionary study.

Host-Plant Resistance to Gallmaker Attack: The Plant-Gallmaker Encounter— The Plant's Perspective

4.1 VARIATION IN HERBIVORE INFESTATION

Variation in the levels of insect infestation on host plants is common among individual plants, plant genotypes, and populations (Fritz and Simms 1992; Weis and Campbell 1992). In many examples, this variation has been shown to be a consequence of genetic differences in the susceptibility of individual plant genotypes to insect-herbivore attack (Strong, Lawton, and Southwood 1984). Indeed, the role of genetic resistance to pests in determining infestation levels has been a major thrust of agronomic research. The preservation of genes from wild relatives that confer insect resistance in crop plants remains one of the most important goals of plant breeding. Resistance to specific pests has been transferred to plants using interspecific hybridization and backcrossing in several crops, including cotton (Knight 1954), potatoes (Ross 1966), and raspberries (Keep and Briggs 1971). Yet in other cases, historical factors such as variation in infestation levels in the previous growing season, host-plant alteration due to the effects of previous attack, or host-plant performance differences due to microsite variation have been implicated as determinants of current levels of attack (e.g., Karban and Carey 1984; Fowler and Lawton 1985; Haukioja and Neuvonen 1985; Karban 1990).

The importance of such historical factors has been illustrated by several studies. Craig, Price, and Itami (1986), for example, reported that attack by the sawfly *Euura lasiolepis* resulted in increased sprouting of new shoots by the willow *Salix lasiolepis*. Susceptibility of the attacked clone was enhanced because of the higher number of actively growing young shoots as compared to clones composed largely of older shoots. Still other studies (e.g., Zangerl 1990) have found an opposite response to herbivore attack in that attacked plants became more, not less, resistant (i.e., induced resistance). Schultz and Baldwin (1982), for instance, reported a decrease in susceptibility to gypsy moths in oak trees that had been previously attacked.

Unfortunately, the roles of historical and environmental factors (i.e., previous attack levels or resource supplies to the host plant) on resistance and susceptibility have received only limited attention in spite of the existence of ecological theories such as the carbon/nutrient balance hypothesis (Bryant et al. 1983; Iason and Hester 1993) and a rich agronomic literature. Agronomists, recognizing the importance of previous attack, have at least partially controlled local herbivore infestation levels by alternating between resistant and susceptible crops (Hull 1974; Russell 1978; Carter and Deeming 1980; Wright 1984). This approach is reasonably effective with insect pests that have limited dispersal or colonizing ability.

It is also clear that environmental factors such as resource supply to the plant can influence insect preferences for or performance on a given host genotype (e.g., Rossi and Strong 1991; Tisdale and Wagner 1991). Plant chemical composition, for example, has been shown to be influenced in fairly predictable ways by resource availability (e.g., Coley, Bryant, and Chapin 1985; Waring et al. 1985; Reichardt et al. 1991), and plant chemical composition in turn can alter herbivore preference or performance (e.g., Scriber 1984; Lincoln 1985; Collinge and Louda 1988; Lincoln and Couvet 1989; Horner and Abrahamson 1992). Oviposition preferences by adult insects as well as performance of their offspring are affected by host plant nutrient, water, and secondary metabolite content (Miller and Strickler 1984). However, as Thompson (1989) and Horner and Abrahamson (1992) have argued, it is likely that the genes involved in host-plant choice by adult females are different from those

103

that influence larval performance on a particular host plant. As a result, the host-plant qualities that may be important to an ovipositing female as she chooses among plant genotypes at best may be only correlated to plant attributes related to her offsprings' performance. Craig, Itami, and Price (1989) are probably correct in their suggestion that ovipositing females may be unable to accurately assess the complex suite of host-plant characters that influence larval success.

4.2 HISTORICAL VERSUS GENETIC FACTORS

Few studies have attempted to address the importance of both historical and genetic factors. Yet infestation levels in non-crop systems are likely to be the result of an interplay between local insect population numbers and the proportion of resistant versus susceptible host genotypes available to attack.

Herbivore Dispersal and Host Infestation

In a series of experiments designed to evaluate the relative contribution of host genetics and previous infestation levels, we framed two alternative hypotheses (McCrea and Abrahamson 1987). First, ball-gallmaker infestation levels are the consequence of genetic differences among host plants for plant traits important to oviposition preference or offspring performance. Second, attack levels are the result of the herbivore's ability to find host plants (i.e., some susceptible genotypes are not found due to limited dispersal).

We followed *Eurosta* infestation levels in more than one hundred small and distinct genotypes of *S. altissima* for three growing seasons. Each genotype was randomly assigned to one of three treatments—"control," in which we simply monitored gall occurrence; "stocked," in which we added galls until there was one gall per ramet; and "no galls," in which we removed all galls within the host-plant genotype. The manipulations were done during each winter when the gallmaker larvae were in diapause within their galls and the host genotypes were recensused each growing season. There were no statistical differences among the treatments, indicating that the stocking or removal of galls in a genotype did not affect the total number of galls nor the number of galls per ramet in the following year. Thus, even though the goldenrod ball gallmaker is reported to make few sustained flights (Uhler 1951), we found no evidence that the variation in ball gall infestation levels was the result of limited dispersal of female flies or proximity to a source of gallmakers. If low dispersal of flies from their point of emergence had been a major factor, then we should have found a higher number of galls present in genotypes that were artificially stocked and fewer galls in those genotypes that had all galls removed. Instead, there were significant, positive relationships for the number of galls and the number of galls per ramet in each genotype between years throughout the 3 years of the experiment. Furthermore, we identified significant differences among genotypes for the number of galls per ramet, the fraction of ramets ovipunctured, and the fraction of ovipunctured ramets that developed galls (fig. 4.1). These results suggest differences among host plants in acceptability to the ovipositing females and/or suitability for the developing larvae.

In another experiment, we asked how local gallmaker abundance and proximity to a source of gallmakers influence the total number of gallmakers present in subsequent growing seasons (McCrea and Abrahamson 1987). After censusing the pattern of gall dispersion within an isolated (by actively tilled agricultural fields) old field, we killed all the ball gallmakers. This was accomplished by mowing the field between the time of oviposition and the time when gallmaker larvae attain full size. We learned that while a major disturbance to a gallmaker population may alter the total number of gallmakers present in the subsequent generation, it did not alter the dispersion pattern of gallmakers within the field (table 4.1). We would expect that if the dispersion pattern among genotypes was due to chance discovery of host genotypes by females, then the dispersion pattern would change over the course of years. Rather, the results of this and the previous experiment suggest that the flies do disperse well under natural conditions.

Frequency of Ovipunctures and Galls

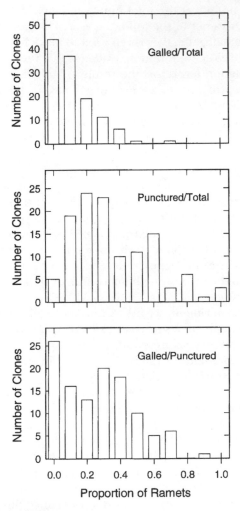

FIGURE 4.1. Frequency distributions of the proportion of ramets galled (*top*), the portion of ramets ovipunctured by *Eurosta* (*middle*), and the proportion of ovipunctured ramets that produced galls (*bottom*). Data are from 117 clones. (Redrawn from McCrea and Abrahamson 1987, "Variation in herbivore infestation: Historical vs. genetic factors," *Ecology* 68:822–827, with kind permission of The Ecological Society of America)

TABLE 4.1. Number of ramets of *Solidago altissima*, number of ball galls, and number of ball galls separated according to content.

| | Number of Ramets | Number of Galls | Gall Contents | | | | # Surviving Insects | % *Eurosta* Surviving |
			Eurosta	*Eurytoma gigantea*	*Eurytoma obtusiventris*	*Mordellistena unicolor*		
			Unmowed					
1982	700	123	36	21	22	10	34	29.3
			Field mowed prior to maturation of *Eurosta*					
1983	703	46	3	6	4	8	25	6.5
1984	585	40	5	5	6	16	7	12.5

Source: From McCrea and Abrahamson 1987.

Note: Counts from 121 0.5 × 0.5 m quadrats in the Elysburg mowing experiment.

The Genetic Component

Even under field conditions, gall infestation rate on golden-rod clones is consistent from year to year (McCrea and Abrahamson 1987; Maddox and Root 1990). Our 3-year study showed that the rank-order correlation among years for the number of galls per ramet varied from 0.50 to 0.64 (McCrea and Abrahamson 1987). To determine if this pattern of gallmaker attack observed in the field experiment was a consequence of genetic differences among host-plant genotypes, we transplanted replicates grown from sections of new rhizomes of each of 30 replicated genotypes (15 genotypes from each end of a resistance-susceptibility continuum) to a common garden (McCrea and Abrahamson 1987). The continuum was defined by the relative rank of the ratio of galled ramets to total ramets for the 129 genotypes of the 3-year marked genotype study. There are potential difficulties with a resistance and susceptibility continuum of host genotypes developed from an existing field pattern. Since such a continuum does not account for all the possible alterations of expression of host phenotypes due to environmental variations, it cannot be a completely reliable indicator of female preference for oviposition. We will discuss this more fully in section 4.3.

Eurosta were provided free access to the garden by placing a large stock of overwintered galls near the common garden. The pattern of attack in the common garden closely followed the pattern in the field (table 4.2). Thus, the same genotypes that had large numbers of galls in the field had high relative numbers in the common garden. The indication is that there is an important genetic component of host-plant resistance to the pattern of herbivore infestation levels seen in nature. However, our common garden experiment does not eliminate the possibility that environmental effects might play a crucial role in resistance since it is likely that some characteristics of ramet performance in the garden were the result of the conditions under which the parent ramet was growing in the previous year.

Our studies have also shown that genetic variation in *S. altissima* strongly influences *Eurosta*'s survivorship by affecting its vulnerability to natural enemies. As we will show in chapters 8 and

TABLE 4.2. Number of ramets of resistant and susceptible *Solidago altissima* clones from the common garden experiment that were punctured or galled.

	Number of Ramets		Number of Punctured Ramets	
	Punctured	Not Punctured	Galled	Not Galled
Resistant	72	99	3	69
Susceptible	94	81	43	51
	$\chi^2 = 4.67$		$\chi^2 = 35.18$	
	$P = 0.03$		$P < 0.001$	

Source: From McCrea and Abrahamson 1987.

9, gallmakers in small galls are more frequently attacked by the parasitoid wasp *Eurytoma gigantea*. This parasitoid penetrates the gall with its ovipositor to lay its egg within the gall's central chamber (Weis, Abrahamson, and McCrea 1985). Consequently, gallmakers in large galls escape the attack of this parasitoid because of the limitations of *E. gigantea*'s ovipositor lengths. *Eurosta* larvae in large galls do not necessarily have the highest survival because larger galls are subject to attack by downy woodpeckers and black-capped chickadees who peck open galls during winter to extract gallmaker larvae (Weis and Abrahamson 1986; Abrahamson et al. 1989; Weis, Abrahamson, and Andersen 1992). Although the insect genotype has a heritable influence on gall size (Weis and Abrahamson 1986; Weis and Gorman 1990), repeated experiments have found that approximately 20% of the variance in gall diameter is due to host-plant genotype. Furthermore, host-plant genotypes vary significantly both in final gall diameter and in gall growth rates (Abrahamson, Anderson, and McCrea 1988).

4.3 ENVIRONMENTAL INFLUENCES ON HOST RESISTANCE

Maddox and Cappuccino's (1986) study of *S. altissima* genotypes showed that resistance to herbivory by the aphid *Uroleucon nigrotuberculatum* was related to host-plant genotype and to water availability in the soil. However, plant genotypes varied in resis-

109

tance to this aphid only when genotypes were well watered. The consequence is that there is little or no genetic variation in resistance to this aphid in some environments. In another study using replicates of host-plant genotypes grown at multiple sites, Maddox and Root (1987) found that *S. altissima*'s resistance to several herbivores varied depending on the field in which the goldenrod genotype grew. These examples clearly indicate that there can be considerable variation in the expression of genetic variation for host-plant resistance.

In a series of experiments (Abrahamson, Anderson, and McCrea 1988; Horner and Abrahamson 1992 and submitted) we have examined the degree to which herbivore resistance is determined by host-plant genotype and by environment. Since plant carbon/nutrient balances have been implicated as an important factor in plant defensive chemistry (e.g., Bryant et al. 1983; Bryant, Chapin, and Klein 1983; Coley et al. 1985; Bryant 1987), we altered the environments of host plants through changes in mineral-element and/or light availability. Subsequently we observed the responses of host plants to the altered conditions while monitoring *Eurosta*'s oviposition preference and offspring performance.

Role of Mineral Element Supply

In an initial experiment, we addressed four questions. First, do goldenrods grow larger under conditions of enhanced mineral-element availability? Second, are host plants grown under conditions of improved mineral nutrition more successfully attacked? Third, is there an increase in gall growth rate or final gall size with increased mineral element availability? Finally, are herbivore defenses modified by resource availability? While we did not directly monitor the degree of defensive chemistry present, we instead inferred changes in resistance or the lack of changes by bioassay of host-plant acceptability to gallmaker attack and suitability for gall and herbivore growth (Abrahamson, Anderson, and McCrea 1988).

The rhizomes used to grow the plants for this study came from six *S. altissima* genotypes from the 3-year field study (McCrea and Abrahamson 1987). Each genotype was grown for two previous years in a common garden at Bucknell University. Three of the genotypes were classified as susceptible and three as resistant (representatives of the extreme ends of the resistance/ susceptibility continuum). Half the replicates of each genotype were fertilized weekly, while the other half of the replicates received no fertilizer. Ramets were provided with water as needed so that water was never a limiting factor. By using the mineral-element concentrations of field-grown ramets (Abrahamson and McCrea 1985), we were able to estimate the K and P contents of the new rhizome pieces of various sizes used in this experiment. Our conclusion was that the unfertilized control ramets in this study were strongly K-deprived and at least partially P-deprived relative to field-grown ramets of similar sizes. This initial experiment assayed offspring performance but did not monitor adult preference. Oviposition was accomplished under no-choice conditions by presenting single ramets, approximately three to four weeks in age, to individual, mated female gallmakers until the fly made at least one ovipuncture in the plant's bud.

A repeated-measures analysis of variance on the week-by-week growth in ramet height found that fertilizer treatment, host-plant genotype, and time were all significant main effects. The addition of fertilizer enhanced growth rates of all genotypes, but genotypes differed in their growth rates throughout the duration of the experiment. All interaction terms were significant, indicating that the genotypes responded differentially to fertilizer treatment and that mineral-element addition did not affect the growth of ramets equally during the experiment (Abrahamson, Anderson, and McCrea 1988). Furthermore, we found that total ramet biomass, as well as the biomass of all component organs, was increased by fertilizer addition but again that the individual genotypes differed in their response to enhanced mineral-element availability.

Importantly, we found no difference in the occurrence of galls by fertilizer treatment regardless of the level examined (i.e., fertilized versus control ramets or resistant versus susceptible genotypes; table 4.3). The latter result suggested that we should compare our experimental results to field data for the same genotypes. There were significant differences with susceptible

111

TABLE 4.3. Gall formation data from a nutrient-manipulation experiment and 1984 field data for the same *Solidago altissima* clones.

Clone	Number of Ramets		Mean Gall Diameter ± SD
	Punctured	Galled	
UNFERTILIZED			
*Susceptible**			
1	34	9	19.8 ± 3.7
2	33	6	16.8 ± 7.4
3	33	1	20.8 ± 1.0
*Resistant**			
4	27	4	18.2 ± 1.1
5	36	0	—
6	32	12	13.5 ± 5.9
All clones	195	42 (21.5%)	17.7 ± 5.3
FERTILIZED			
*Susceptible**			
1	34	12	21.2 ± 3.2
2	34	13	17.0 ± 8.5
3	32	8	21.8 ± 2.1
*Resistant**			
4	30	2	13.6 ± 1.5
5	34	0	—
6	31	9	13.3 ± 7.0
All clones	195	44 (22.6%)	

1984 FIELD DATA

	Total Initially Examined	Punctured	Galled
*Susceptible**			
1	86	53	35
2	43	33	17
3	89	43	28
*Resistant**			
4	55	9	
5	49	4	
6	78	0	

Source: From Abrahamson, Anderson, and McCrea 1988.
*Based on the occurrence of galls in the field.

clones for field-grown genotypes versus the same genotypes grown under experimental conditions, and ramets with and without galls. One resistant genotype exhibited a very different galling pattern in our experiment and in the field. In the field, we found that the genotype did not produce galls simply because it was not ovipunctured; apparently it was unattractive to ovipositing females. However, under our experimental no-choice, do-or-die conditions, fully one-third of the ramets of this genotype formed galls. This latter result suggests that there may be a number of reasons why a particular host-plant genotype is or is not resistant to herbivore attack, including that some genotypes are unacceptable to most gallmakers in spite of being highly susceptible. On the other hand, some genotypes may be resistant and form no galls in spite of being oviposited.

The similarity of gall occurrence rates in the treatment groups of this experiment could result from two possibilities. First, that our alteration of the mineral-element environment of the host plant had little effect on the goldenrod's resistance in spite of marked changes in the phenotypic expression of plant chemical defense against the ball gallmaker. If this was true, our result runs contrary to the predictions of the carbon/nutrient balance hypothesis and to findings with many other plant-herbivore systems (e.g., Bryant et al. 1983; Bryant, Chapin, and Klein 1983; Waterman, Ross, and McKey 1984; Gershenzon 1984; Bryant 1987).

Even though appreciable changes in host-plant growth were induced by the fertilizer treatment, there is a second interpretation of the similarity of gall occurrence rates between fertilized and control treatments. Since we did not measure plant defensive chemistry directly, it is possible that the carbon/nutrient balance was unaffected by the addition of fertilizer. If host-plant growth and defensive chemistry changed in proportion to each other, we would observe no net change in herbivore-infestation levels.

Interactions of Mineral Element and Light Supplies

To examine this possibility, we performed a second experiment in which we grew replicates of four genotypes of *S. altissima* (two each from the extreme ends of the resistance/susceptibility

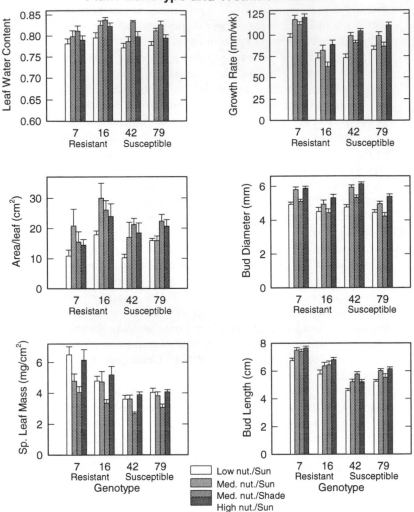

FIGURE 4.2. The effects of plant genotype and treatment on plant traits. Two genotypes, 7 and 16, represent clones resistant under natural conditions, and two genotypes, 42 and 79, represent clones susceptible under field conditions. Means ± standard errors of five to six ramets for leaf water content (as a proportion of fresh mass; *top left*), leaf areas (*middle left*), and specific leaf masses (*bottom left*), and of 38–48 ramets for growth rates (*top*

continuum) under low, medium, or high levels of nutrient supply in full sun or with medium levels of nutrients in 50% shade (Horner and Abrahamson 1992). By varying the resource supply to the host plant through altering both light and nutrient availability, we can change host-plant growth and host-plant defensive chemistry in nonproportional ways. Replicates of each genotype were attacked in the greenhouse by mass-released flies when ramets were at the same stage of plant development as when field attack occurs. Shaded replicates were removed from the shade treatment just before gallmaker oviposition since the purpose of this experiment was to determine the relative contribution of host-plant genotype and resource supply to the host plant, not whether this herbivore prefers light versus shaded environments.

This experiment found that a variety of host-plant characteristics including leaf-water content, average leaf area per leaf, specific leaf mass, bud diameter and length, and plant growth rate varied as a result of plant genotype, resistance level, nutrient supply, and/or light level (fig. 4.2; Horner and Abrahamson 1992). While the number of plants ovipunctured was affected by host-plant genotype, and there was a significant effect of the interaction between plant genotype and nutrient supply (i.e., one resistant and one susceptible genotype were most preferred), there were no significant main effects of resistance level, nutrient supply, or light level (fig. 4.3). An important point to remember is that the resistance or susceptibility of host-plant genotypes was defined by their relative rank of galled ramets to total ramets under field conditions. Thus, an unacceptable genotype that was highly suitable would be classified as resistant. Yet under no-choice or limited-choice ovipositional conditions such a genotype may become susceptible.

---

right), bud diameter (*middle right*), and bud length (*bottom right*). (Redrawn from Horner and Abrahamson 1992, "Influence of plant genotype and environment on oviposition preference and offspring survival in a gallmaking herbivore," *Oecologia* 90:323–332, with kind permission of Springer-Verlag)

Ovipuncture by Treatment

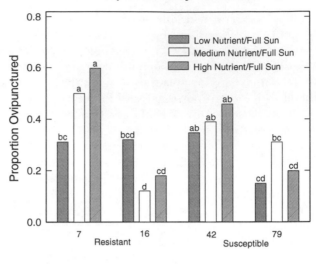

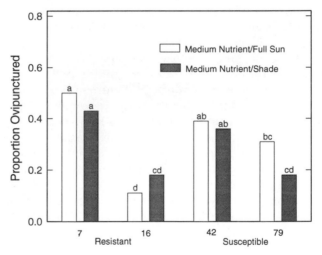

FIGURE 4.3. Distribution of the proportion of ramets ovipunctured among genotypes grown under different conditions of nutrient supply (*top*) and light levels (*bottom*). Two genotypes, 7 and 16, represent clones resistant under natural conditions, and two genotypes, 42 and 79, represent clones susceptible under field conditions. Bars with common letters are not significantly different (<0.05). (Redrawn from Horner and Abrahamson 1992, "Influence of plant genotype and environment on oviposition preference and offspring survival in a gallmaking herbivore," *Oecologia* 90:323–332, with kind permission of Springer-Verlag)

A regression of the percentage of plants ovipunctured on various physical attributes of the host plant (e.g., growth rate, height, bud length and diameter, water content, leaf area, and specific leaf mass) found that bud diameter, leaf area, and water content explained about 74% of the variation in the number of plants ovipunctured (Horner and Abrahamson 1992). Since bud diameter was strongly correlated with ramet growth rate during oviposition in this experiment, we suspected and confirmed that female *Eurosta* were preferentially ovipositing in plants with higher growth rates. This preference for host plants with high growth rates has been recorded in other gallmakers as well as in other of our experiments with *Eurosta* (Craig, Price, and Itami 1986; Craig et al. 1988; Craig, Itami, and Price 1989; McCrea and Abrahamson 1987; Price, Waring, and Fernandes 1987). This preference for rapidly growing ramets may originate from their need of undifferentiated host tissue (Abrahamson and Weis 1987).

This experiment also found that the number of surviving larvae was significantly different among the four host-plant genotypes than would be expected from a random distribution among the ovipunctured plants. As anticipated, the susceptible genotypes formed the expected number of or more galls, while the resistant genotypes formed fewer than predicted. However, when we analyzed gall formation and larval survival for the effects of nutrients and light, we found no significant effect of treatment. Again, this result emphasizes the importance of plant genotype and the somewhat limited role of plant environment in determining resistance to gallmaker attack, the same inference reached in the initial study (Abrahamson, Anderson, and McCrea 1988; but see Obeso and Grubb 1994, and Hartvigsen, Wait, and Coleman 1995). This is a noteworthy conclusion because enhanced resource availability did have the expected effects on plant factors such as growth rate. Host-plant growth rate is likely a critical characteristic as there are indications that host growth rate is positively related to larval survival in several gallmakers (Price, Waring, and Fernandes 1987; Anderson et al. 1989; Craig, Itami, and Price 1989).

Resistance level of the genotypes used in this experiment was not a reliable indicator of female preference for oviposition since the more preferred genotypes came from both the resistant and the susceptible groups (Horner and Abrahamson 1992). However, nutrient supply to the host plant did influence adult preference for at least the more-preferred genotypes. The number of plants attacked tended to increase with enhanced fertilizer treatment among ramets of the more-preferred genotypes. Our conclusion was that the relative attractiveness of at least the more-preferred host genotypes can be influenced by the availability of nutrients.

Water Supply Effects on Preference

In a third experiment, we examined the influence of host-plant genotype and early-season water deficits on oviposition preference and offspring performance (Horner and Abrahamson, submitted). This study addressed a number of questions including whether transient environmental conditions, such as early-season water deficit, can alter the suitability of a host genotype. To answer this question, we grew replicates of eight genotypes of *S. altissima* representing the range of resistance to *Eurosta* under greenhouse conditions. Replicates were provided either a control (i.e., sufficient water to remain turgid between waterings) or low-water regime (i.e., half the volume supplied to control plants, which resulted in wilting between waterings). These treatments were initiated after replicates were established (3 weeks before oviposition) and were continued for 4 weeks after oviposition. Earlier studies (Abrahamson, Anderson, and McCrea 1988; Anderson et al. 1989) had indicated that early larval establishment is a particularly vulnerable period, and early-season water deficits occur with some regularity in the field (Sumerford, Abrahamson, and Weis, unpub. data). After manipulating insect emergence times by incubating galls in an environmental chamber, we mass released sixty females and sixty males at the stage of ramet development typical during the oviposition period in the field.

The early-season water deficit treatment had the expected effects on host plants. Low-water treatment ramets had significantly more negative xylem pressure potentials and exhibited reduced ramet height, ramet growth rate, bud length, bud diameter, leaf water content, and average area per leaf at the time

118

of oviposition. These host-plant characters were also affected significantly by plant genotype. However, a number of these characters, including ramet height, bud diameter, leaf water content, average area per leaf, and specific leaf mass, were not significantly influenced by the interaction of plant genotype and treatment. Thus, for these characteristics, genotypes were responding to water deficits in similar ways. However, other traits (e.g., ramet growth rate, bud length, and gall formation) potentially related to infestation levels among genotypes were affected significantly by the interaction of host genotype and treatment. For example, while a greater proportion of control ramets formed galls in some genotypes, in other genotypes the low-water treatment ramets developed more galls. Finally, the interaction between plant genotype and treatment did not significantly influence oviposition preference.

The conclusion of this water-deficit study parallels those of earlier studies (Abrahamson, Anderson, and McCrea 1988; Anderson et al. 1989; Horner and Abrahamson 1992): that plant genotype strongly influences the occurrence of *Eurosta* galls through its effects on oviposition preference and the proportion of ovipunctured plants that form galls. However, even though early-season water deficits affected oviposition preference among host genotypes, this transient environmental effect did not influence host genotypes differentially. Ovipositing females preferred control plants over those in the low-water regime regardless of their genotype (Horner and Abrahamson, submitted).

Taken together, the Abrahamson, Anderson, and McCrea (1988) and Horner and Abrahamson (1992 and submitted) studies indicate that environmental influences, such as enhanced mineral-element availability or early-season water deficits, can influence host-plant traits. However, in spite of significant alterations to numerous plant characteristics, host-plant resistance in *S. altissima* appears to have a strong genetic component.

Host Resistance to Ball Gallmakers and Other Herbivores

Maddox and Root (1987) examined the genetic and environmental components of *S. altissima*'s resistance to sixteen diverse

119

species of herbivorous insects and found that the observed phenotypic variation in insect attack levels was correlated with underlying genetic variation in resistance for about two-thirds of the species examined. Using a quantitative genetic analysis of heritability in the narrow sense (i.e., the proportion of phenotypic variance that is from additive genetic variation), Maddox and Root (1987) examined genotype resistance to herbivory as the number of herbivores attacking the genotype relative to other genotypes. A heritable difference among host-plant genotypes would indicate the degree to which *S. altissima*'s resistance to herbivores is genetically based.

These authors found that the differences among host-plant genotypes were based on genetically correlated aspects of plant phenotypes (table 4.4). However, the correlations were less than 1.0, indicating that other factors, including microsite differences and insect behavior, were also contributing to insect-attack patterns. There were significant differences in herbivore attack levels among host plants for resistance to fifteen of sixteen commonly recorded herbivores. Interestingly, the degree of differences in herbivore attack depended on insect species. For example, infestation levels by stem-galling insects such as *Eurosta, Epiblema scudderiana* (a moth that creates stem galls), and *Rhopalomyia solidaginis* (a midge that forms a rosette gall at the goldenrod's apex; Raman and Abrahamson 1995) were more strongly correlated with plant genotype than was attack by free-feeding herbivores (table 4.5). Genetic correlations of insect resistance between years in a garden ranged from 0.79 to 0.92 for *Eurosta,* from 0.75 to 0.90 for *Epiblema,* and from 0.60 to 0.81 for *Rhopalomyia.* Such patterns indicate that successful gall development depends on very specific and well-tuned responses to host-plant variation in traits (Strong, Lawton, and Southwood 1984; Abrahamson and Weis 1987; Maddox and Root 1987).

More recently, Maddox and Root (1990) investigated the responses of a diverse set of seventeen insect herbivores to genetic variation among host-plant genotypes. The four herbivore groups shown in figure 4.4, which Maddox and Root (1990) termed "herbivore suites," emerged from a cluster analysis on the genetic correlations among the herbivores' responses. These herbivore suites consisted of insect species that were

TABLE 4.4. Analysis of variance results, Spearman rank correlations of genotypes between years, and rank correlations of genotypes between Hamilton Reserve (NY) and the parental garden for resistance to seven insects common at Hamilton Reserve.

Species	R^2 1982	R^2 1983	Rank Correlations 1982–1983	Field–Garden
Philaenus	0.354**	0.396**	0.567*	0.140
Uroleucon nigrotuberculatum	0.653**	0.601**	0.492*	0.880**
Trirhabda	0.927**	0.899**	0.796**	0.767**
Asteromyia	ND	0.286*	—	0.429*
Rhopalomyia	0.230*	0.368**	0.435*	0.509**
Eurosta	0.705**	0.697**	0.826**	0.899**
Ophiomyza/Phytomyza	0.160	0.230	0.183	0.332

Source: From Maddox and Root 1987.
*$P < 0.10$, ** $P < 0.05$, ND = no data

TABLE 4.5. Genetic correlation of insect resistance of *Solidago altissima* between years in a parental garden.

Insects	Genetic Correlations 1982–1983	1983–1984	1984–1985
Lygus	—	—	0.48*
Corythuca	0.16	−0.20	0.08
Philaenus	0.33	0.67*	0.50*
Uroleucon caligatum	0.39	0.60*	0.55*
U. nigrotuberculatum	—	—	—
Exema	—	—	0.45*
Microrhopala	—	0.15	0.75*
Ophraella	—	—	0.30
Trirhabda	0.53*	0.49*	0.61*
Epiblema scudderiana	0.75*	0.83*	0.90*
Epiblema spp.	—	—	0.68*
Asteromyia	—	0.56*	0.44*
Rhopalomyia	0.60*	0.79*	0.81*
Eurosta	0.88*	0.92*	0.79*
Ophiomyza/Phytomyza	0.41*	0.52*	—

Source: From Maddox and Root 1987.
Notes: Since these are clones, correlations are of total genotypic effects. *Slaterocoris* did not occur in the parental garden. *Uroleucon nigrotuberculatum* occurred only in 1985. *$P < 0.05$

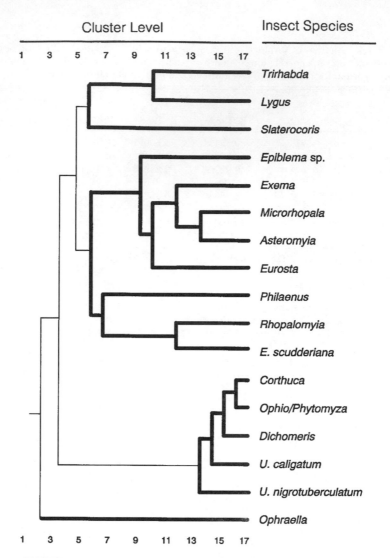

Cluster Level   Insect Species

1   3   5   7   9   11 13 15 17

Trirhabda

Lygus

Slaterocoris

Epiblema sp.

Exema

Microrhopala

Asteromyia

Eurosta

Philaenus

Rhopalomyia

E. scudderiana

Corthuca

Ophio/Phytomyza

Dichomeris

U. caligatum

U. nigrotuberculatum

Ophraella

1   3   5   7   9   11 13 15 17

FIGURE 4.4. Cluster diagram based on the genetic correlations among herbivore responses to genotypes of *Solidago altissima* growing in two gardens at Whipple Farm near Ithaca, N.Y. Four herbivore suites emerged from a diverse set of seventeen herbivores examined. The herbivores in a suite were attracted or repelled by similar host genotypes. (Redrawn from Maddox and Root 1990, "Structure of the encounter between goldenrod [*Solidago altissima*] and its diverse insect fauna," *Ecology* 71:2115–2124, with kind permission of The Ecological Society of America)

attracted or repelled by similar plant genotypes. Whereas free-feeding herbivores tended to fall into most herbivore suites (leaf chewers in all four suites and suckers/raspers into three suites), mining and gallmaking herbivore species tended to have similar responses to the same host plant genotypes. All four gallmaking species examined in their study (i.e., *Eurosta, Rhopalomyia, Epiblema,* and the Cecidomyiid leaf gallmaker *Asteromyia carbonifera*) exhibited strong affinity with the same host genotypes. Among these four gallmakers, the three stem gallmakers were more correlated to one another in genotype preference than they were to the leaf gallmaker (Maddox and Root 1990). However, the gallmaker suite also contained non-gallmakers, including two leaf miners, a leaf chewer that lives in a case built of its frass, and a spittlebug that taps xylem fluids. Yet most of the species in this suite share the trait of feeding on internal host tissues. The Maddox and Root (1990) study extends the conclusions of our investigations (McCrea and Abrahamson 1987; Abrahamson, Anderson, and McCrea 1988; Horner and Abrahamson 1992 and submitted) in showing that resistance in *S. altissima* to a variety of herbivores is genetically based and, to a lesser extent, phenotypically based. An important corollary, however, is that no one host genotype is resistant to all herbivore species. This makes frequency dependent selection of host-plant genotypes a possible consequence of variation in herbivore abundance over time.

The existence of heritable variation for resistance to several herbivores indicates that a host plant such as *S. altissima* is at least capable of an evolutionary response to selection by herbivores (Maddox and Root 1987). Whether a host-plant genotype responds to selection depends not only on the presence of heritable variation in the host plant but also on the intensity and duration of the herbivore impacts.

4.4 MECHANISMS OF HOST-PLANT RESISTANCE

The nonrandom distribution of *Eurosta* galls between and within field sites is striking to anyone who has ever noticed these conspicuous stem galls. Within fields, for example, some *S. altissima* genotypes have no galls while others may support a gall on nearly every ramet (Hess, Abrahamson, and Brown, 1996). Among those host-plant genotypes that support *Eurosta* galls, some hosts have consistently smaller galls while others have galls larger than average (Weis and Abrahamson 1986). Such observations have led us to explore the mechanisms behind such distribution patterns.

Numerical Relations of Host and Gallmaker

A considerable portion of the observed field-level variation in *Eurosta* gall occurrence is due to the density of host-plant ramets (Abrahamson, Armbruster, and Maddox 1983). Our study of numerical relations among trophic levels found that all measures of occurrence of *S. altissima,* including ramet density, dominance, and frequency, were strongly and positively correlated with the abundance of *Eurosta*. A regression of ramet densities at twenty field sites explained over 77% of the variation in *Eurosta* gall density (Abrahamson, Armbruster, and Maddox 1983). However, within fields, the same simple numerical relations between host-plant ramet density and *Eurosta* gall density do not explain the patterns of gall occurrence. Host-plant genotypes with a large number of ramets are not necessarily more attacked than genotypes with a small number of ramets (R. L. Walton, unpub. data).

There are a variety of possibilities besides historical and numerical factors for the nonrandom, host-genotype use just described. These include variation in host-plant characters such as defensive chemistry (Cooper-Driver and LeQuesne 1987; Bosio et al. 1990; Abrahamson, Anderson, and McCrea 1991), physical defenses (Anderson et al. 1989), or genetically determined reactions of the host plant that are lethal to the gallmaker (Anderson et al. 1989).

Host-Plant Defensive Chemistry

While *S. altissima* is attacked by a variety of herbivorous insects from many feeding guilds (e.g., gallmakers, leaf chewers and miners, stem borers; Maddox and Root 1987, 1990), the majority are specialists that either restrict their attack to the genus

TABLE 4.6. Effects of *Solidago altissima* ($X \pm$ SD) on the growth and development of *Trichoplusia ni* (Lepidoptera: Noctuidae).

	Larval Width Day 7—mm	Pupal Mass g	n	# Eggs Laid	n
Normal diet	1.65 ± 0.344	0.240 ± 0.020	119	978.9 ± 211.6	15
Leaf powder	0.52 ± 0.114	—	—	—	—
Hexane extract	0.48 ± 0.286	—	—	—	—
Residue	1.08 ± 0.295	0.206 ± 0.020	29	959.0 ± 126.2	15
Cellulose control	1.08 ± 0.275	0.212 ± 0.015	9	—	—

Source: From Bosio et al. 1990.

Solidago or infest only specific species within the genus. The lack of appreciable numbers of generalist insect herbivores on *S. altissima,* for instance, implies that this host plant possesses defensive chemistry that either repels or kills polyphagous herbivores that could potentially infest it.

To evaluate the potential defensive role of *S. altissima* chemistry against generalist herbivores, we bioassayed both live plants and chemical extracts of plant leaves for their effects on survivorship and development of larval cabbage loopers (Noctuidae: *Trichoplusia ni*), a generalist moth (Bosio et al. 1990). Larvae raised on intact *S. altissima* plants had very low survivorship (about 11% to pupation and less than 6% to adulthood) relative to larvae fed on a normal control diet (87%; see Nitao and Berenbaum 1988 for details of this diet). Furthermore, development times were extended and the pupal masses were reduced in the surviving *T. ni* on intact tall goldenrods (table 4.6). Our observation that leaf damage on living host plants increased throughout the experiment suggested that the effect of the plant defensive chemistry was to poison the moth larvae rather than to deter their feeding (Bosio et al. 1990).

This study included a second series of experiments in which we separately incorporated lyophilized *S. altissima* leaves, leaf-powder residue (the powder was extracted with a series of solvents to remove chemicals with varying polarities), and the resultant hexane, methanol, and water extracts into an artificial diet. As illustrated in figure 4.5, moth larvae survived better on

125

Survivorship of a Generalist Herbivore

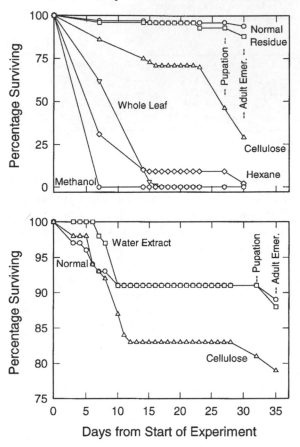

FIGURE 4.5. Survivorship of *Trichoplusia ni* (Lepidoptera: Noctuidae, cabbage looper) larvae in Run 1 (*top*) and Run 2 (*bottom*) fed artificial diets containing normal, cellulose control, whole-leaf powder of *Solidago altissima*, hexane, methanol, water extracts (extractions were done in the order listed), and leaf-powder residue after extraction. Leaf powder was extracted with different solvents to remove compounds of different polarities. It was not possible to run all the extract treatments at once, so the water extract was run separately in Run 2. In Run 1, 50 g of whole-leaf powder, the cellulose control, the hexane, and methanol extract-treated cellulose powders, and postextraction residues were incorporated into separate portions of the modified diet. Run 2 consisted of the water extract, cellulose control, and

the leaf-powder residue and water-extract treatments than on the cellulose control. This pattern indicates that *S. altissima* does contain nutrients that are usable by this generalist herbivore. However, larvae reared on either the whole-leaf powder (i.e., unextracted) or the hexane-extract treatments were less than one-third the size (i.e., larval width) of larvae living on the normal control diet after only one week. The mortality rates of moth larvae on the whole-leaf powder, methanol-extract, and hexane-extract diets were much higher (100%, 100%, and 97%, respectively) than for larvae reared on the cellulose-powder control diet (17%). Observations of the larvae that were fed methanol-extract diets showed that the diet was barely disturbed by the larvae, suggesting that they were not feeding. This result indicates that the methanol fraction contains some deterrent compound or compounds. On the other hand, larvae offered the hexane-extract diet did feed and a few did survive to pupate. This outcome suggests that the high mortality observed in this treatment was due to the presence of toxic chemicals in this fraction.

We can conclude from these experiments (Bosio et al. 1990) that *S. altissima* contains chemicals that both repel and poison the generalist *T. ni*. Furthermore, the increase in larval mortality that we measured is likely the consequence of more than one class of chemical compound since effects were found in both the methanol and hexane fractions.

The Compositae is a family well known for its distinctive chemistry involving numerous terpenoid compounds. Of particular interest in *Solidago* species are diterpenoids that have been shown to have a wide range of biological activities including the inhibition of insect larval growth and feeding (Cooper-Driver and LeQuesne 1987). Many of the diterpenes and polyacetylenes that have been isolated from *Solidago* species are toxic

---

normal artificial diet treatments. Otherwise, Run 1 and Run 2 were conducted identically. (Redrawn from Bosio et al. 1990, "Defense chemistry of *Solidago altissima*: Effects on the generalist herbivore *Trichoplusia ni* [Lepidoptera: Noctuidae]," *Environmental Entomology* 19:465–468, with kind permission of The Entomological Society of America)

to specialist herbivores (LeQuesne et al. 1986; Cooper-Driver and LeQuesne 1987; and Lu et al. 1993). However, much work remains to be done before we will understand the defensive chemistry of *Solidago*.

Host-Plant Physical Defenses

To assess the role of physical and/or mechanical defenses in resistance to *Eurosta*, we measured trichome densities and lengths for the undersides of leaves (the leaf undersides face *Eurosta* as they evaluate host-plant buds) of constant age from both susceptible and resistant host genotypes growing in a common garden (Anderson et al. 1989). While host-plant genotypes did vary significantly for both trichome density and length, trichome characteristics were not related to resistance to *Eurosta*. Thus trichome characteristics do not appear to be related to defense against *E. solidaginis*.

We examined other morphological characteristics of the host plant in a greenhouse experiment using both resistant and susceptible host-plant genotypes (Horner and Abrahamson 1992). We found that the percentage of plants oviposited by *Eurosta* was related to several physical attributes of the host, including bud diameter, leaf area, and leaf water content. Each of these characters significantly contributed to the percentage of ramets ovipunctured. Bud diameter and leaf water content both had positive relationships, while leaf area had a negative relationship with the percentage of ramets ovipunctured. Taken together, these variables explained approximately 74% of the variation in ovipuncture occurrence. It is not clear, though, how any of these characteristics would be related to host defense. In fact, in two more recent greenhouse experiments (How, Abrahamson, and Craig 1993; Horner and Abrahamson, submitted), ramet bud diameter was not related to oviposition occurrence. To date, we have not identified any morphological characters that are directly related to defense.

Necrosis and Larva-Killer Host Plants

We conducted several related investigations to explore the failure of some goldenrod genotypes to develop galls in spite of being ovipunctured and oviposited (Anderson et al. 1989). In a field experiment, we determined the percentage of ramets ovipunctured, the number of unhatched and hatched eggs, and larval survival in thirty-eight large genotypes (i.e., >300 ramets each). For the purposes of this experiment, we defined and examined three principal steps leading to gall formation. First, the female *Eurosta* must locate and ovipuncture an appropriate host-plant bud. Second, once the ovipositor is inserted into the host's bud, she must determine whether to inject an egg (i.e., the ramet's bud must be acceptable). Third, the larva must succeed in stimulating the host plant to develop a gall (i.e., the ramet must appropriately react to the larva's stimulus to be a suitable host).

We learned that the primary reason ramets did not form galls was that they were never ovipunctured (47%; fig. 4.6, top left). A smaller fraction of ramets (18%; fig. 4.6, bottom left) had an egg injected but still did not develop a gall, and an even smaller proportion was ovipunctured but did not receive an egg (11%; fig. 4.6, top right). We suspect that this ranking also reflects the relative importance of these three stages in the host plant's resistance to *Eurosta* attack (Anderson et al. 1989).

To examine this ranking, we used path analysis to determine the relationships among this set of hierarchical variables. Path models are useful because they demand explicit decisions concerning the structure of the causal relations among variables, and once the structure is established path models allow the decomposition of correlations into direct and indirect effects. Our path analysis confirmed that the most significant factor in determining gall occurrence within host genotypes was variation in the percentage of ramets ovipunctured (fig. 4.7). Our path model further suggested that the factors controlling ovipuncturing were the most crucial. That is, the decision of whether to oviposit an egg into the bud after the bud was ovipunctured was relatively unimportant in determining the percentage of ramets galled within a host plant. This result was the consequence of *Eurosta* females injecting an egg into the majority of buds they ovipunctured (77%; fig. 4.6, bottom right). The variation in the percentage of ramet buds that received an egg was so small that this variation did not appreciably add to the variation in the

129

Eurosta Oviposition

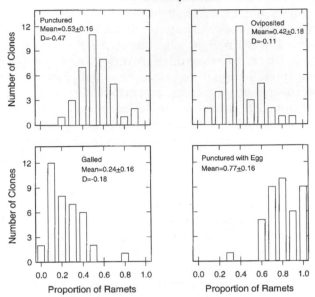

FIGURE 4.6. Results of an adult-choice study expressed as the proportion of ramets within each clone. Means ± standard deviations and frequency distributions of the proportion of ramets ovipunctured (*top left*), the proportion of ramets in which oviposition occurred (*top right*), the proportion of ramets galled (*bottom left*), and the proportion of ovipunctured ramets into which an egg was injected (*bottom right*). D values represent the change in the proportion of the total ramets reaching each stage of gall formation (e.g., in *top left,* the proportion of ramets not producing galls because they were not ovipunctured). (Redrawn from Anderson et al. 1989, "Host genotype choice by the ball gallmaker *Eurosta solidaginis* [Diptera: Tephritidae]," *Ecology* 70: 1048–1054, with kind permission of The Ecological Society of America)

percentage of ramets of a host genotype that developed galls (fig. 4.7).

However, we were left wondering why some host-plant genotypes that are extensively ovipunctured rarely develop galls. In an attempt to explore more thoroughly the relationship among ovipuncturing, egg injection, and larval survival, we dissected ovipunctured buds at weekly intervals following oviposition from four of the thirty-eight host-plant genotypes used in the field

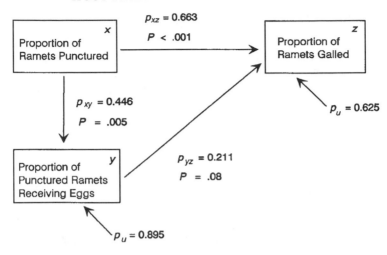

FIGURE 4.7. Path model for the stages of gall development. Values on arrows connecting boxes are the standardized partial regression coefficients and their significances. Values on arrows not originating on boxes represent residual errors. The model explained a total of 61% of the variation observed in the percentage of ramets in clones that developed galls. (Redrawn from Anderson et al. 1989, "Host genotype choice by the ball gallmaker *Eurosta solidaginis* [Diptera: Tephritidae]," *Ecology* 70: 1048–1054, with kind permission of the Ecological Society of America)

study (Anderson et al. 1989). To maximize our chances of finding differences among host plants, we used the two most resistant and two most susceptible host genotypes. Ramets of each host genotype were examined and marked daily for ovipuncture scars (see fig. 2.6) throughout the oviposition period.

These bud dissections showed that all the larvae found in the folded bud leaves of the susceptible genotypes were alive whereas only two-thirds of the larvae detected in the two resistant host plants were alive (table 4.7). More significant, however, was the finding that only about 27% of the larvae that had chewed to the meristem regions of resistant genotype buds were still living, while fully 96% of the larvae were alive in the meristematic regions of susceptible-genotype buds. In approximately 72% of the cases of larval death, the plant tissues surrounding the larva were necrotic, and this necrosis appeared to be related

TABLE 4.7. Mortality of *Eurosta solidaginis* larvae in ramets of *Solidago altissima*.

A. Mortality of *Eurosta* larvae in the developing bud leaves or meristem of resistant or susceptible ramets.

	Resistant Ramets		Susceptible Ramets	
	Dead	Alive	Dead	Alive
Bud leaves	3	6	0	21
Meristem	30	11	2	51

B. Two-tailed probabilities from Fisher's Exact Test of larval mortality in bud leaves versus meristem (tissue) within resistance levels and resistant versus susceptible ramets within each tissue.

Mortality vs.:	Within	Probability
Tissue	Resistant ramets	0.047
	Susceptible ramets	n.s.
Resistance level	Bud leaves	0.021
	Meristem	$\ll 0.001$

Source: From Anderson et al. 1989.

to the larva's death. Even more striking was that all but one case of necrosis took place in the two resistant host-plant genotypes. The reactions of the two resistant genotypes in our study suggest an important host-plant resistance mechanism. Through necrotic, hypersensitive reactions in response to gallmaker larval stimulation, the host plant can save an appreciable amount of plant tissue that would otherwise be directed away from functions that directly benefit the plant. The necrotic region, surrounding the very young first-instar larva, consists of highly pigmented, dark-brown cells. The larva dies very quickly after this hypersensitive reaction occurs and well before the development of a gall. Since the initial observation of a necrotic reaction related to *E. solidaginis* stimuli, we have recorded a similar reaction in other genotypes (How, Abrahamson, and Craig 1993; Hess, Abrahamson, and Brown 1996).

Such a hypersensitive response could potentially kill the herbivore in either of two ways. The insect could die from either contact with or ingestion of chemical poisons associated with the necrosis. Alternatively, the larva could simply die from starvation since the hypersensitive response effectively separates the larva from its food source. This type of resistance to gallmakers is particularly efficient because the gallmaker is mobile only within the gall and cannot move to other parts of the host plant, and the tissue sacrificed by the host plant is bound to develop into abnormal gall tissues that do not support the plant (Abrahamson, Anderson, and McCrea 1991).

The Role of Phenolics in Resistance

Resistance to herbivore attack due to hypersensitive, necrotic reactions has been identified in a number of gallmaker/host-plant interactions (e.g., Fernandes 1990; Westphal et al. 1992; Bentur and Kaslode 1996). Furthermore, several studies have suggested a link between necrotic reactions and phenolic compounds in the plant's defense against both herbivores and pathogens (e.g., Kosuge 1969; Westphal, Bronner, and LeRet 1981; Bazzalo et al. 1985). Rohfritsch (1981), for example, found that the spruce *Picea excelsa* produced a necrotic response to the stimuli created by the gallmaking aphid *Chermes abietis*. In this spruce-aphid system, the hypersensitive plant tissue reaction was accompanied by the accumulation of phenolic compounds. Rohfritsch (1981) suggested that the phenol accumulation inhibited aphid feeding.

These reports from other gallmaker/host-plant systems led us to hypothesize that phenolic defenses were part of *S. altissima*'s resistance to *Eurosta* gallmakers. However, as we showed in chapter 3, our experiments (Abrahamson, Anderson, and McCrea 1991) resulted in the rejection of this hypothesis. We found that the phenolic compounds in both the water and acetone extractions from bud leaves and meristematic tissues of normal, unpunctured, and ungalled hosts increased over time as the host plants grew. Furthermore, the phenolic level of water-extracted bud leaves and meristems of ungalled ramets of susceptible genotypes were significantly higher than the levels for the same tissues of resistant host genotypes. In addition, the initiation of normal gall development increased, not decreased, the phenolic concentrations (approximately 2- to 5-fold, depending on the stage of gall development) of both the water-

and acetone-extracted meristematic tissues relative to ungalled tissues. Finally and most important to our hypothesis, buds with a hypersensitive, necrotic reaction to gallmaker presence did not exhibit elevated levels of total phenolics (fig. 3.1). Thus, the hypothesis that the hypersensitive response would raise total phenolic levels was not supported by our results. These data suggest that it is unlikely that phenolic defenses are involved in *S. altissima*'s necrotic reaction to *E. solidaginis*.

It remains a possibility, however, that individual phenolic compounds were increased or decreased in response to necrosis even though total phenolic levels did not increase with necrosis. Our analysis examined total phenol levels and did not investigate the individual phenolic compound concentrations. It is also possible that the Folin-Ciocalteau method may have detected qualitative as well as quantitative differences in phenolic composition. This would happen if a specific set of strongly reacting (with the Folin-Ciocalteau reagents) phenols were present (Abrahamson, Anderson, and McCrea 1991).

Our study is not the first to report enhanced phenolic levels associated with normal gall development. Hartley (1992) surveyed total phenolic levels in ungalled and galled tissues associated with a wide array of gallmakers. She reported that phenol levels in gall tissues were either higher or not significantly different from phenolic levels in ungalled tissues in the vast majority of cases. However, elevated levels of phenolics have not consistently been found in gall tissues. Purohit, Ramawat, and Arya (1979) found that while mite-caused (*Eriophyes prosopidis*) leaf galls on the legume *Prosopis cineraria* had higher than normal phenolic levels, stem galls on this legume induced by an unknown chalcid had lower than normal phenolic amounts. Lower than normal leaf phenolic levels were also reported by Joshi, Tandon, and Rajee (1985) in leaf-roll galls on *Camellia sinensis* and *Elaeocarpus lancifolius* that were induced by "some aphids." Additional examples of reduced phenolic levels in gall tissues are summarized in Hartley's (1992) study.

Phenolic compounds associated with many galls are probably not related to host-plant resistance to gallmakers. Instead, we can offer several alternative hypotheses to explain why the overwhelming majority of the galls examined have elevated total phenolic concentrations associated with normal gall development. First, the phenolic increases are involved in the protection of the gallmaker from its natural enemies (e.g., parasitoids) or pathogens such as fungi that invade the gall chamber (Taper and Case 1987). According to this hypothesis, gallmaking insects stimulate their host plant to produce phenolic compounds for their own protection (Hartley 1992). If this were the case and if phenolics negatively affect gallmaker feeding, we might expect that elevated phenolic concentrations would only be found in the outer portion of the gall. The phenolic levels of the inner, nutritive zone of the gall would likely have normal, or lower-than-normal phenol concentrations. This is not the case, however, in the *Eurosta* gall. We found elevated levels of phenolics in the host meristematic tissues next to the developing larva (Abrahamson, Anderson, and McCrea 1991).

The more probable alternative explanation for *E. solidaginis* is that phenols are involved in the normal development of galls. For *Eurosta*, we found increased levels of phenolic compounds in galls of both susceptible and resistant host genotypes; however, the susceptible genotypes had the higher endogenous concentrations of phenols (Abrahamson, Anderson, and McCrea 1991). It may be that the gallmaking larva's ability to manipulate its host is at least partially dependent of the host's ability to synthesize phenolics. Susceptible host plant genotypes may, for example, require less stimulation for normal gall development than a host with a lesser propensity for phenol production. An additional possibility is that the higher phenolic concentrations of the susceptible host genotypes make them more attractive and hence more acceptable to ovipositing *Eurosta* (Abrahamson, Anderson, and McCrea 1991).

4.5 THE HIERARCHY OF PLANT RESISTANCE

At a minimum, a plant's resistance to insect herbivory results from traits that influence host acceptability (i.e., herbivore preference) and those that affect host suitability (i.e., herbivore growth and survival) in supporting the herbivore's feeding or its offspring's development. Acceptability of a host plant to a gallmaker likely involves chemical and/or physical traits that are used in determining whether oviposition will occur. Herbivore preference, a measure of acceptability, is usually measured by the frequency of oviposition on a particular host relative to other plants. Suitability of a plant for larval development, on the other hand, is related to host traits such as growth rate or reactivity to gallmaker stimuli. Host-plant suitability can be measured by herbivore growth, survival, and fecundity after attack (Anderson et al. 1989; Abrahamson, Anderson, and McCrea 1991; Horner and Abrahamson 1992).

Successful exploitation of a host plant by a gallmaker requires that the adult female first recognize and accept a given plant for oviposition, and second, once accepted, that the host plant be suitable for the offspring's development (Horner and Abrahamson 1992 and submitted). Thus, host-plant resistance is hierarchical with varying degrees of acceptability and suitability. The consequence of these two steps and their interaction is a continuum of resistance levels from very resistant (e.g., a host that is both unacceptable and unsuitable) to very susceptible (a plant that is both very acceptable and very suitable) host plants. The outcome of this hierarchy is that host-plant resistance to a herbivore occurs whenever either acceptability or suitability fails.

While there has been a long-standing interest in the factors that influence the resistance of plants to herbivores, only recently has attention shifted to the relative contribution of plant genotype, environment, and their interaction (e.g., Craig, Itami, and Price 1989; Fritz 1990; Strauss 1990; Fritz and Simms 1992). Horner and Abrahamson's (1992 and submitted) studies of the influence of *Solidago* genotype and environment on *Eurosta* oviposition preference and offspring performance are among the few studies that have directly assessed the relative contribution of these factors at more than one level in the resistance hierarchy. Our results corroborate the importance of plant genotype in the *Solidago-Eurosta* system for both acceptability and suitability. What is more important, our results indicate that the relative contribution of plant genotype and environment can differ between levels in the hierarchy of resistance. We will more fully develop this concept of hierarchical relations of genetic and environmental contributions and the intermediate and emergent phenotypes in the final chapters.

Host-Plant Choice

The relationship between adult oviposition preference for particular host-plant species, phenotypes, or genotypes and the performance of offspring on chosen host plants is central to our understanding the evolution and ecology of phytophagous insects. As Thompson (1988) pointed out, this relationship (or lack thereof) between adult and larval characteristics influences how shifts onto novel hosts occur and, consequently, how insect species distribute among plant species over evolutionary time. Unfortunately, preference and performance relationships are influenced by numerous factors (e.g., breadth of feeding niche, competition, natural enemies), and the number of studies examining this relationship is limited (Tisdale and Wagner 1991; Burstein and Wool 1993; Fox and Lalonde 1993; Hanks, Paine, and Millar 1993; Nylin and Janz 1993; Heard 1995a,b; Larsson and Ekbom 1995; Preszler and Price 1995; Price and Ohgushi 1995). As a consequence, ecologists have a number of hypotheses about how the preference and performance relationship can vary, but we have only limited data available to help us discriminate among our hypotheses.

According to the convention established by Singer (1986) and Thompson (1988), adult "preference" refers to the hierarchical ordering of the choices available to an ovipositing female; or stated as a working definition, adult preference is the nonrandom oviposition on plant resources offered simultaneously or sequentially. Consequently, a preferred oviposition site is one that is attacked at a higher rate than expected if attacks were random (Craig, Itami, and Price 1989). Thompson (1988) defined offspring "performance" as a composite term to include survival of all immature stages (egg, larva, pupa), larval growth and efficiency, pupal mass, and resultant adult fecundity and longevity. Subsequent studies (e.g., Dodge et al. 1990; Rossi and

Strong 1991; Kouki 1993; Leddy, Paine, and Bellows 1993) have largely adopted these definitions, although in practice offspring performance typically is estimated by measuring only specific components of the composite "performance" term defined by Thompson (1988).

In our work with *E. solidaginis,* we have begun to explore both the genetic and nongenetic mechanisms that influence adult oviposition preference and offspring performance. In this chapter, we will first consider the host-plant traits that correlate with host-plant choice by *E. solidaginis* and subsequently discuss the relationship (or lack of relationship) between adult oviposition preferences and offspring performance for this gallmaker.

5.1 HOST-PLANT CHOICE

In the previous chapter we saw that the level of *E. solidaginis* infestation among host-plant genotypes is dependent in large part on genetic variation in the host's resistance (McCrea and Abrahamson 1987; Abrahamson, Anderson, and McCrea 1988; Anderson et al. 1989; Horner and Abrahamson 1992 and submitted). Differences in infestation levels occur among clones under field, common-garden, and greenhouse conditions. Furthermore, experiments that alter gallmaker distributions (McCrea and Abrahamson 1987) or resource supplies (i.e., nutrients, light, water) to *S. altissima* host plants show that host genotype is a significant determinant of oviposition and galling levels (Abrahamson, Anderson, and McCrea 1988; Horner and Abrahamson 1992 and submitted). These studies have also shown that *S. altissima*'s resistance to *E. solidaginis* is hierarchical with varying degrees of acceptability (preference) and suitability (performance). This hierarchy creates a continuum of host-plant resistance levels from very resistant (e.g., a host that is both unacceptable and unsuitable) to very susceptible (a plant that is both very acceptable and very suitable). The hierarchy also means that there are multiple ways for genotypes to be resistant. For example, a host genotype is resistant if it is either unacceptable and suitable or if it is acceptable but unsuitable. Since the degrees of acceptability and suitability vary by genotype, a field of goldenrods presents to *E. solidaginis* a continuum of clones of varying degrees of resistance or susceptibility. The critical question is whether *E. solidaginis* females prefer those host genotypes that are most suitable for their offspring. The nonrandom distribution of galls on clones of any goldenrod-dominated old field suggests that *E. solidaginis* females do discriminate among host plants and/or goldenrod clones vary strongly in their suitability for larval development. A critical question, then, is whether female preference is tightly coupled to offspring performance.

Choice among Solidago *Taxa*

We explored *E. solidaginis* host-plant choice in a series of studies (Abrahamson, McCrea, and Anderson 1989; Anderson et al. 1989; Horner and Abrahamson 1992 and submitted; Craig et al. 1993; Abrahamson et al. 1994; Craig et al., unpub. data; Itami et al., unpub. data) ultimately aimed at relating adult oviposition preference to the performance of their offspring.

In our initial study (Abrahamson, McCrea, and Anderson 1989), we measured the degree of host specificity at the host-species level using central Pennsylvania *E. solidaginis* populations. We offered one ramet of each of five *Solidago* taxa (i.e., *S. altissima, S. gigantea, S. canadensis,* and two ploidy levels of *S. rugosa*) in random order to mated females derived from galls of *S. altissima* to determine if females would ovipuncture and oviposit into the buds of taxa other than *S. altissima*. Some 50% of the *S. altissima* ramets tested were ovipunctured; however, only one ramet each (5%) of *S. gigantea, S. canadensis,* and the diploid *S. rugosa* were ovipunctured. The one hundred ramets tested received a total of eighty-seven ovipunctures, eighty of which (92%) were in buds of *S. altissima* (table 5.1). We grew the ovipunctured ramets to maturity to ascertain if galls would develop on non-host taxa. Six of the ten ovipunctured *S. altissima* ramets and the single ovipunctured *S. canadensis* ramet developed a gall. The fate of the larva in the *S. canadensis* gall was especially important because galls do not occur on this taxon in central Pennsylvania. After overwintering, an approximately one-half size female emerged from the galled *S. canadensis* ramet. However, we were unsuccessful in our attempts to mate this female with *E. solidaginis* males emerged from *S. altissima* ramets. We

TABLE 5.1. Results of a greenhouse, timed, no-choice experiment to determine whether mated central Pennsylvania *Eurosta solidaginis* females would ovipuncture and oviposit into the buds of taxa sympatric with *Solidago altissima.*

	Total Punctures	Percentage of All Punctures	Plants Punctured	Percentage of Tested Replicates
S. altissima	80	92.0	10	50
S. gigantea	2	2.3	1	5
S. canadensis	2	2.3	1	5
S. rugosa (tetraploid)	0	0	0	0
S. rugosa (diploid)	3	3.5	1	5
Totals	87		13	

Source: From Abrahamson, McCrea, and Anderson 1989.
Note: $\chi^2 = 281.8$ (total punctures), $P \ll 0.005$.

can conclude from these oviposition preferences that central Pennsylvania *E. solidaginis* prefer *S. altissima.*

We also used a no-choice, untimed experiment to learn if central Pennsylvania *E. solidaginis* females would inject eggs into the buds of taxa other than *S. altissima* (Abrahamson, McCrea, and Anderson 1989). The buds of all four non-*S. altissima* taxa used in the timed experiment were ultimately ovipunctured. However, most of these ovipunctures only penetrated the outermost bud leaf. We found eggs only in *S. gigantea* (two eggs from the same ovipuncture probe that penetrated the fifth bud leaf) and *S. canadensis* (one egg from five deep and nine shallow ovipunctures) (table 5.2). While these results establish that central-Pennsylvania *E. solidaginis* will oviposit on taxa other than *S. altissima* under no-choice conditions, the limited ovipuncture penetration and infrequent egg injection in non-*S. altissima* taxa indicate that *E. solidaginis* females accurately discriminate among host-plant options.

Choice among Genotypes of S. altissima

While host monophagy is well developed in central Pennsylvania *E. solidaginis* for *S. altissima* (Abrahamson, McCrea, and Anderson 1989), there is remarkable variation in infestation levels among *S. altissima* clones (McCrea and Abrahamson 1987). One can readily observe heavily galled clones next to ungalled

TABLE 5.2. Results of a greenhouse, untimed, no-choice experiment to determine if mated central Pennsylvania *Eurosta solidaginis* females would ovipuncture and inject eggs into the buds of taxa sympatric with *Solidago altissima.*

	Ramets Punctured	Ovipunctures	Eggs Laid
S. gigantea	4	13	2
S. canadensis	4	27	1
S. rugosa (tetraploid)	3	11	0
S. rugosa (diploid)	5	22	0
Totals	16	73	3

Source: From Abrahamson, McCrea, and Anderson 1989.

clones in virtually any *S. altissima*-dominated old field. To understand this pattern of variation in infestation among clones, we carefully examined *E. solidaginis* host choice among thirty-eight naturally occurring clones of *S. altissima* at the end of the spring oviposition period (Anderson et al. 1989). We determined the percentage of ramets ovipunctured by randomly selecting one hundred ramets from each clone and examining each for ovipuncture scars (see fig. 2.6). To determine the number of unhatched eggs, hatched eggs, and larvae for each bud, we collected a sample of twenty ovipunctured buds from each clone for laboratory dissection. Finally, to ascertain the percentage of ovipunctured ramets that eventually formed galls, we marked at least forty ovipunctured ramets in each clone and in late July we recensused these marked ramets for gall presence.

Solidago altissima clones differed significantly in (1) the percentage of ramets in which oviposition is attempted by *E. solidaginis*, (2) the percentage of ovipunctured ramets with *Eurosta* eggs, and (3) the percentage of ramets with ovipunctures that became galled (see figs. 4.6 and 4.7; Anderson et al. 1989). Ramets did not form galls for three reasons: (1) they were never ovipunctured (47%); (2) buds were ovipunctured but no egg was injected (11%); and (3) no gall developed in spite of the bud receiving an egg (18%). A female's choice of whether to ovipuncture is the most important factor in determining the occurrence of galls among goldenrod clones. The differential attack of clones by *Eurosta* females suggests that females perceive

141

some *S. altissima* genotypes as more acceptable (having the proper cues to stimulate oviposition) or more suitable (capable of supporting offspring development) than other genotypes. However, this field study cannot provide a completely unbiased evaluation of female oviposition preference since host-plant resources cannot be offered to *Eurosta* females simultaneously or sequentially under field conditions (see Singer 1986).

Consequently, we also compared the field susceptibility of *S. altissima* genotypes with their common-garden susceptibility. As described in the previous chapter, we replicated thirty *S. altissima* clones with fifteen representing genotypes on the resistant end and fifteen characterizing the susceptible end of the resistance-susceptibility continuum developed from our field observations (Anderson et al. 1989).

Analysis of the ramet growth rates and heights in this common-garden study showed that ovipunctured ramets were growing significantly faster and were taller at the time of oviposition than unpunctured ramets (fig. 5.1; Anderson et al. 1989). Furthermore, susceptible host genotypes as a group grew more rapidly and received more ovipunctures than resistant genotypes. However, as galls developed on ovipunctured ramets, their growth rates slowed (presumably due to resource limitations to the meristem caused by gall growth) such that by the time of maximum gall growth (weeks 4 and 5 after oviposition), ovipunctured ramets were growing significantly more slowly than unpunctured ramets (fig. 5.1).

Eurosta's preference to oviposit on taller ramets and/or on more rapidly growing ramets was also established under experimental conditions (Walton, Weis, and Lichter 1990; Horner and Abrahamson 1992). Walton, Weis, and Lichter (1990) offered tubs of 4–10 genetically identical ramets of *S. altissima* to *E. solidaginis* females to determine their choice of ramet heights for oviposition. Seventy-three of 242 (30.2%) ramets were ovipunctured at least once. Ovipunctured ramets were significantly taller on average than unpunctured ramets (64 cm versus 45 cm), and were of higher relative rank height (9.0 cm versus 6.4 cm). A discriminant function analysis suggested that absolute and relative ramet heights were nearly equal in their importance in predicting whether a female oviposited into a given ramet.

Growth Rates and Oviposition

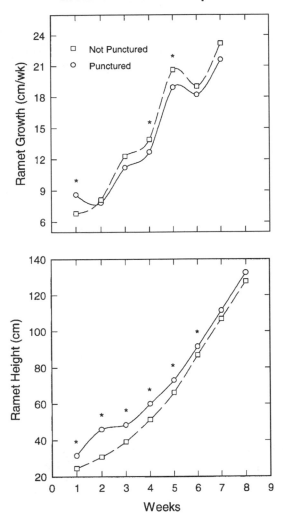

FIGURE 5.1. Growth rates (*top*) and heights (*bottom*) of ramets not ovipunctured or ovipunctured by *Eurosta solidaginis* in a common garden experiment. Weeks are according to time after oviposition. *Indicates a significant difference ($P < 0.05$, one-way analysis of variance) between ovipunctured and unpunctured ramets during that week. (Redrawn from Anderson et al. 1989, "Host genotype choice by the ball gallmaker *Eurosta solidaginis* [Diptera: Tephritidae]," *Ecology* 70:1048–1054, with kind permission of The Ecological Society of America)

To determine how ramet height and growth rate affected female oviposition preference, we randomized replicated, potted ramets of twenty-one *S. altissima* genotypes in a grid enclosed by a large outdoor cage (Craig et al., unpub. data). One-half of the replicates of each genotype was elevated 20 cm by placing their pots on another inverted pot while the remaining replicates remained at ground level. Five mated females were released in the cage, and censuses were made over the next 2 days at 2-hour intervals for the number of ovipuncture scars on each bud. At the conclusion of the trial, buds were dissected for *Eurosta* eggs. Our results indicate that female preference is dependent on the actual ramet (bud) height rather than the apparent height. An analysis of covariance using height treatment (i.e., pot elevated or not) as a factor and actual ramet height as a covariate found that the actual ramet height was significant ($F_{1,180} = 9.97$, $P = 0.002$) but that height treatment (i.e., elevated or not elevated) was not significant.

We examined the relationship between genotype growth rate and ovipuncture preference in a common garden planted with ramets from field clones of known *Eurosta* attack and emergence rates (Craig et al., unpub. data). Fifteen replicates of each fifteen *S. altissima* clones were monitored for weekly height growth and were enclosed in a large nylon-mesh cage. Five mated females were allowed to oviposit freely over a 48-hour period. A one-way analysis of variance showed that clones were growing at significantly different rates ($F_{14,199} = 5.03, P < 0.001$). Furthermore, a regression of mean growth rate for each clone and the number of ovipunctures that each clone received was highly significant (fig. 5.2). The results of these experiments suggest that (1) females use some cue(s) to discriminate between rapidly growing ramets and the ramets that had artificially high buds, and (2) that ramet height is a good indicator of rapidly growing ramets because ramet height is strongly correlated with ramet growth rate ($r^2 = 66\%$, Craig et al., unpub. data).

Oviposition Deterrent

Preferential attack of the taller, more rapidly growing ramets of any host-plant population could lead to appreciable intraspecific competition among *Eurosta* larvae unless females avoid ovi-

Ovipunctures by Ramet Growth Rate

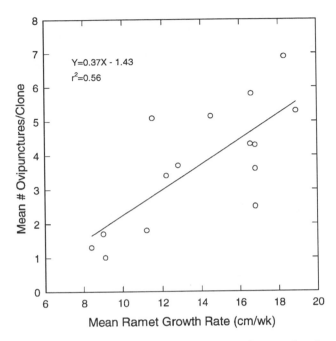

FIGURE 5.2. The mean number of ovipunctures per clone as a function of mean ramet growth rate. Fifteen replicates of fifteen clones were grown in a common garden and exposed to ovipuncturing by five mated female *Eurosta solidaginis*. (From Craig et al., unpub.)

positing on previously attacked buds. However, we have observed numerous instances of females ovipuncturing previously ovipunctured buds. Under natural field conditions, multiple-galled ramets occur as frequently as expected by the occurrence of galled and ungalled ramets (Abrahamson, unpub. data). Walton, Weis, and Lichter (1990), for instance, concluded that a female's acceptance or rejection of a ramet was much more strongly influenced by ramet height than by the presence of previous ovipunctures. These observations suggest that *Eurosta* lack the oviposition-deterring pheromones that are reported in some other tephritids (Averill and Prokopy 1987).

Craig et al. (in prep.) examined this possibility by testing *E.*

solidaginis preference for ovipuncturing and/or ovipositing on buds of *S. altissima* with and without previous ovipunctures. Mated, naive females were placed sequentially on an unpunctured and a previously ovipunctured ramet, with equal numbers of females randomly assigned to begin either on the unpunctured or on the ovipunctured ramet. The likelihood that *E. solidaginis* females would ovipuncture a bud was not affected by the presence of previous ovipunctures or by the experience of the fly. Virtually the same proportion of females ovipunctured previously attacked buds as unattacked buds and there was no significant difference in the mean number of ovipunctures made on previously attacked and unattacked buds (paired t-test, $P < 0.10$, 47 df). Neither the order in which females encountered unpunctured and previously ovipunctured buds nor the previous experience of the female influenced oviposition preference.

For these trials, the number of ovipunctures per visit to a bud varied from 0 to 12 with a mean of 3.6 ± 3.1 (SD) for all visits that resulted in ovipuncturing (fig. 5.3). Ovipunctures were typically made in clusters during an ovipuncturing bout (fig. 5.4; see fig. 2.6 for an illustration of oviposition clusters); however, females often made more than one cluster of ovipunctures during a single bout ($\overline{X} = 1.6 \pm 0.9$, SD). Thus, counting the number of ovipunctures or the number of clusters of ovipunctures does not accurately reflect the number of female visits to a bud. Furthermore, neither the number of ovipunctures nor the number of ovipuncture clusters could predict the number of eggs oviposited by a female during an oviposition bout. Of all the factors examined, only the time spent ovipuncturing was a significant, albeit inexact, predictor of the number of eggs oviposited (fig. 5.5).

Resultant Intraspecific Competition

The willingness of *Eurosta* females to oviposit into previously oviposited buds creates the potential for intraspecific competition among *E. solidaginis* larvae. Our experiments (Abrahamson et al. 1994; Hess, Abrahamson, and Brown 1996; Craig et al., unpub. data) have shown that larvae occupying a single bud do strongly compete with one another. In one of these experiments

Ovipunctures per Oviposition Bout

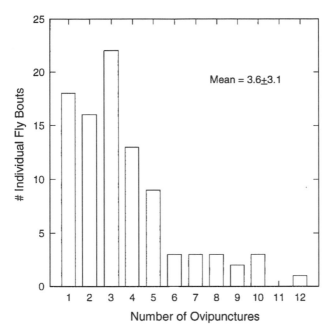

Mean = 3.6±3.1

FIGURE 5.3. Number of ovipunctures resulting from a single oviposition bout on a *Solidago altissima* bud. Mean ± standard deviation is given for forty-nine females that were individually monitored in two oviposition bouts, one on a previously ovipunctured ramet and one on an unpunctured ramet, for a total of ninety-eight oviposition bouts. (From Craig et al., unpub.)

(Hess, Abrahamson, and Brown 1996), we exposed over six hundred replicates of a single susceptible *S. altissima* genotype to attack by *E. solidaginis* females under greenhouse conditions. Samples of one hundred buds were dissected immediately after oviposition and at 3 weeks following oviposition. The remaining ramets and galls were allowed to grow to maturity. The number of ovipunctures per attacked bud ranged from five to forty-one while the number of eggs found in a single bud varied between one and eleven (fig. 5.6). Although the number of eggs oviposited increased significantly with the number of ovipunctures ($Y = 0.20 \ X + 1.59$, $r^2 = 0.61$), within 3 weeks of oviposition

147

Oviposition Clusters per Bout

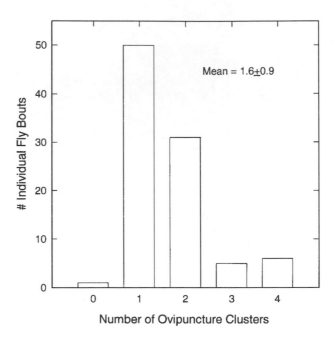

FIGURE 5.4. Number of clusters of ovipunctures resulting from a single oviposition bout on a *Solidago altissima* bud. Mean ± standard deviation is given for forty-nine females that were individually monitored in two oviposition bouts, one on a previously ovipunctured ramet and one on an unpunctured ramet, for a total of ninety-eight oviposition bouts (from Craig et al., unpub.). See figure 2.5 for an illustration of ovipuncture clusters.

there was no significant variation in the number of larvae per bud regardless of the number of ovipunctures. Thus, larval mortality during this pre-gall stage increased as the number of larvae per bud increased. The principal sources of mortality were the loss of apical growth in buds damaged by extensive ovipuncturing and bud necrosis induced at high levels of ovipuncturing. At 3 weeks after oviposition, the incidence of larval mortality due to necrotic reactions (see chapter 4.4) and bud damage was only 8% in buds receiving the lowest numbers of ovipunctures (5–9), but was 38% in heavily attacked buds (>15 ovipunctures). This

Oviposition Time and Egg Deposition

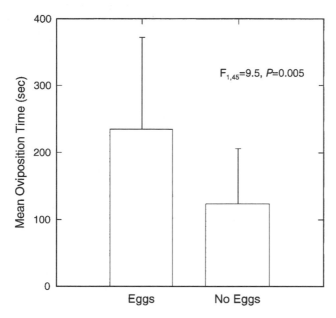

FIGURE 5.5. Mean oviposition time ($\pm$ standard deviation) in seconds when no eggs were oviposited and when eggs were oviposited. Forty-nine females were individually monitored in two oviposition bouts, one on a previously ovipunctured ramet and one on an unpunctured ramet, for a total of ninety-eight oviposition bouts. (From Craig et al., unpub.)

striking increase in the frequency of necrosis among intensely ovipunctured buds suggests that such necrosis can be induced even in highly susceptible genotypes like the one used in this experiment. At 9 weeks after oviposition, a census of the remaining 412 ramets found significantly more bud damage in heavily attacked buds (39%) than in less ovipunctured buds (24%).

The results of a second larval competition experiment were similar (Craig et al., unpub. data). In this experiment, we exposed ninety replicate ramets of one *S. altissima* genotype to *Eurosta* attack. Ramet buds were checked for oviposition scars every two hours and buds with >10 ovipunctures were removed. The buds of thirty ramets were dissected for eggs and larvae imme-

Eggs or Larvae and Ovipunctures

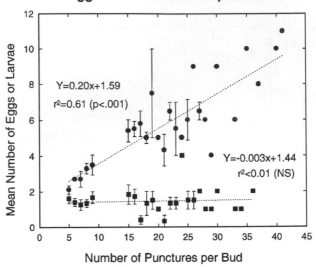

FIGURE 5.6. Mean number ± standard error of eggs (circles) or of larvae (squares) as a function of the number of ovipunctures for ramets attacked by Pennsylvania *Eurosta solidaginis* in a low-attack treatment (5–9 ovipunctures/bud) or high-attack treatment (>15 ovipunctures/bud). Regression analyses were performed on the number of eggs or larvae for each bud, not on the illustrated means. (Redrawn from Abrahamson et al. 1994; Hess, Abrahamson, and Brown 1996)

diately following ovipuncturing. Thirty additional ramets were dissected at one week and thirty more replicates at two weeks after oviposition. The mean number of living eggs and/or larvae significantly declined over the two-week post-oviposition period (fig. 5.7; $F_{2,83} = 42.5$, $P < 0.001$). Many buds dissected the first week contained multiple eggs or larvae. However, after weeks 1 and 2 the number of buds with multiple larvae declined sharply. A primary cause of this decline was the shedding of eggs placed in the outermost bud leaves as leaves expanded and unfolded.

We also have indirect evidence of intraspecific competition among *Eurosta* larvae in natural field populations. Hess, Abrahamson, and Brown (1996) found that the buds of some highly acceptable *S. altissima* genotypes under field conditions are both frequently (up to 100% of the ramets of a clone) and heavily

Decline in Living *Eurosta* over Time

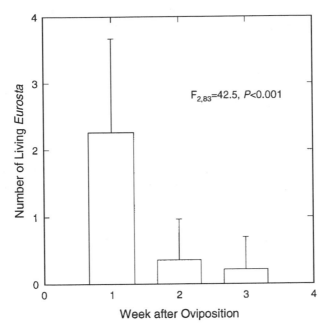

FIGURE 5.7. Mean number of living *Eurosta solidaginis* ($\pm$ standard deviation) in buds dissected the first, second, and third weeks following oviposition. (From Craig et al., unpub.)

ovipunctured (up to forty-two ovipunctures per bud). Yet in spite of such intense oviposition, galls with multiple larvae are exceedingly rare.

This lack of avoidance of previously oviposited buds, and as a consequence of intraspecific competition, contrasts with the pattern reported for other gallmaking insects (Whitham 1978, 1980, 1986; Craig, Itami, and Price 1990). There are probably several reasons why *E. solidaginis* has not evolved behaviors to avoid intraspecific competition (Craig et al., unpub. data). First, there might be no reliable indicators of the presence of eggs. Because many ovipunctured buds have not been oviposited (see fig. 4.7), females that shun buds with ovipunctures would miss many suitable opportunities for oviposition. Furthermore, even

151

if eggs could be detected, they may not represent potential competitors if previous eggs were located in sites where larvae have a low probability of tunneling to the meristematic region. In some instances, it might be advantageous for females to oviposit multiple eggs in a single bud, as females may then have a higher probability of one offspring surviving. It is also possible that *Eurosta* females may be time limited rather than egg limited. Because *Eurosta* are active under only a narrow range of environmental conditions (Walton, Weis, and Lichter 1990), are very short-lived as adults (Uhler 1951; Hess, Abrahamson, and Brown 1996), and suffer high rates of parasitism and predation (Weis and Abrahamson 1985), females may rarely have the time to oviposit all of their eggs. If the probability of a female ovipositing all of her eggs is low, then it is to a female's advantage to oviposit eggs at any potential site regardless of site quality.

5.2 HOST RECOGNITION AND PREFERENCE

Knowing that females preferentially attack taller and faster-growing *S. altissima* ramets does not answer the question of which host-plant traits are critical for *Eurosta*'s recognition of more and less suitable hosts. The behavior of females before and during ovipuncturing bouts gives us some clues about the evaluation process. Before ovipuncturing, a female typically climbs to the bud's apex and rapidly rubs the apex with her forelegs, occasionally pulling the tip of the bud to her mouth parts. If she is not discouraged by whatever information this behavior elicits, the fly appears to evaluate the bud as she walks around the base of the bud. It seems that she is searching for a suitable oviposition site (Abrahamson, McCrea, and Anderson 1989). Females can spend an appreciable amount of time walking over a ramet's bud, stem, and adjacent leaves, presumably evaluating potential ovipuncture sites (Walton, Weis, and Lichter 1990). Once a female initiates ovipuncturing, she appears to gain information from her ovipositor as ovipuncturing flies may quickly conclude the oviposition bout without penetrating the bud, ovipuncture only the outermost bud leaves, or ovipuncture through four or five layers of the bud leaves surrounding the apical meristem (Abrahamson, McCrea, and Anderson 1989).

Furthermore, if a female ovipunctures a bud, she rarely leaves that bud after only one ovipuncture. Instead, females typically leave a number of ovipunctures in varying numbers of clusters on a given bud (Walton, Weis, and Lichter 1990; Craig et al., unpub. data; fig. 5.4). Females appear to test many bud locations as they attempt to find appropriate egg deposition sites.

Physical Cues

Eurosta solidaginis females can discriminate among genotypes of *S. altissima* under field, common-garden, and greenhouse conditions (McCrea and Abrahamson 1987; Anderson et al. 1989; Horner and Abrahamson 1992 and submitted). Ramet height and growth rate are both important traits to such discrimination, but are there other host-plant traits, both correlated and uncorrelated with ramet height and growth, that influence *Eurosta* oviposition behavior? We gained at least a partial answer to this question from experiments aimed at examining environmental influences on host-plant resistance within a species (Horner and Abrahamson 1992 and submitted). In the earlier experiment, replicates of four genotypes of *S. altissima* (two each from the ends of the resistance and susceptibility continuum) grown under three levels of nutrient supply in full sun or with medium nutrient supply in 50% shade varied for an assortment of host-plant characteristics according to plant genotype, resistance level, nutrient supply, and/or light level (see fig. 4.2). A regression of the percentage of ramets of each genotype ovipunctured on various physical traits of the host plant found that bud diameter, leaf area, and leaf water content explained about 74% of the variation in the number of ramets ovipunctured (Horner and Abrahamson 1992). Bud diameter and leaf water content both had positive relationships, while leaf area had a negative relationship with the percentage of ramets punctured. Since bud diameter was strongly correlated with ramet growth rate during oviposition in this experiment, we suspected and confirmed that female *Eurosta* were preferentially ovipositing in plants with higher growth rates. According to these results, any female that can accurately assess a bud's diameter should be able to correctly identify the more rapidly growing ramets. However, the degree of correlation between bud diameter and growth rate likely varies among plant genotypes, since in other experiments using different sets of host-plant genotypes we found no correlation between bud diameter and ramet growth rate (How, Abrahamson, and Craig 1993; Horner and Abrahamson, submitted).

In the second greenhouse experiment, replicates of eight genotypes of *S. altissima* were subjected to either early growing season water deficits or a normal watering regime (Horner and Abrahamson, submitted). While oviposition preferences were significantly affected by plant genotype (categorical analysis $\chi^2 = 23.2$, df = 7, $P = 0.002$) and watering regime ($\chi^2 = 4.3$, df = 1, $P = 0.04$), ramet height and growth rate at the time of oviposition were the only physical attributes that were significantly correlated with oviposition preference (r = 0.73, $P = 0.001$, and r = 0.52, $P = 0.04$, respectively). Thus, ramet height and growth rate appear as the only physical host characteristics we have consistently identified as related to *Eurosta* host preference.

Eurosta's preference for host plants with high ramet growth rates is not unusual; such preference is well developed in several other species of gallmakers (Craig, Price, and Itami 1986; Craig, Itami, and Price 1988, 1989; Abrahamson and Weis 1987; Price, Waring, and Fernandes 1987; Price 1994; Weis, Walton, and Crego 1988). A preference for rapidly growing tissues may be the rule among gallmakers since gall initiation and development are dependent on the presence of rapidly growing, undifferentiated host-plant tissue (see chapter 3; Abrahamson and Weis 1987; Weis, Walton, and Crego 1988). Furthermore, host-plant growth is positively related to larval survival in several gallmakers (Price, Waring, and Fernandes 1987; Anderson et al. 1989; Craig, Itami, and Price 1989).

Preference for tall and rapidly growing hosts means that *Eurosta* host choice could shift according to the phenotypic expressions of host-plant genotypes. Our results show that the relative attractiveness of at least the more-preferred host genotypes of *S. altissima* can be strongly influenced by the availability of nutrients or water (Horner and Abrahamson 1992 and submitted). Since resistance is hierarchical (see chapter 4), environmental conditions that affect traits related to acceptability but not suitability or vice versa will alter the relative preference rankings observed. Consequently, a suitable host genotype that is highly unacceptable could appear highly resistant under some environmental conditions. Yet under environmental conditions that influence the expression of traits related to acceptability, this genotype could become acceptable and thus be highly susceptible to ovipositing females.

Cues from the Ovipositor

We investigated how females discriminate among potential *Solidago* hosts by testing whether *E. solidaginis* females could distinguish host species or host genotypes that varied in resistance with their ovipositor (Abrahamson, McCrea, and Anderson 1989). Newly unfolded bud leaves from four clones of *S. altissima* ramets were carefully wrapped around either the intact buds of *S. rugosa* (a non-host; see tables 5.1 and 5.2) or the intact buds of the same clone of *S. altissima*. The wrapped leaves were held in place with very fine, varnish-coated, copper wires (there was no discernible effect of the wire on *Eurosta* behavior). The *S. altissima* clones were known to differ in their resistance to *E. solidaginis* (McCrea and Abrahamson 1987) and included a clone with (1) low frequency of ovipuncture and egg injection (i.e., most resistant), (2) low frequency of ovipuncture but relatively high occurrence of egg injection, (3) high ovipuncture occurrence but low frequency of egg injection, and (4) a high ovipuncture occurrence coupled with frequent egg injection (i.e., most susceptible). Ramets with wrapped buds were placed into an observation chamber with an individual, mated female for 45 minutes. Randomly selected bud-wrap combinations were offered so that nineteen females were exposed to all eight possible combinations (each of the four resistance levels of *S. altissima* around *S. rugosa* and the four resistance levels of *S. altissima* around itself). Each bud was examined to determine the number of ovipunctures, depth of ovipunctures, and presence of eggs.

Eurosta females were able to discriminate between *S. altissima* and *S. rugosa* with only the information available from their

TABLE 5.3. Results of a laboratory bud-wrap experiment to test whether mated *Eurosta solidaginis* females reared from *Solidago altissima* can distinguish host species with their ovipositor. Significance levels are based on binomial probability for equal puncturing and egg injection.

| | Bud | | |
	Solidago altissima	Solidago rugosa	Significance
Number of ramets	76	76	—
Punctured ramets	26	25	n.s.
Punctures into wrap leaf	79	95	n.s.
Punctures into wrapped bud	24	1*	<0.001
Eggs injected	8	0	<0.005

Source: From Abrahamson, McCrea, and Anderson 1989.
*Ovipunctured only through outermost leaf of bud.

ovipositors. Twenty-five out of the seventy-six *S. rugosa* buds wrapped with *S. altissima* bud leaves were ovipunctured. These twenty-five buds received a total of ninety-five ovipunctures, but only one of these penetrated beyond the *S. altissima* wrap leaves (table 5.3). In this case, the ovipositor pierced only the first *S. rugosa* bud leaf and no egg was injected. Twenty-six of the seventy-six *S. altissima* buds wrapped with *S. altissima* bud leaves were ovipunctured, receiving a total of seventy-nine ovipunctures. In contrast to the *S. rugosa* bud trials, twenty-four of the ovipunctures to the *S. altissima* buds penetrated beyond the wrap leaves and eight eggs were injected. According to the null hypothesis that the buds of both species were equally likely to be ovipunctured, the probability that the twenty-five buds ovipunctured beyond the wrap leaves would occur in the ratio of one *S. rugosa* and twenty-four *S. altissima* buds is less than 0.001. Likewise, the probability that all eight eggs injected would be deposited in *S. altissima* is 0.004.

However, there were no differences among the four resistance levels of *S. altissima* used in the bud-wrap experiment (Abrahamson, McCrea, and Anderson 1989). Although information from the ovipositor alone enabled females to distinguish between two *Solidago* taxa, females did not discriminate among the resistance levels of *S. altissima*. The information gained from the ovipositor may need to be combined with information available from other sensory organs. We suspect that preoviposition evaluation is important in enabling discrimination among host-plant genotypes.

Chemical Cues

Many studies (e.g., Shelley, Greenfield, and Downum 1987; Feeny et al. 1989; Mitter, Farrell, and Futuyma 1991) have documented the important role of host-plant chemical cues to host choice in free-feeding herbivores, but few studies have considered the role of host chemistry to gallmaker host choice. Yet host-plant chemical cues should play an especially crucial role in host selection by gallmaking insects because of the intimacy of the interactions between gallmaking insects and their host plants (Abrahamson et al. 1994). The results of the bud-wrap experiment described above (Abrahamson, McCrea, and Anderson 1989) implicate host-plant chemical cues in host selection by *E. solidaginis*. To investigate this possibility, we determined female responses to plant extracts applied to artificial buds (a strip of sponge ≈0.5 × 1 × 6 cm wrapped with polyester/rayon gauze) (Abrahamson et al. 1994). We used extracts or stems from greenhouse-grown *S. altissima* and *S. gigantea* to test if Pennsylvania, *S. altissima*-origin *Eurosta* (1) could distinguish between real and artificial buds, (2) would accept artificial buds containing a crude methanolic extract of *S. altissima*, and (3) could distinguish between artificial buds containing *S. altissima* extracts, *S. gigantea* extracts, or a combination of both extracts. *Eurosta* ovipositional behavior includes walking on the bud, abdomen arching, and finally ovipositor insertion. However, abdomen arching constituted bud acceptance for purposes of this experiment.

Eurosta can discriminate between real and artificial buds. Eight females that initially rejected distilled-water-treated artificial buds accepted real buds, and all eleven females that initially accepted real buds never accepted distilled-water-treated artificial buds ($\chi^2 = 19$, $P < 0.001$, for the hypothesis that females equally accept the two bud types). However, when we applied a crude methanolic extract of *S. altissima* (10 ml of 50% methanol homogenized with each gram of fresh bud mass, centrifuged at 3000 rpm for 5 minutes) containing such potential cues as flavonoids, terpenoids, and simple phenolics to an artificial bud (standard amount being 25% of the total amount of extract per bud), eleven of thirty-six (31%) females tested accepted an artificial bud. Even though physical cues may be altered, *Eurosta* females do respond to chemical cues in crude extracts of *S. altissima*.

In a subsequent experiment, we learned that the relative concentrations of these chemical cues are important. Some 45% of the eleven females tested accepted artificial buds with low concentrations (15% of the total amount of extract per bud) of *S. altissima* extract, but 100% accepted artificial buds when the concentration was raised to the standard amount (table 5.4; Abrahamson et al. 1994). Furthermore, all eleven Pennsylvania *Eurosta* females tested rejected artificial buds that were painted with standard amounts of *S. gigantea* extracts. Even though the rejection rate decreased when the concentration of *S. gigantea* extract was lowered, the decline was not significant.

Another series of experiments was conducted to examine whether the addition of extracts of *S. gigantea* would function as a deterrent to oviposition by Pennsylvania *E. solidaginis*. *Eurosta solidaginis* in the mid-Atlantic region typically attack only *S. altissima* in spite of sympatric populations of *S. gigantea* and other potential *Solidago* hosts (see chapter 2). Of nine females that originally accepted an artificial bud treated with *S. altissima* extract, none would do so once a standard concentration of *S. gigantea* extract was applied (table 5.4; Abrahamson et al. 1994). This rejection rate did decrease, although not significantly, if a low concentration of *S. gigantea* extract was used. We could lower the rejection rate of artificial buds with standard concentrations of *S. gigantea* extract by superimposing standard concentrations of *S. altissima* extract. The highest acceptance rate of buds initially painted with *S. gigantea* extract resulted from low-concentration *S. gigantea* extract buds to which we applied standard concentrations of *S. altissima* extract ($\chi^2 = 5.1$, $P < 0.025$, compared to buds treated with only *S. gigantea* extract).

These results show that the chemicals present in or absent from *Solidago* buds are important in eliciting behavioral responses that result in ovipuncturing by *E. solidaginis*. However,

TABLE 5.4. Acceptance and rejection of artificial buds treated with *Solidago altissima* and *S. gigantea* extracts in low (15% of the total extract per bud) and standard (25% of total extract per bud) concentrations by Pennsylvania *Eurosta solidaginis* infesting *S. altissima*.

Extract Type	Number Accepting	Number Rejecting
S. altissima only		
Low	5	6
Standard	11	0
	$\chi^2 = 8.3; P < 0.01$	
S. gigantea only		
Low	3	8
Standard	0	11
	$\chi^2 = 3.5; P < 0.10$	
S. altissima [standard] with *S. gigantea*		
Low	2	7
Standard	0	9
	$\chi^2 = 2.3; P < 0.20$	
S. gigantea [standard] with *S. altissima*		
Low	0	9
Standard	5	4
	$\chi^2 = 6.9; P < 0.01$	

Source: From Abrahamson et al. 1994.
Note: Chi-square goodness of fit tests were performed to test the hypothesis that females would accept artificial buds equally, regardless of extract concentration.

these preliminary experiments also suggest that both the absolute amount and the ratio of stimulant to deterrent compounds may be important. Our results imply that host choice is based on a complex interaction of stimulatory to deterrent cues since ovipositional stimulants were effectively masked by deterrents and vice versa (Abrahamson et al. 1994).

5.3 ADULT PREFERENCE AND LARVAL PERFORMANCE

Performance on Tall and Short Ramets

Results from a number of experiments (Anderson et al. 1989; Horner and Abrahamson 1992 and submitted; How, Abrahamson, and Craig 1993) indicate that ramet height and growth rate are important determinants of oviposition preference. Walton, Weis, and Lichter (1990) specifically examined the relationship between ramet height and offspring performance measures in a greenhouse experiment where females were given plants of varying heights in a no-choice situation. Mated females were placed individually into cages (20 cm diameter, 30 cm height) that enclosed only the terminal bud and the first few unfolded leaves of 120 ramets from ten different clones. This design offered no means for the fly to determine a ramet's height before oviposition. Ramet heights were measured the day following oviposition and the diameters of the resultant galls were measured approximately 5 weeks later at gall maturity. In spite of *E. solidaginis'* preference for taller and/or more rapidly growing ramets in field and greenhouse environments, ramet height was at best a weak predictor of host quality in this experiment. Although the probability of gall induction rapidly increased as ramet height increased from approximately 4 to 18 cm, the probability of gall formation leveled off at about 60% for ramets greater than 18 cm (fig. 5.8). The rank corelation between ramet height and the probability of gall induction only approached significance ($r_s = 0.58$, $P = 0.064$) and taller ramets did not produce larger galls.

However, *E. solidaginis'* preference for tall and rapidly growing ramets may have some survival advantages for its offspring because ramets that are relatively small in the early season are less likely to survive than are larger ramets. Cain's (1990b) study of *S. altissima* ramet growth and mortality found that large size inequalities exist among *S. altissima* ramets in the spring, but that these inequalities decrease over the course of the growing season due to the mortality of small ramets. An analysis of the ramets that survived the growing season showed low correlation between early and late-season ramet heights. However, correlations between early and late-season ramet heights were much stronger when ramets that died were included. Cain (1990b) interpreted these results to suggest that a herbivore could use early season ramet height to discriminate those ramets with a high probability of death from others with a high probability of survival. The consequence is that *E. solidaginis* may avoid ramets that are likely to die by assessing their early season height.

Galled Ramets by Stem Height

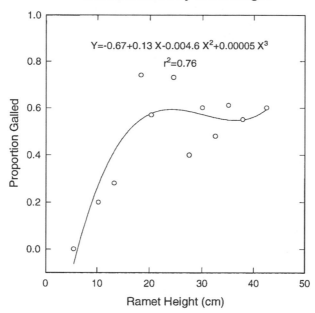

FIGURE 5.8. Proportion of ramets galled as a function of ramet height. The curve is a third-order regression of the points representing the proportion of ramets galled in groups of approximately ten ramets each in graduated size classes. (Redrawn from Walton, Weis, and Lichter 1990, "Oviposition behavior and response to plant height by *Eurosta solidaginis* Fitch [Diptera: Tephritidae] on *Solidago* [Compositae]," *Annals of the Entomological Society of America* 83:509–513, with kind permission of The Entomological Society of America)

Correlation of Preference and Performance

A field experiment suggested that *E. solidaginis* oviposition preferences were reasonably stable over time (McCrea and Abrahamson 1987). There were significant positive Spearman rank correlations among our 130 clones for the number of galls and the number of galls per ramet in each clone over 3 years (table 5.5). A common-garden experiment using fifteen replicates of thirty *S. altissima* genotypes also suggested a relationship between preference and performance for *E. solidaginis* (McCrea and Abrahamson 1987). In this experiment *Eurosta* strongly preferred to oviposit in those genotypes classified as susceptible in

TABLE 5.5. Spearman-rank correlations between years from 1982 to 1984 for the number of galls per clone and galls per ramet for marked clones along Pennsylvania Route 147.

	Number of Galls		Galls per Ramet	
	1982	1983	1982	1983
1983	0.58	—	0.47	—
1984	0.50	0.64	0.34	0.58

Source: From McCrea and Abrahamson 1987.
Notes: All correlations are significant at $P < 0.001$. Nonparametric correlations were used because the data were not normally distributed.

the field while infrequently ovipositing in those genotypes classified under field conditions as resistant (see table 4.2). Susceptible clones were significantly taller than the resistant clones at the time of oviposition (fig. 5.9), which agrees with *Eurosta*'s preference for more vigorous clones. As they did in the field, the susceptible genotypes had a high frequency when exposed to *Eurosta* in the garden, while most resistant genotypes failed to produce galls. Overall, the proportion of the fifteen replicates for each clone to form galls was positively correlated with the clonal mean stem height ($r = 0.59$, $P < 0.001$). Within each class, there were also positive correlations between height and proportion of galled replicates (resistant, $r = 0.39$, n.s.; susceptible, $r = 0.47$, $P = 0.08$), although neither were significant with sample sizes of only fifteen clones per category. Taken as a whole, the evidence points to correlations between plant vigor and insect preference and host suitability.

As we pointed out in the introduction to this chapter, it is often assumed that ovipositing females can and do evolve mechanisms to place their offspring in locations that yield the greatest fitness in their progeny through enhanced larval survival and/or reproduction. Furthermore, such mechanisms are presumed to be especially well developed in herbivores with immobile larvae (like gallmakers) since such larvae cannot compensate for poor host quality by moving to a new host. Unfortunately, there have been few studies of the preference and performance relationships of gallmaking insects (excepting the Price group's

Early Stem Height and Susceptibility

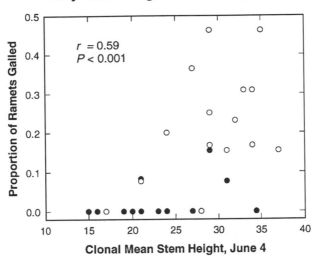

FIGURE 5.9. Proportion of ramets galled as a function of early season (June 4) ramet height in a common-garden experiment using fifteen replicates of thirty *Solidago altissima* genotypes. Clones that were classified as susceptible in the field are illustrated with open circles, while those genotypes classified under field conditions as resistant are shown as filled circles. Susceptible clones were significantly taller than the resistant clones at the time of oviposition, which agrees with Eurosta's preference to oviposit in more vigorous clones (see table 4.2). (Data from McCrea and Abrahamson 1987)

findings of strong preference-performance correlation for gall-making sawflies attacking willows; Fritz et al. 1987; Craig, Itami, and Price 1989; Price 1991; Price and Ohgushi 1995) in spite of the expectation that gallmakers as a group should have strong positive correlations between oviposition preference and off-spring performance. Strong correlation for gallmakers would be an expected outgrowth of (1) the immobile nature of their larvae (hence making adult host choice crucial to offspring performance), (2) the intimate association of the gallmaking larvae with the tissues of their host plant (larvae must be able to stimulate their host plant to initiate and develop a gall), and (3) the necessity of temporal synchronization of herbivore and host plant (larvae must stimulate their host tissues at a time when these tissues are appropriately reactive). Larval immobility, intimate association, and synchronization have the potential to create strong selection pressures for adaptation to the host-plant species or even to clusters of similar host-plant genotypes that share crucial characteristics. We might further expect a gallmaker like *Eurosta* to adapt to specific host genotypes because of the strongly clonal nature of its host plant (*S. altissima* creates sizable collections of identical ramets) and the variation among clones for susceptibility and resistance to gallmaker attack. The weak relationship found between *Eurosta*'s oviposition preference for tall ramets and the probability of gall induction in the no-choice situation (Walton, Weis, and Lichter 1990) contrasts with the slightly stronger preference-performance correlation expressed in the garden experiment (fig. 5.9). This second experiment indicates that plant genotypes of lower vigor are less likely to sustain gallmaker attack.

Plant Genotype, Environment, and the Correlation between Preference and Performance

We examined the degree of overlap between traits related to preference and to performance in two greenhouse studies that manipulated the availability of resources to the host plants. The first study (Horner and Abrahamson 1992) found that both oviposition preference and larval survival were significantly affected by host genotype; however, only preference was influenced by the interaction of genotype and nutrient supply to the host plant. This study did find that a bud characteristic (the ratio of bud diameter to bud length) was positively related to the percentage of ovipunctured ramets that formed galls. Such a result suggests that accurate placement of eggs near the apical meristem may be easier in short, thick buds. Most importantly, there was no significant correlation between preference and larval survival (i.e., performance) at the population level. The results of this experiment indicated that the host traits affecting *Eurosta* oviposition preference have a greater magnitude of phenotypic plasticity than those affecting larval performance. Furthermore, the degree of plasticity in host traits affecting *Eurosta* oviposition preference differs among *S. altissima* genotypes.

A subsequent greenhouse study (Horner and Abrahamson, submitted) that manipulated the early-season water supplies to host plants found that ramet height and growth rate at the time of oviposition were significantly correlated with oviposition preference. The occurrence of galls on ovipunctured ramets and the proportion of galled ramets from which adult offspring emerged (measures of performance) were significantly affected by plant genotype and the interaction between plant genotype and early season water deficits. Furthermore, plant genotype significantly affected the mass of adult offspring, a trait that is highly correlated with potential fecundity in females. However, we found no significant correlations when relationships between these offspring performance measures and indices of host vigor were examined. Most important, there was no correlation between oviposition preference (i.e., the proportion of ramets ovipunctured) and offspring performance as measured by either gall formation or the emergence of adult offspring ($r = 0.28$, $P = 0.29$, and $r = 0.05$, $P = 0.85$, respectively).

An examination of *Eurosta*'s oviposition preference for and offspring performance on individual genotypes of *S. altissima* in two garden experiments produced complicated results. At our Bucknell University garden, we constructed 1 m × 1 m plots separated by 45 cm deep aluminum flashings (to prevent interdigitation of clones). One replicate of each of twenty *S. altissima* genotypes of known resistance (Anderson et al. 1989) was randomly planted into each of four blocks to make a total of eighty plots. At our Cedar Creek Natural History Area garden near Bethel, Minnesota, we established a second set of 272 1 m × 1 m plots (also divided by flashing extending to a depth of 45 cm) into which we randomly planted four replicates each of thirty-four genotypes of *S. altissima* and of thirty-four genotypes of *S. gigantea*. We monitored the height and growth rate of a sample of ramets in each plot and recorded the occurrence of ovipunctures and galls over two years in each garden. *Eurosta* females expressed clear oviposition preferences for particular host genotypes in each year (Craig et al., unpub. data). However, these preferences for host genotypes were different in each of the 2 years. Most importantly, *Eurosta*'s preferences based on rates of ovipuncture did not correlate with off-

165

spring performance (as measured by gall initiation, gall size, or peak larval mass) in either of the years nor with the oviposition preferences that we had observed in the field (Anderson et al. 1989).

The reasons for these apparently contradictory results are likely related to differences in the questions asked in the early versus recent experiments (and hence in the analyses applied) and in our experimental designs. The preference pattern we saw in our field survey (McCrea and Abrahamson 1987) was based in part on genetic differences among host clones for traits like height or growth rate. However, clonal variation in the field is confounded by environmental variation, and environmental variation masks some of the effects of genetic variation (Horner and Abrahamson 1992 and submitted). Consequently, clones growing at microsites with highly favorable growing conditions may be preferred over those at poor microsites regardless of their genotype.

In addition, our early garden experiment (McCrea and Abrahamson 1987) only estimated preference and performance for two groups. It compared resistant genotypes and susceptible genotypes. The estimates of oviposition preference and performance (galling) were based on 225 single-ramet replicates for each category. Consequently that experiment found large differences between resistant genotypes and susceptible categories. Our recent garden experiments (Craig et al., unpub. data; Itami et al., unpub. data) were designed to determine preference at the much finer host-genotype level but could do so with only a small number of replicates since each replicate was a clonal fragment. It is entirely possible that the genotypic component of host variation is masked by the extensive variation we found within individual plots as well as among the clonal fragments of each genotype. We do know those host-plant traits important to preference rankings (e.g., ramet height and growth rate) are under considerable environmental control (Horner and Abrahamson 1992 and submitted). Environmental variation in such traits will obscure the preference and performance relationship even under the relatively uniform common-garden conditions if the genetic influence is relatively weak.

166

5.4 THE PARADOX: THE WEAK RELATIONSHIP BETWEEN PREFERENCE AND PERFORMANCE

It seems intuitive that ovipositing females should evolve mechanisms to place their offspring in locations that yield the greatest fitness in their progeny through enhanced larval survival and/or reproduction (Dodge et al. 1990; Price 1991; Whitham 1992; Hanks, Paine, and Millar 1993; Preszler and Price 1995; Price and Ohgushi 1995). As a consequence, we might anticipate a strong correlation between female oviposition preference and larval performance in herbivores such as gallmaking insects. However, the results for E. *solidaginis* show that there is only a weak relationship of adult oviposition preference with offspring performance.

The reasons for poor correspondence between preference and performance are potentially numerous and complex (Rausher 1985; Courtney and Kibota 1990; Burstein and Wool 1993; Fox and Lalonde 1993). For introduced plants or for novel host plants, an often-cited explanation is that the herbivore has not had sufficient time to adjust its oviposition behavior to the new host. Alternative explanations have suggested that the respective densities of high- and low-quality hosts influence oviposition behavior and that herbivore fecundity is constrained by the length of a female's life (Rausher 1985; Larsson and Ekbom 1995). Furthermore, Fox and Lalonde (1993) argued that there are bounds to the extent of discriminatory behavior in ovipositing females. Consequently, two host species that have similar cues but different suitabilities for offspring survival could create oviposition confusion for females. Such confusion could result in both plants being retained as oviposition sites because discriminating against the less-suitable host would also exclude the more-suitable host.

Larsson and Ekbom (1995) used a simulation model based on data from gallmaking cecidomyiids to explore the circumstances under which host plants of low suitability can be retained among those selected for oviposition. Their findings suggest that females that can discriminate among plants differing in suitability for their offspring have little advantage over nondiscriminating ("confused") females if the proportion of less-suitable hosts is

167

large and the time available for oviposition is short. Furthermore, the advantage of host discrimination is negligible when nondiscriminating females oviposit on their eclosion plant (a suitable host).

Possibilities for Eurosta solidaginis

There are a number of reasons, including those described above, why *E. solidaginis* lacks a strong correlation between preference and performance. One may be the non-overlap of the traits influencing host plant choice with the traits related to larval survival (Horner and Abrahamson 1992 and submitted). Ovipositing females may be unable to accurately assess the complex of plant traits that influence offspring performance. Instead, females may use correlates of these traits (Craig, Itami, and Price 1989; Horner and Abrahamson 1992 and submitted). Our data imply that the phenotypic correlations between those traits influencing oviposition preference and those influencing offspring performance differ under different environments and/or for different host genotypes. Furthermore, the environmental conditions that exist during oviposition may be transient, causing the genotypes that appear highly suitable at oviposition to become less suitable during larval development.

An additional possibility is the artificially complex host plant distribution offered under our experimental conditions (Horner and Abrahamson 1992 and submitted). *Solidago altissima* is strongly clonal so that in the field each genotype occurs as a clumped resource for ovipositing *Eurosta*. Under natural conditions, a female emerges within the clone on which she fed as a larva (a suitable clone) and may ovipuncture some ramets of the eclosion clone (some reproduction in the natal clone is expected if only for bet hedging). If ovipositing females are time limited, it would be to their advantage to assess only the patch rather than each ramet because the host plant occurs in clonal patches. If this is so, then the experimental conditions we used to assess preference could result in no apparent correlation when one does exist under natural conditions.

An important additional point is that our experiments have measured only the mean preference for a population of ovipositing *Eurosta* females. By ignoring the behavior and preference of individual females and measuring only the population's preference, it is possible that an existing correlation was obscured (Thompson 1988; Thompson and Pellmyr 1991; Horner and Abrahamson 1992 and submitted). Indeed, Via (1990) detected a significant genetic correlation between oviposition preference and larval performance at the individual level that was not detectable at the population level (see also Via 1991; Fox 1993). If individual females have widely varying host preferences, estimates of population-level preference will not correlate with offspring performance estimates.

Still, it may be that a positive relationship between preference and performance simply does not exist in *E. solidaginis*. We know, for instance, that the availability of preferred hosts under field conditions drops rapidly during the oviposition period (Anderson et al. 1989). We also suspect that the relationship between preference and performance is influenced by differential levels of attack by natural enemies on preferred and nonpreferred clones (Abrahamson et al. 1994; Brown et al. 1995; Hess, Abrahamson, and Brown 1996). *Solidago* clones should have differential natural-enemy attack given the variation in gall size among clones. The size of a ball gall is strongly related to survival of its larva (Weis and Abrahamson 1985; Abrahamson et al. 1989; Weis, Abrahamson, and Andersen 1992) such that it is better for a *Eurosta* to develop in a larger gall rather than in a small gall because of parasitoid attack. However, developing in the biggest galls is not always best because of bird attack (Weis, Abrahamson, and Andersen 1992; see chapter 8). Thus, preference for hosts that react to *Eurosta* stimulation to develop very small or very large galls could be counteradaptive.

In addition, the time that has been available to evolve strong preference and performance correlations may affect the apparent correlations or non-correlation. The distributions and abundances of both *Solidago* and *Eurosta* have markedly increased over the past two centuries. Before the colonization of North America by Europeans, old-field species like *S. altissima* and *S. gigantea* were likely restricted to river floodplains, areas recovering from forest fires, and canopy openings in the extensive forest cover of eastern North America (the center of richness for

Solidago species). It is probable that many if not most *Solidago* and *Eurosta* populations existed as small patches (by today's standards) that were partially isolated from other such patches. Any adaptations of fly genotypes for preference of or performance on particular host clones could have been lost as the abundance of both host and herbivore rapidly expanded following the massive disturbance of the North American landscape.

Studies have also shown that factors such as intense larval competition on highly preferred hosts (Hanks, Paine, and Millar 1993), host-plant water stress (Tisdale and Wagner 1991; Horner and Abrahamson 1992 and submitted), and the availability of host-plant tissues of appropriate age (Weis and Abrahamson 1985; Kouki 1993) can also markedly alter the preference and performance relationship.

Finally, it may be that there is simply no advantage to *Eurosta* females that discriminate relative to nondiscriminating ("confused") females. Larsson and Ekbom (1995) have argued that the insects most likely to possess a nondiscriminatory oviposition behavior are those that have (1) an intimate relationship with predictable host plants highly variable in suitability, (2) a short time available for oviposition, and (3) a limited capacity for directed flight. Goldenrod clones have remarkable longevity and do differ markedly in their suitability for offspring. Consequently, these clones are predictable and variable. Furthermore, adult female *Eurosta* are short lived (e.g., ≈10 days; Uhler 1951) and consequently may not be able to afford a highly selective host-plant choice strategy. Finally, even though *Eurosta* adults are capable of strong and rapid flight, they typically make short erratic flights. Even when opportunities avail, *Eurosta* adults typically do not make sustained flights (Uhler 1951; pers. obs.).

Preference and Performance Correlation in Other Herbivores

In spite of what we know, we still lack a set of general conclusions about the degree of correlation between oviposition preference and offspring performance among herbivorous insects. Although the results of several studies indicate a positive relationship between oviposition preference and offspring performance regardless of larval mobility (e.g., Nylin 1988; Thompson

1988; Craig, Itami, and Price 1989; Dodge et al. 1990; Price 1991; Rossi and Strong 1991; Brody 1992; Whitham 1992; Hanks, Paine, and Millar 1993; Kouki 1993; Leddy, Paine, and Bellows 1993; Nylin and Janz 1993; Heard 1995b, Larsson and Ekbom 1995; Preszler and Price 1995; Price and Ohgushi 1995), other studies show either weak correspondence or no relationship (e.g., Courtney 1981, 1982; Karban and Courtney 1987; Thompson 1988; Roininen and Tahvanainen 1989; Valladares and Lawton 1991; Horner and Abrahamson 1992 and submitted; Krebs, Barker, and Armstrong 1992; Larsson and Strong 1992; Burstein and Wool 1993; Fox 1993; Leddy, Paine, and Bellows 1993). Such variation among herbivores is not surprising given that oviposition preference by adult insects as well as performance of their offspring are affected by environmental factors including host-plant nutrient, water, and secondary metabolite content (Miller and Strickler 1984; Horner and Abrahamson 1992; Burstein and Wool 1993). Furthermore, it is likely that the genes involved in host-plant choice by adult females are different from those that influence larval performance on a particular host plant (Thompson 1989; Horner and Abrahamson 1992 and submitted). As a result, the host-plant qualities that may be important to an ovipositing female as she chooses among plant genotypes may be correlated only to plant attributes that are loosely related to her offspring's performance. Ovipositing females may be unable to accurately assess the complex suite of host-plant characters that influence larval success. In addition, it is possible that there are only minor penalties if an herbivore retains a non-discriminating oviposition strategy (Larsson and Ekbom 1995).

Attempts to understand the preference-performance relationship will be more successful if we are careful to discriminate the context in which the preference-performance relationship evolved from its context today. For example, it is likely that the populations of *Eurosta* and *Solidago* are markedly more abundant today than in prehistoric time. Changes in the landscape mosaic, brought about by European colonization of eastern North America, created a substantial increase in habitat appropriate for these organisms. However, the rapid expansion of both herbivore and host-plant populations may have altered the match between adult preference and offspring performance as particular fly and plant genotypes expanded at differenial rates. More time may be required to reestablish the equilibrium between fly genotypes and host-plant genotypes.

Furthermore, a clearer understanding of preference-performance patterns may result if we more carefully distinguish correlations at various levels of discrimination. For example, *E. solidaginis* can discriminate between susceptible host-plant genotypes and resistant genotypes when offered only the extreme clones of the susceptible-resistant continuum (McCrea and Abrahamson 1987). However, adult females of *Eurosta* do not appear particularly good at predicting the best host genotype for their offspring when quite similar susceptible clones are tested. This could be a consequence of their life-history attributes. For example, factors such as adult longevity may alter preference-performance patterns by influencing whether it is worthwhile to discriminate host-plant quality at the finest level. It may be better for *Eurosta* females to leave ten offspring on mediocre host plants than only five offspring on the highest-quality host plants. The time (cost) that it takes to discriminate the best host-plant genotypes from mediocre host genotypes will determine this trade-off. Herbivores with short adult longevity may gain fitness by being less fussy rather than more fussy. As a consequence, flies may discriminate among potential host-plant species, and even among subsets within a host-plant species, but not distinguish among genotypes that vary only slightly in suitability.

CHAPTER SIX

The Gall as *Eurosta*'s Extended Phenotype

6.1 THE ISSUE OF INSECT GENES IN THE EVOLUTION OF GALL PHENOTYPE

Near the end of the last century, the elegant structure of plant galls suggested to some that they are of adaptive value to the plant that produces them; by encapsulating its enemy in a gall, the plant restricts damage to a nonvital structure (Cockerell 1890). Others suggested that galls were parasitic manipulations of plant development, induced by insects that have been selected to force the plant to contribute to their survival and reproduction (Romanes 1889). This brief debate arose when Mivart (1889), Darwin's most nettlesome critic, suggested that plant galls seemed to provide the fatal example of a "structure of any one species . . . formed for the exclusive good of another species" which Darwin (1859) admitted would "annihilate" his theory. Interestingly, in anticipation of the "panglossian paradigm" debate (Gould and Lewontin 1979) almost a century later, Mivart was sure that the Darwinians would concoct some sort of adaptive explanation for gall formation.

In the previous chapters, we reviewed available evidence to determine if goldenrods benefit by producing a gall when attacked by *Eurosta,* and found that gall production is at very best neutral. In subsequent chapters, we will present evidence that gall structure provides benefits to the insect beyond provision of food. Thus, in concurrence with Price, Waring, and Fernandes (1987), we conclude that the adaptive benefits of gall formation go to the insect and not the plant. The focus of this chapter is on one of the fundamental assumptions underlying the notion that galls can evolve as insect adaptations.

Romanes's suggestion (1889) that galls evolve through selection acting on the insect has been echoed in this century (Mayr

Genetic and Environmental Influences on Gall Development

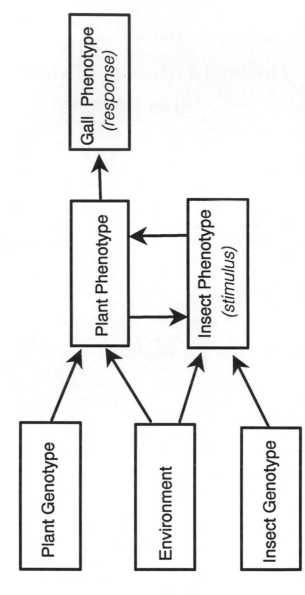

FIGURE 6.1. A developmental-genetic model for gall growth. (From Weis and Abrahamson 1986, "Evolution of host-plant manipulation by gall makers: Ecological and genetic factors in the *Solidago-Eurosta* system," *American Naturalist* 127:681–695; with permission)

1970; Päclt 1972; Cornell 1983; Price, Waring, and Fernandes 1987). Dawkins (1982) proposed that plant galls can be viewed as the insect's "extended phenotype"—an alteration of the environment caused by a gene, which ensures that gene's propagation. However, before we can conclude that selection on the insect can cause evolution of this phenotype, it must be shown that the insect genome exerts some control over the phenotype's development. In a larger sense, it is obviously true to say that insect genes contribute to gall formation. In *Eurosta*'s case, insect genes carry the instructions for neural and muscular structures that cause the female to deposit her eggs into the plant bud, and the instructions that cause the resulting larvae to burrow into the stem just below the apical meristem and begin feeding. Without these behaviors, made possible by expression of insect genes, no galls will form. However, these types of behaviors are, at most, minute variations on those seen in numerous other phytophagous insects that do not induce galls, including other tephritid flies. Modifications of the preferences within these behavioral sequences make specialization on goldenrod possible, but are not key to the actual process of gall formation. To determine the manner of genetic control that insects exert over gall induction it would be helpful if a more complete knowledge of the insect's stimulus were known, but it is not (chapter 3). However, the product of that stimulation is available for study.

In figure 6.1, we present a simple developmental-genetic model for gall induction and growth. Plant genotype interacts with the environment to develop the plant phenotype, which includes the stem where galls can be induced. Similarly, insect genotype interacts with the environment in the development of the insect phenotype, which includes the stimulus it applies to the plant. When the normal stem phenotype is subjected to the insect's stimulus, a phenotypic response, the gall, develops. This model suggested an experimental approach for determining if insect genes have an influence over gall development which involved searching for naturally occurring genetic variation in the stimulus. The basic logic behind this approach is that if there are segregating alleles that contribute to the stimulus phenotype, then insect groups that share an allele should induce similar galls, and that their galls will differ from those induced by groups with other alleles. In other words, if gall phenotype covaries when insect genotype is systematically varied, the existence of genetic variation in the stimulus can be inferred, and as Mendel showed, demonstrating the existence of genetic variation demonstrates the existence of genetic control. We used quantitative genetic analysis to search for insect genetic variance in gall size. Before describing these experiments, we will review some of the principles on which they are based.

6.2 SOME BASICS OF QUANTITATIVE GENETIC ANALYSIS

In a foundational paper, R. A. Fisher (1918) showed how resemblance among relatives could be statistically analyzed to reveal the relative contributions of genes and environment to quantitative characters, such as height (and in the process invented analysis of variance). Fisher's approach starts with the assumption that a character's development can be influenced by a large number of loci, that alleles are segregating at many of these loci, and that each allele adds or subtracts a small amount from the character's phenotype. In the development of an individual, the phenotype of a quantitative trait, P, will be the sum of contributions from genes, G, and contributions from environment, E, or

$$P = G + E. \tag{6.1}$$

When many polymorphic loci are involved, individuals will vary for their genetic contributions, and G will have a continuous probability distribution within a population. Similarly, when many small environmental factors influence phenotypic development, a continuous distribution can be expected for E. This will usually result in a continuous distribution for P that is either normal, or transformable to a normal (see Falconer 1989). Thus the variance in the phenotype, V_P, will be the following:

$$V_P = V_G + V_E, \tag{6.2}$$

assuming that G and E are uncorrelated. Experiments can then be designed to determine if V_G is greater than zero.

Suppose an experimenter randomly draws a very large number of individuals from a population, produces many replicates of each by cloning, allows the replicates to develop at randomly selected points in the environment, and then measures some character on each. The variance in the distribution of those measurements, the phenotypic variance, can then be partitioned into genetic and environmental components. The variance among the individuals within each clone will be due to environmental effects, since the replicates are genetically identical. The variance among the clones will be due to genetic differences since the environmental influences are randomized with respect to genotype. If the clonal means differ, they will have some distribution, and the variance of that distribution will be V_G. On the other hand, if all clones have the same mean value for the measured character, then $V_G = 0$, and one can assume that the clones were drawn from a population that is genetically uniform for the character. The potential for natural selection to change a character depends on its genetic variance. This is clear from the limiting case in which all individuals in a population are genetically identical—no matter how strong selection may be on such a population, the individuals in the next generation will be genetically identical to the previous generation.

For many types of organisms it is impossible to clone individuals. However, because relatives share genes through common descent, family groups will be partial replicates of inherited allelic combinations. For instance, an offspring inherits half the genes of each parent. Thus phenotypes of offspring will be correlated to their parents because of inherited genetic differences. The magnitude of that correlation, when measured from many parent-offspring groups sampled from a breeding population, is proportional to V_G. By the same logic, full-sibs will on average share half their genes, and so the variance among full-sibships will equal half of V_G, assuming completely additive gene effects.

Just as phenotypic variance can be decomposed into genetic and environmental variances, genetic variance can be decomposed into several components. The most important of these for understanding natural selection is the additive component, V_A. The magnitude of the additive genetic variance in a population measures how faithfully offspring resemble parents, and determines the rate of response to selection (Fisher 1918; Falconer 1989; Houle 1992). If all genetic variance is due to different alleles making additive contributions to the phenotype, then V_A is equal to V_G. However, alleles do not always make additive contributions to a trait, as is the case when a dominant allele masks the effect of a recessive. Allelic differences at one locus can also be masked by the action of other loci, a phenomenon known as *epistasis*. Such interactions among alleles can distort the phenotypic resemblance among relatives (Falconer 1989). Quantitative geneticists have designed breeding experiments that can be used to measure the amount of genetic variance that is additive, and the amount due to distortions caused be dominance and epistasis (Falconer 1989). Often the degree of additive genetic variation in a trait is expressed as the heritability, that is,

$$h^2 = V_A / V_P, \tag{6.3}$$

which is scaled between 0 and 1.0. Having sketched the basic principles of quantitative genetics, we can go on to our application of them to determine if gall size is controlled by insect genes.

6.3 INSECT GENES AND GALL DIAMETER, I: HOMOGENEOUS PLANTS

Galls are plant tissue that develops under the influence of an insect's stimulus. Thus, to partition phenotypic variance in gall characters into genetic and environmental components, the contribution of both insect and plant genetic variance has to be considered. A simple quantitative genetic model for a gall character, one that assumes that all genetic variance is additive, would be

$$V_P = V_{Gi} + V_{Gp} + V_{(Gi \times Gp)} + V_E, \tag{6.4}$$

where V_{Gi} is the insect genetic variance, V_{Gp} is plant genetic variance, $V_{(Gi \times Gp)}$ is the variance due to the interaction among genotypes, and V_E is the residual variance, including environmental effects on gall development. In an experiment that seeks to determine if V_{Gi} is or is not greater than zero, it would be advisable to eliminate as many of the other sources of phenotypic variance as is possible, and to reduce the impact of those that cannot be eliminated. This was the approach in our first attempt to determine if gall size can be a genetically heritable character of the insect.

In this first experiment (Weis and Abrahamson 1986), thirteen singly mated females were given greenhouse-grown goldenrod plants for oviposition. The galls resulting from the offspring of a single female thus are produced by a full-sibship, and so thirteen sibships were included in the analysis. Each sibship consisted of 5–11 offspring. The among-sibship variance in gall diameter will be proportional to V_{Gi} for that character. We took two steps that allowed us to reduce other sources of phenotypic variance in gall size. First, the plants given to females were cloned from a single goldenrod genet. This eliminated both the plant genetic and the "plant genotype × insect genotype" interaction effect components of variance. Second, by performing the experiment in the greenhouse, we reduced environmental heterogeneity that could have contributed to phenotypic variance.

The experiment indicated significant genetic variation in gall size (fig. 6.2). Diameter was measured at three times during development, and at all three times sibships were significantly different (Weis and Abrahamson 1986). Growth rate also differed among sibships early in development, but not later as the gallmakers approached their mature size. Heritability of gall diameter at maturity was estimated at 0.40. This value falls within the range typical for morphological measurements (Mousseau and Roff 1987), and would suggest that gall size could evolve rapidly under moderate selection. However, this value must be interpreted with great caution.

Although the results of this experiment point to the qualitative conclusion that gall size develops under the genetic influence of *Eurosta,* the experiment was performed under conditions that tend to overestimate heritability. First, the use of full-sibs to estimate additive genetic variance is troubled by the fact that part of the variance due to dominance is included in the variance among sibships (Falconer 1989). Second, since full-sibs share the same mother, differences among sibships could

179

Gallmaker Sibships Differ in Gall Size

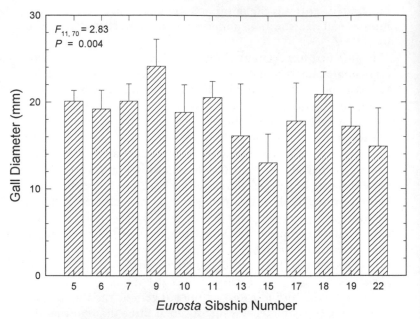

FIGURE 6.2. Mean diameters of galls induced by the offspring of twelve different, single-mated females. All plants were cloned from a single genotype. The significant variance among sibships indicates genetic variance in the *Eurosta* population for gall size.

be in part due to differences in the environment provided by the different mothers. This would be a serious concern if *Eurosta* galls were initiated by substances injected by the mother into the plant, as occurs with galls caused by sawflies (Price 1992). However, *Eurosta* galls form away from the oviposition site and about a week after the egg is deposited (chapter 2). Neither of these problems are likely to have changed the qualitative result of the experiment. Both problems would have been eliminated were we able to use alternate breeding designs, such as half-sib families or father-offspring regression (see Falconer 1989); but unfortunately *Eurosta* is not nearly so easy to culture as *Drosophila*, and attempted experiments with these more complex designs never achieved adequate sample sizes. The third condition that would tend to overestimate heritability is more serious. The denominator for this heritability estimate was the phenotypic variance in gall diameter, but efforts were made to eliminate or reduce all causes of gall phenotypic variance except for insect genotype. By reducing plant genetic variance and environmental variance, insect genetic variance may have constituted a greater proportion of the remaining variance. Under less proscribed conditions, heritability may have been lower. These three conditions would bias the estimate of heritability, even if the additive genetic variance actually expressed by the experimental larvae was exactly the same as the population additive genetic variance. However, another difficulty could have caused the additive genetic variance expressed in this situation to differ from the population variance expressed under typical conditions.

The use of a single plant genotype in this experiment potentially biased the estimate of insect genetic variance. If the interaction effect between insect and plant genotypes is a substantial determinant of gall size in natural populations of *Eurosta* and goldenrod, then V_{Gi} for gall size could be different on different plant genotypes (Weis and Abrahamson 1986; Weis and Gorman 1990). The insect genetic variance estimate from this experiment could be either higher or lower than what would be found when V_{Gi} was measured against a random background of plant genotypes. Thus, although we can say that this experiment established in principle that gall phenotype is influenced by insect genotype, it is impossible to say from these data alone whether the potential for the evolution of gall size in natural environments is high or low, or, for practical purposes, nonexistent. Before considering a more complex experiment on insect genetics, we will turn to information on the plant's contribution to gall size.

6.4 PLANT INFLUENCES ON GALL SIZE

When walking through a goldenrod population, one is often struck by the fact that goldenrod genets differ not only in the percentage of galled ramets, but in the size of the galls they pro-

Plant Clones Differ in Gall Size

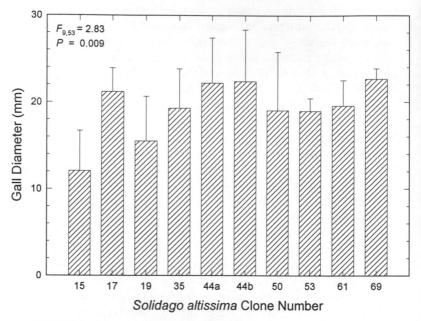

FIGURE 6.3. Mean gall diameter of galls induced on potted plants cloned from ten different *S. altissima* genotypes. Gallmaker females were randomized over plant clones. Significant differences indicated a plant genetic variation for gall size.

duce. In a series of genets examined at the Bucknell University Natural Area in 1983, mean gall size ranged from a low of 14.2 to a high of 23.9 mm (Hollenbach 1984). The plant's genetic contribution to gall size variation was confirmed in a greenhouse experiment. Cloned replicates of ten *S. altissima* were presented to randomly chosen *Eurosta* females for oviposition. The genotypes differed significantly in the sizes of the resulting galls (fig. 6.3). We became curious as to how plant variation, genetic and otherwise, may affect the expression of *Eurosta*'s genetic variance for gall size.

In 1982 we started a field study to determine if the date of gall

Growth Curves for Age Cohorts of Galls

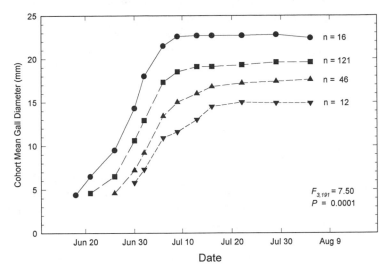

FIGURE 6.4. Growth curves for four successive age cohorts of gall. Data were collected at the Bucknell University Natural Area in 1983. (From Weis and Abrahamson 1985, "Potential selective pressures by parasitoids on the evolution of a plant-herbivore interaction," *Ecology* 66:1261–1269; with permission)

appearance had any effect on gall size and *Eurosta*'s survival. We marked several hundred plants with oviposition scars and then returned every 2–3 days to see when new galls appeared. After new galls stopped appearing, these visits were continued in order to measure the growth of the developing galls. This was repeated the next year. We found that the late-starting galls matured at significantly smaller sizes than early ones (fig. 6.4). On one hand, this suggested that plants lose their reactivity to the stimulus as they age. However, in the process of collecting these data, we got the impression that on any one day new galls showed up in clusters, and that cluster boundaries seemed roughly to correspond with genet boundaries. This type of spatio-temporal clustering could have two simple explanations. First, each cluster could represent the offspring of a single fe-

183

male who attacked on a single day, with the early clusters being deposited by early emerging females and the late clusters by late emergers. If so, late starters may have induced smaller galls because all genets lose reactivity with age. Second, clusters are the product of variation in plant reactivity such that in some genets, gall growth starts later and proceeds more slowly than in others. But it was also possible that the explanation was not simple, and that both oviposition time and variation in plant reactivity contributed to the pattern.

We set up an experiment to determine whether plant age at gall appearance, plant genotype, or both influenced gall growth (Weis and Abrahamson 1985). Clones, produced from cuttings from each of four plant genotypes, were started at three dates, separated by about 12 days. These three cohorts were then exposed to ovipositing females on the same date. The time from oviposition to gall appearance, which we called "lag time," was measured, as was gall diameter at four dates through development. We found that both plant clone and plant age affected gall size, as revealed in a repeated measures analysis of variance (table 6.1). No interaction between the clone and plant age was found. The analysis further showed that the shape of the gall growth curve differed among plant genotypes (the Plant Clone × Gall Age interaction was significant), but not among plant ages (the Plant Age × Gall Age interaction was not significant). This suggested that some clones produce slower-growing and smaller-maturing galls than others, and that all clones reduce growth rate and size by the same amount as they mature. What did not show up in this experiment was a significant effect of clone on lag time ($F = 2.09$, df $= 3, 32$, $P = 0.12$). In a larger experiment, presented in the next section, we discovered that some plant genotypes do take longer to respond to the gall-inducing stimulus than others. Before we deal with that result, however, we want to offer an idea as to why late-starting galls are smaller.

Goldenrod stems continue to grow throughout the season, reaching their maximum height only when they produce the last of their flower buds in late August or September. However, growth rate in stem height starts to decline in mid-July (fig. 6.5). Before this decline newly developing nodes behind the apical

TABLE 6.1. The effect of plant age at time of oviposition, and of plant genotype, on gall growth.

A. Mean gall diameter at four times during development, by plant age at oviposition.

Plant Age at Oviposition (days)	Mean Gall Diameter (mm)			
	Apr 24	May 1	May 10	May 21
56	4.38	8.50	14.59	14.79
44	6.27	13.11	18.77	19.00
33	6.88	13.24	17.53	17.87

B. Repeated measure analysis of variance.

Source of Variation	df	MS	F	P
Within galls				
Plant clone	3	1.034	3.60	.023
Plant age	2	1.156	4.02	.027
Plant clone × Plant age	6	0.253	0.88	.519
Residual	34	0.287	—	—
Among galls				
Gall age	3	8.15	254.02	<.001
Gall age × Plant clone	9	0.106	3.29	.002
Gall age × Plant age	6	0.039	1.22	.302
Gall age × Plant age × Plant clone	18	0.017	0.53	.935
Residual	102	0.032	—	—

Notes: Plants were initiated at three dates. Gall diameter was measured at four times during growth.

bud produce only leaves, and frequently the leaves mature at successively larger sizes as the stem grows. After this point, however, many plants begin also to produce side branches from the newly developing nodes (these branches eventually bear inflorescence), and in all plants, the newly produced leaves become progressively smaller. This slowing of stem elongation coincides with the end of gall growth (fig. 6.5). We suggest that the stem undergoes a physiological change at this time that reduces reactivity to the gallmaker. Late-starting galls may fail to reach their potential size because the plant stops reacting before growth is

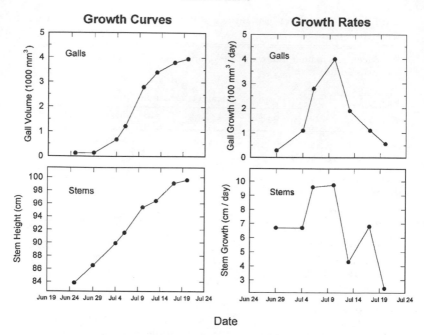

FIGURE 6.5. Growth curves and growth rates for stems and galls at the Bucknell University Natural Area during 1983.

complete. However, the effect of a late start may not be the same for all *Eurosta* larvae, as we shall show next.

6.5 INSECT GENES AND GALL DIAMETER, II: HETEROGENEOUS PLANTS

In our first experiment to test for *Eurosta*'s genetic variance in gall size, the plants we provided for gall development were all cloned from a single genotype. We selected this genet because in the previous year it had produced an abundance of average-sized galls, and in the greenhouse we obtained evidence for a contribution of insect genes to gall size. Our choice of plants was perhaps a lucky one. In a subsequent experiment, we obtained evidence that the level of insect genetic variance for gall size that gets expressed probably changes with plant genotype.

To explain how genetic variance can change, we present a modification of a model devised by G. de Jong (1990a,b) that examines the consequences for expressed genetic variance when two alleles at a locus show different sensitivity to the environment. Assume we have an annual plant that flowers at the end of the season, and that the height of the flowering stalk is influenced by a locus with two alleles, G_1 and G_2, both with a frequency of 0.5. For simplicity we will make the contributions for these loci additive, such that heterozygotes are intermediate. Also assume that an environmental factor, such as rainfall, also contributes to height in a linear fashion; in wet years stalks grow taller than in dry years. We can write a set of equations that predict the phenotype for plants of each genotype across the rainfall gradient as follows:

$$G_1G_1 \quad P_{11} = 2a_1 + 2b_1R$$
$$G_1G_2 \quad P_{21} = (a_1 + a_2) + (b_1 + b_2)R$$
$$G_2G_2 \quad P_{22} = 2a_2 + 2b_2R,$$

where R is the deviation of seasonal rainfall away from an average year, a_1 and a_2 are the contributions to height of alleles G_1 and G_2, respectively, to stalk height during an average year, and b_1 and b_2 are the changes in height per unit change in rainfall for alleles G_1 and G_2 respectively. Since alleles act additively, homozygotes receive the allelic contribution from both copies of the allele, and the heterozygote receives the sum of the two different allelic contributions. Each of these equations predicts stalk height as a function of rainfall. Functions such as these are called *reaction norms*, in the parlance of quantitative genetics (Lewontin 1986; Stearns 1988; Gomulkiewicz and Kirkpatrick 1992; Weis 1992; Gavrilets and Scheiner 1993a,b). In essence, a reaction norm depicts the phenotype expected by a given genotype across an array of environments. Parameters within the function that determine its slope or curvature can describe the genotype's response to the environment.

The item of importance in this example is that b_i, the reaction norm slope, describes the sensitivity of allelic expression to the environment, and variation in b_i will affect how much of the underlying genetic variation gets expressed in a given environment. If b_1 equals b_2 then all genotypes will react to changes in rainfall equally (fig. 6.6a). The three genotypes will differ by the same amount regardless of rainfall, and so the expressed genetic variance is constant. However, if $b_1 \neq b_2$, then the three genotypes have different sensitivity to rainfall, and thus the difference among the phenotypes they produce changes with rainfall. Figure 6.6b shows an example where the expressed genetic variance will be high in years of low rain and zero in years of moderately high rain. We emphasize that the underlying genetic variability is the same at all points on the rainfall gradient (the frequency of both alleles is always 0.5), but the phenotypic variance that is caused by genetic factors is not. An experimenter that measured genetic variance in stalk height at only one rainfall level could get a high or a low estimate, depending on which level was chosen—and each estimate would be correct, but each would pertain only to the rain level experienced during the experiment.

We realized that *Eurosta* may show a similar pattern, that is, some combinations of insect alleles may respond differently to the variation in plants during gall induction and growth than others. This led to an experiment we hoped would detect differences in sensitivity (Weis and Gorman 1990) by looking at gall-size reaction norms. We grew members of sixteen full-sib families of gallmakers on genetically diverse arrays of goldenrod plants, measured gall diameter at maturity, and also measured a number of other plant traits as the gall grew. The reasoning behind the design was this: some measures of plant growth are likely to reflect the quality of the plant as an environment for gall growth and development, and if so, we could test whether the gall sizes of the full-sib families responded differently to these environmental quality measures. Thus each reaction norm measured in this experiment would predict gall size for members of a family as a function of some plant character.

Each singly mated *Eurosta* female received a series of plants for oviposition. These were drawn at random from an array of seventy-two plant genotypes, each genotype cloned multiple times. These plants were clones of individuals used in an unrelated experiment on genetic and maternal effects on seedling growth (Weis, Hollenbach, and Abrahamson 1987) that used a plant full-sib mating design. Because we knew the origin of each plant, we could also estimate the genetic variance for the plant

Reaction Norms

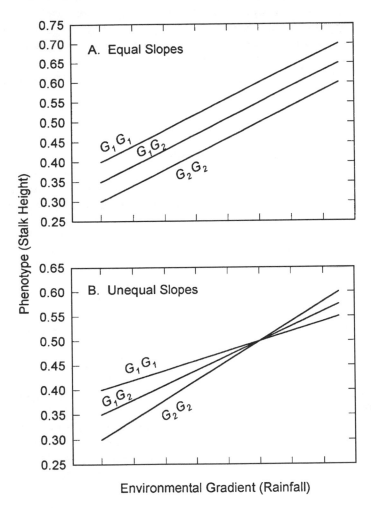

FIGURE 6.6. Hypothetical reaction norms for the three genotypes derived from alleles G_1 and G_2. See text.

characters used as environmental quality measures. As potential measures of environmental quality, we determined stem height and stem width several times during the experiment, calculated the growth rate in each, and in addition calculated lag time as the number of days between oviposition and gall appearance. To determine which of these traits actually affect gall size, we performed stepwise multiple regression to find the combination of plant characteristics that best predicted gall size (Weis and Gorman 1990). In the end, only lag time explained significant variance in gall size.

Intuitively, lag time should be strongly influenced by insect genes. The time between oviposition and gall appearance will depend on the time it takes the egg to hatch, the larva to bore into and settle inside the stem, and then perhaps on the strength of the larva's initial stimulus. However, our analysis showed no signs of variation among insect full-sib families for this trait ($F = 0.90$, df $= 15, 120$, n.s.). On the other hand, lag time did differ significantly among the plant full-sib families ($F = 7.51$, df $= 8, 9, P = 0.008$). Thus, lag time was used as a measure of environmental quality for the gallmaker.

To estimate the reaction norms, we performed a linear regression of gall size over lag time for each of the sixteen full-sib families. For reasons that will be made clear directly, we recorded each gall's lag time as its deviation from the population mean lag time. We also tried including a quadratic term in the family regressions, but the term was marginally significant in only one of the families, and so we proceeded with the assumption that linear functions were adequate descriptors of gall-size reaction norms. The slope of a family's regression line thus indicates its sensitivity to lag time. Because we used the deviation from mean gall size as the independent variable, the y-intercept of the reaction norm regression takes on the meaningful interpretation of being the expected gall size if a family member gets an average plant. This is equivalent to saying it is the expected gall size on a plant drawn at random from the population, and so is the expectation independent of information on lag time. We called the intercept term the reaction norm "elevation" to indicate when elevation is high, the family makes large galls, on average, and small galls when elevation is low.

We used a mixed model analysis of covariance (Henderson 1982) to test for among-family variance in gall size, including variance that could be explained by differences in sensitivity to lag time. The model is expressed in the equation

$$D_{ij} = \mu + A_i + \beta L_{ij} + \beta_i L_{ij} + \epsilon_{ij} \qquad (6.6)$$

where D_{ij} is the diameter of the gall induced by the jth offspring in the ith full-sib family, L_{ij} is the lag time of the plant it occupies, μ is the mean gall size, A_i is the random effect of the ith full-sib family on gall size, β is the fixed effect of the regression of diameter over lagtime (i.e., the mean regression slope), β_i is the deviation of the family-specific regression slope of the ith family from the mean regression slope, and ϵ_{ij} is the residual error. A significant A_i term would indicate that the families differed in the mean size of the galls, after accounting for the effects of lag time—or, in other words, families different for reaction norm elevation. This variance component is one-half the insect additive genetic variance for gall size (assuming negligible dominance and maternal effects; see above). A significant β_i term would indicate that families differ in the linear component of the response to lag time, or less formally, their reaction norm slopes differ. This variance component is equivalent to a "genotype × environment" interaction term in a two-way analysis of variance, or at least its linear component. It constitutes one-half of the additive genetic variance for sensitivity to lag time (see Scheiner and Goodnight 1984).

All three variance terms were significantly greater than zero (table 6.2). The variance among full-sib families explained about one-tenth of the phenotypic variance in gall size, averaged over all lag times. Since the siblings are related by one-half, this proportion is multiplied by two to get the heritability of gall size. This calculation then indicates that 20% of the variance in gall size is attributable to insect genetic variance after accounting for sensitivity to lag time. This is half as large as was estimated by the "single plant genotype" experiment. By contrast, approximately 56% of the variance in gall size can be attributed to insect genetic variance in sensitivity to plant lag time.

The individual family reaction norms show that although the overall trend is for smaller galls to be produced on slow-reacting

TABLE 6.2. Analysis of covariance of gall size by gallmaker full-sib family and plant lag time.

Source	MS	S of S	df	F-ratio	F	Variance Component	Proportion Variance
Family	M4	3.44	15	M4/M1	2.14*	1.688	0.103
Lag time	M3	258.70	1	M3/M2	10.02***	1.082	0.066
Family × Lag	M2	387.02	15	M2/M1	2.40**	4.395	0.269
Residual	M1	1107.77	103				

$*P < .05, **P < .01, ***P < .001.$

Gall Size Reaction Norms

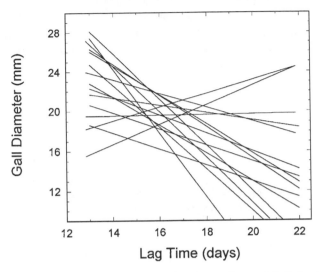

FIGURE 6.7. Gall-size reaction norms for sixteen *Eurosta* full-sib families, inducing galls on a genetically diverse array of *S. altissima*. The overall height of the line indicates the family average gall size, and the slope of the line indicates the family-specific sensitivity to plant lag time (number of days between oviposition and gall initiation). (From Weis and Gorman, 1990, "Measuring canalizing selection on reaction norms: an exploration of the *Eurosta-Solidago* system," *Evolution* 44:820–831, with permission)

plants, there is much heterogeneity in the response (fig. 6.7). Sample size was limited (~eight sibs per family), and so none of the individual family reaction norms were measured with comfortably narrow confidence limits. Yet the analysis of covariance showed that the variances in elevation and slope are significantly greater than the random expectation, and so the "pile of lines" in figure 6.7 should give an impression of what the distribution of reaction norms really is like. Some families showed large decreases in gall size over slight increases in lag time, others produced intermediate galls on all plants, and one showed an increase in gall size with lag time. What is interesting is that the reaction norms tend to converge near the mean lag time of 16.2 days. As a result, the expressed genetic variance in gall size is the least in the average environment.

This convergence of reaction norms is not due to any mathematical necessity, but simply follows from the fact that the variance in elevation is less than the variance in slope. But this then raises the question of why one source of variance should be larger than the other. The answer to that question must involve, among other things, the rate at which natural selection is acting on the two reaction norm parameters, because as selection acts, it erodes genetic variance. The strength of selection depends on what is the optimal reaction norm, that is, one that yields the maximal fitness at all points of the environmental gradient. Further, it depends on the rate of fitness loss as genotype deviates further from the optimum. And finally, it depends on whether a deviation from optimal elevation causes a more or less severe fitness loss than a similar deviation in slope. These issues are taken up again in chapter 9.

6.6 ARE THERE GALL-SPECIFIC PLANT GENES?

The developmental genetic model presented in figure 6.1 indicates that plant genotype contributed to gall phenotype. This opens the possibility that gall phenotype can evolve in response to the selective pressure the gallmaker places on the plant. Although we concluded in chapter 3 that *Solidago* is under weak selection, at best, to resist *Eurosta* attack, it is still interesting to consider generally what sorts of genetic constraints plants would face when gallmakers exert strong selection.

Plant reactivity to the gallmaker may be strongly influenced by genes with important roles in normal developmental and physiological processes. Rohfritsch and Shorthouse (1982) summarized many histochemical studies by stating "there are no essentially new physiological processes nor any new metabolic pathways in the insect gall tissue." Simply, gall development may be governed by normal plant processes, but ones that are turned on at atypical times and at atypical rates. The fitness contribution an allele makes through decreased reactivity to a gallmaker could easily be outweighed by the negative pleiotropic effects it would have on normal growth and development. To take an extreme case, a locus coding for a dysfunctional auxin receptor would render the plant unable to respond when the gallmaker's stimulus boosts auxin levels, but a plant with dysfunctional auxin receptors would be unable to do anything at all.

The involvement of normal plant genes in gall development suggests that plant reactivity to gallmakers evolves as a correlated response to the many other selection pressures that act on plant growth and development. This will severely constrain the evolution of plant resistance through non-reactivity. However, if a plant gene for reactivity was expressed only in the gall itself, selection on lowered reactivity would be free of constraints. Duplication of functioning genes through proliferation of gene families or through polyploidy could provide a source of loci that are silent except when stimulated by a gallmaker. However, Weis, Walton, and Crego (1988) argued that insects which induce galls by turning on such gall-specific genes would be at a disadvantage. Low-reactivity alleles at such loci would be brought quickly to fixation since their positive effects on plant fitness would have no negative pleiotropic effects. Zero-reactivity alleles that put the gallmaker "out of business" would offer complete resistance. The seeming stability of plant-gallmaker interactions argues that gall-specific genes are likely to be of no more than minor importance, if any, in the evolution of plant-gallmaker interactions.

Host Specificity and Herbivore Speciation

Host specificity in phytophagous insects varies from taxa that are broadly polyphagous (e.g., gypsy moth) to those that are oligophagous or monophagous. While there has been some debate over whether generalist or specialist phytophagous insects are more likely to radiate new species (Futuyma 1987), it is clear from several examples that specialist taxa can be composed of host-associated populations or host races, each of which exploit different host plant species (e.g., Bush 1969a,b, 1975a,b, 1994; Knerer and Atwood 1973; Wood 1980; Menken 1981; Scriber 1983; Tauber and Tauber 1989; Craig et al. 1993; Berlocher 1995). A host race has been defined by Diehl and Bush (1984) as "a population of a species that is partially reproductively isolated from other conspecific populations as a direct consequence of adaptation to a specific host." The existence of such closely related, host-associated populations or host races implies that the availability of host-plant resources has the potential to influence the patterns of herbivore differentiation. However, evolutionary biologists have differed in their interpretation of such variation in host utilization by herbivores and consequently they continue to debate the means by which a single species divides into multiple species (Futuyma 1987; Rice and Hostert 1993; Bush 1994, 1995; Claridge 1995). In spite of the attention these issues have received, the fundamental issue of determining the requisite processes of speciation still remains essentially unresolved.

One focus of our research group's studies has been the understanding of these requisite processes of host-race formation and speciation. There is evidence that *E. solidaginis* is actively undergoing speciation through host-plant association. The following discussion is intended to synthesis the results currently available for *E. solidaginis*. This synthesis draws heavily from unpublished and/or ongoing studies by several of our coworkers, including Jonathan Brown, Timothy Craig, John Horner, and Joanne Itami. We greatly appreciate their willingness to share their insights into the *S. altissima/E. solidaginis*/natural-enemy interaction.

7.1 ALLOPATRIC AND SYMPATRIC SPECIATION

At the center of this enduring discussion about the processes and modes of speciation is the means by which reproductive isolation is attained among differentiating populations. Some authors contend that speciation requires that populations be geographically separated to undergo genetic differentiation (Mayr 1942, 1963, 1988; Carson 1975, 1989; Mayr and Provine 1980; Futuyma 1986). Such geographic models of speciation argue that populations must be spatially separated for a sufficient time to enable populations to genetically differentiate through genetic drift and/or adaptation to a new environment. Geographic or allopatric models are of two broad types: (1) those in which genetic changes gradually accumulate as populations adapt to different environmental conditions, and (2) models in which small, isolated populations experience rapid genetic revolution (Futuyma 1986).

Others have argued that under some circumstances speciation processes may operate among populations in sympatry (e.g., Bush 1975a,b, 1994; Diehl and Bush 1984; Rice 1984; Feder 1995; Johnson et al. 1996). Because of their short generation times and close associations with host plants, phytophagous insects are seen as likely to experience relatively rapid speciation using simple modes of speciation (Bush 1969a,b, 1975a,b, 1994, 1995). Insect herbivores could undergo such sympatric speciation by shifts in host-plant utilization and subsequent adaptation to a novel host plant (Bush 1974, 1975a,b, 1994; Price and Willson 1976; Wood 1980; Wood and Guttman 1983; Rice 1987; Tauber and Tauber 1989; Feder 1995; Feder et al. 1995; Payne and Berlocher 1995). Bush (1974, 1975a,b), for example, maintained that there must be at least two steps in a herbivore's successful shift to a novel host, both of which require genetic

197

change. If sympatric speciation is to occur according to this model, genetic changes are necessary first in the herbivore's host-recognition process and second in traits that affect survival on the new host. More recently Craig et al. (1993) argued that host-race formation in a phytophagous insect minimally requires (1) alteration of the insect's preferences for feeding and/or oviposition, (2) appropriate physiological adaptation to the new host plant, and (3) assortative mating to maintain genetic changes important to the use of the new host.

The concept of sympatric speciation through host-race formation has been highly controversial because of the restrictive conditions necessary for the required genetic changes (Tauber and Tauber 1989). For example, if herbivore traits for host recognition, physiological adaptation to a new host, and assortative mating are mediated by different gene loci, then the likelihood of sympatric speciation through host-race formation is extremely low. Genetic linkage of these loci would be necessary to prevent recombination from breaking up coadapted combinations of these traits at a rate faster than selection could reinforce these combinations (Felsenstein 1981; Craig et al. 1993). This constraint has led to arguments that sympatric speciation is unlikely (Futuyma 1983, 1986; Futuyma and Peterson 1985).

However, a number of authors have maintained that this constraint can be relaxed if mate choice and host choice are coupled (Bush 1975a,b; Diehl and Bush 1984, 1989; Rice 1984, 1987; Craig et al. 1993; Johnson et al. 1996). Such coupling is quite possible in herbivorous insects since many species mate on their host plants (Price 1980). The consequence is that mate choice and host choice can proceed as correlated characters (Rice 1984, 1987; Rice and Salt 1990; Johnson et al. 1996).

Unfortunately, examples of host-race formation are rare (Bush and Howard 1986; Tauber and Tauber 1989; Johnson et al. 1996) and the precise mechanisms of host shifts remain unclear (including the necessity of genetic change in host recognition and the requirement of geographic reproductive isolation). However, Bush and his colleagues, in their studies of *Rhagoletis pomonella* (Diptera: Tephritidae), have convincingly demonstrated that (1) host-race formation can occur, (2) host shifts can occur rapidly, and (3) that host races can serve as an intermediate step in speciation (McPheron, Smith, and Berlocker 1988; Diehl and Bush 1989; Feder and Bush 1989a,b, 1991; Feder, Chilcote, and Bush 1989a,b, 1990a,b; Feder 1995; Feder et al. 1995; Berlocher 1995; Bush 1994). Our group's studies of *Eurosta solidaginis* provide another example of host-race formation by way of a host shift and thus further document both its occurrence and the ecological, behavioral, and genetic conditions necessary.

7.2 THE STUDY OF SPECIATION MECHANISMS

Narrowly oligophagous or monophagous insects are excellent candidates for studies of speciation mechanisms since these organisms complete their life cycles on one to several individual plants of only a limited number of related host-plant species. The consequence of narrow host association is the possibility that herbivorous insects may speciate if some subset of the original insect population shifts to a novel host plant and subsequently becomes reproductively isolated from the original population (Craig et al. 1994). Studies utilizing herbivorous insects to investigate the relative roles of adaptation, behavior, and gene flow in the speciation processes began in the 1960s (e.g., Ehrlich and Raven 1964) and have continued to the present (Wood 1980; Bush 1982; Futuyma and Peterson 1985; Bush and Howard 1986; Tauber and Tauber 1989; Waring, Abrahamson, and Howard 1990; Craig et al. 1993). Many of these studies have explored the assumptions of various models of speciation (reviewed in Tauber and Tauber 1989; Via 1990) and have provided valuable description of the current ecological and genetic factors that promote speciation. If present-day factors are representative of past conditions, we will be able to understand the patterns of diversification among herbivores.

Unfortunately, we can never be absolutely certain that contemporary conditions match past circumstances nor that present population-level situations can be extended to explain larger scale, phylogenetic patterns (Rausher 1988a). This difficulty occurs in part because speciation is a historical process (Abrahamson et al. 1994). Studies of present-day behavioral, ecological, and genetic characteristics of recently speciated or speciating taxa can only suggest the most parsimonious explanation of how speciation has proceeded (Wood and Keese

1990). We will likely never know all the details of a specific speciation event. However, studies of recently speciated or currently speciating taxa can provide crucial insights into the processes of speciation since many of the characteristics that cause speciation still exist in such taxa.

Additional difficulties in understanding the processes of herbivore speciation arise from the variety of associations that phytophagous feeding guilds have with their host plants. Borers, miners, and free feeders, for instance, have distinct constraints placed on them by their hosts and, because of unique lines of descent, have differing adaptive potentials for the exploitation of their host plants (Weis and Berenbaum 1989; Abrahamson et al. 1994). The consequence is that modes of speciation have likely varied among herbivore guilds and lines of descent. However, by careful examination of the speciation processes in herbivores that have an intimate and specific association with their host, we should be able to understand the conditions that promote or inhibit speciation.

7.3 GALLMAKERS AS MODELS OF SPECIATION PROCESSES

Gallmaking insects are excellent model systems for exploration of speciation processes. Gallmakers typically have a very narrow host range and an intimate association with host-plant tissue, both of which can create strong selection pressures for adaptation to their host (Abrahamson et al. 1994; Craig et al. 1994; see chapter 2). Furthermore, since gall characteristics are influenced by both the insect's and the host plant's genotype (Weis and Abrahamson 1986), gall characters that affect, for instance, the emergence time of a gallmaker might be altered simply by ovipositing on a different host (Craig et al. 1993). Such alterations of adult emergence, mating, or oviposition times could markedly reduce gene flow between populations associated with different host plants and could promote speciation, possibly under sympatric conditions (Abrahamson et al. 1994).

Furthermore, gallmakers that oviposit on a closely related but novel host will likely develop in a gall of different size and/or morphology. Such modification of an insect's gall could alter adult performance, emergence times, or other life-history traits, and subsequently facilitate reproductive isolation or change in the susceptibility of the gallmaker to its natural enemies (Abrahamson et al. 1994; Brown et al. 1995; Sumerford and Abrahamson 1995). The potential for complete escape from or reduction of natural-enemy attack would markedly enhance gallmaker survival on a novel host plant. As we have argued earlier, galls are typically apparent, predictable resources for their natural enemies (Price et al. 1980; Abrahamson, Armbruster, and Maddox 1983; Abrahamson and Weis 1987) and consequently gallmakers support diverse communities of parasitoids, inquilines, and predators (Mani 1964; Askew 1975; Abrahamson, McCrea, and Anderson 1989; Weis, Abrahamson, and Andersen 1992). While some natural enemies are known to assail fully formed galls and likely cue on gall traits, other natural enemies attack before gall formation and appear to utilize host-plant cues to locate their prey. The consequence is that a novel host plant may not be recognized by one or more of the natural enemies. Furthermore, many features of the gall phenotype are likely to influence natural-enemy attack including gall rigidity, size, or the presence of allelochemicals (e.g., Washburn and Cornell 1979; Weis 1982a,b; Weis and Abrahamson 1985; Abrahamson et al. 1991; Jones and Lawton 1991; Hartley 1992; Hartley and Lawton 1992). The net result is that a shift to a novel host with its resultant altered gall phenotype may affect infestation rates by natural enemies.

In the following sections we will examine evidence that *E. solidaginis* has undergone a host shift with resultant host-race formation. In addition, we will explore some of the factors that may have facilitated the acquisition of new hosts by this gallmaker and consequently enhanced its opportunity for speciation. Specifically, we will investigate if and how factors have influenced the potential for a host shift and subsequent speciation.

7.4 GEOGRAPHIC DISTRIBUTIONS OF *EUROSTA SOLIDAGINIS*

Eurosta solidaginis ranges from New Brunswick to Florida and Texas to British Columbia (chapter 2). While the taxonomic relationships of *Eurosta* appear to be reasonably straightforward

(Ming 1989; Brown, Abrahamson, and Way 1996), the identification of *Solidago* species is difficult and their taxonomy is controversial. Consequently, *E. solidaginis* has been reported to infest as many as seven *Solidago* species, including *S. altissima, S. gigantea, S. canadensis, S. rugosa, S. graminifolia, S. serotina,* and *S. ulmifolia* (chapter 2.3; Wasbauer 1972; Ming 1989). It is likely, however, that the difficulty of identification and the sheer number of *Solidago* species has created confusion about the host-plant associations of many *Eurosta* species (Abrahamson, Armbruster, and Maddox 1983; Abrahamson, McCrea, and Anderson 1989; Abrahamson et al. 1989). As a consequence, *Eurosta* species are probably more narrowly oligophagous than the literature would lead us to believe. While *E. solidaginis* infests only *S. altissima* within the mid-Atlantic portion of its distribution, this species commonly infests both *S. altissima* and its abundant sympatric congener *S. gigantea* in the northern and western portions of *S. gigantea*'s distribution (from New England across the northern tier of states and the southern provinces to the north-central U.S.) (Lichter, Weis, and Dimmick 1990; Waring, Abrahamson, and Howard 1990; Abrahamson et al. 1994; Brown et al. 1995; Sumerford and Abrahamson 1995; Brown, Abrahamson, and Way 1996).

7.5 GENETIC EVIDENCE FOR A *EUROSTA* HOST SHIFT

Our work on host-plant recognition and preference suggested that *E. solidaginis* existed as two or more host-associated populations, host races, or sibling species (Abrahamson, McCrea, and Anderson 1989). To investigate these possibilities, we initially used horizontal starch-gel electrophoresis to analyze protein variation in twenty-one sympatric and allopatric *Eurosta* populations infesting *S. altissima* ($n = 9$) and *S. gigantea* ($n = 12$) over a broad geographic range (i.e., New England to north-central U.S.; Waring, Abrahamson, and Howard 1990). The relatively broad geographic area examined and the comparison of both allopatric and sympatric host-plant populations permitted us to discriminate between host-plant related variation and geographic variation. We were consistently and reliably able to score variation in twenty-one enzymes.

The results of the allozyme study showed that average heterozygosity was higher for *E. solidaginis* populations from *S. altissima* (0.028 ± 0.016 SD) than for populations from *S. gigantea* (0.009 ± 0.010) (Waring, Abrahamson, and Howard 1990). Such a pattern of genetic variation plus the more limited range of *E. solidaginis* populations on *S. gigantea* (i.e., *E. solidaginis* is associated with *S. altissima* rather than *S. gigantea* throughout most of its range) suggested that *S. altissima* was the ancestral host plant.

This study also found that genetic variation at most loci conformed to Hardy-Weinberg expectations (122 of 126 χ^2 tests). Three of the four populations were out of Hardy-Weinberg equilibrium at the HBDH (D-β-hydroxybutyrate dehydrogenase) locus and occurred at sites with both host-associated *Eurosta* populations present. These three populations had a higher than expected number of homozygotes characteristic of populations from the alternate host plant. While it is difficult to know the origin of these homozygotes for low frequency alleles, Waring, Abrahamson, and Howard (1990) offered that it is possible that they either represent offspring from ovipositions on the alternate host plant or are the consequence of scoring errors (e.g., mistaking a heterozygote for a homozygote) at the HBDH locus.

Not only did populations over the wide geographic area exhibit conspicuous and significant allele-frequency differentiation at five polymorphic loci (table 7.1), but populations infesting different host plants within the same field showed statistically significant heterogeneity at both the HBDH and PGM

TABLE 7.1. Wright's F_{ST} estimates and χ^2 analysis of variable loci for all populations of *Eurosta solidaginis*.

Locus	F_{ST}	Approx. χ^2	N
GAP-1	0.105	233.1**	555
TPI-1	0.114	126.3**	554
IDH-1	0.020	44.4	555
HBDH-1	0.623	1,308.3**	525
GPD-1	0.109	240.7**	552
PGM-1	0.139	306.9**	552
Mean	0.438	—	—

Sources: Workman and Niswander 1970; Waring, Abrahamson, and Howard 1990.
**$P < 0.005$.

TABLE 7.2. Allele frequencies and G-tests of frequency of heterogeneity at HBDH and PGM among *Eurosta solidaginis* populations on both host plants at all sites, within host plants, and on sympatric host plants.

	HBDH$^{1.00}$	P	PGM$^{1.00}$	P
All sites				
S. altissima	84.3		77.8	
S. gigantea	13.0	**	98.1	**
Among S. altissima sites				
All populations	84.3	**	77.8	**
New England	89.6	n.s.	77.8	**
New England (w/o NE12)	87.0	n.s.	72.2	n.s.
North-central	81.0	**	79.0	n.s.
Among S. gigantea sites				
All populations	13.0	**	98.1	*
New England	2.8	**	99.5	n.s.
New England (w/o NE12)	0.01	n.s.	99.4	n.s.
North-central	43.3	*	93.7	n.s.
Between S. altissima and S. gigantea at sympatric sites				
NE1				
S. altissima	90.0		81.2	
S. gigantea	19.5	**	100.0	*
Cedar Creek				
S. altissima	57.0		82.0	
S. gigantea	40.0	n.s.	94.0	*
Illinois				
S. altissima	87.0		75.0	
S. gigantea	61.0	**	89.8	n.s.

Source: Waring, Abrahamson, and Howard 1990.
Notes: Average frequencies are presented for all sites and among sites.
$*P < 0.05$, $**P < 0.005$.

(phosphoglucomutase) loci (table 7.2). Furthermore, *Eurosta* populations associated with *S. altissima* were significantly heterogeneous at both the HBDH and PGM loci due to geography. Allele frequencies among populations in New England were more similar to one another than were frequencies within the other populations. Similarly, heterogeneity among populations associated with *S. gigantea* also varied geographically. North-central populations contained more variation at the HBDH locus than other populations. The facts that these allele-frequency differences existed over a broad geographic area and persisted in host-associated populations that occur sympatrically suggest that the pattern of genetic variation is related to discrepancy in host-plant association and is not attributable solely to geographic variation (Waring, Abrahamson, and Howard 1990).

An unweighted pair group (UPGMA) clustering of Nei's unbiased genetic distance (D) separated *E. solidaginis* populations from *S. altissima* and all but one of the twelve *S. gigantea* on different branches (fig. 7.1). This result suggests that an initial host shift onto *S. gigantea* took place in a single geographic area. If shifts had occurred independently in various geographic regions, populations would cluster by region rather than by host association. This fact and the absence of *E. solidaginis* populations on *S. gigantea* in areas where it commonly attacks *S. altissima* (e.g., mid-Atlantic region) support the notion that a successful host shift occurred in a single geographic area with the resultant *S. gigantea*-infesting population becoming the progenitor for present-day *S. gigantea*-associated populations. We can surmise from these results that host shifts in this gallmaker have been extremely uncommon events (Waring, Abrahamson, and Howard 1990; Abrahamson et al. 1994).

Recently we examined the phylogenetic relationships and geographic distributions of mitochondrial haplotypes of the *E. solidaginis* host-associated populations (Brown, Abrahamson, and Way 1996). We sequenced 492 bp from the 3' ends of the mitochondrial cytochrome oxidase I and II subunits from a single individual from ten *S. altissima*- and eight *S. gigantea*-associated populations and from two outgroup species, *E. comma* and *E. cribrata*. The haplotypes fell into two groups (termed East and West clades) that differed by four substitutions. One of these substitutions occurred within the recognition site of the DdeI restriction enzyme, making it possible for us to use the presence or absence of the restriction site to survey a number of individuals from eleven *S. gigantea* and twenty *S. altissima* host-associated populations. All *S. gigantea*-associated populations were of the E-clade haplotype regardless of their geographic origin. However, flies from *S. altissima* differed depending on their geographic origin. *Solidago altissima* flies east of Michigan were consistently the E-clade haplotype, while those west of Michigan

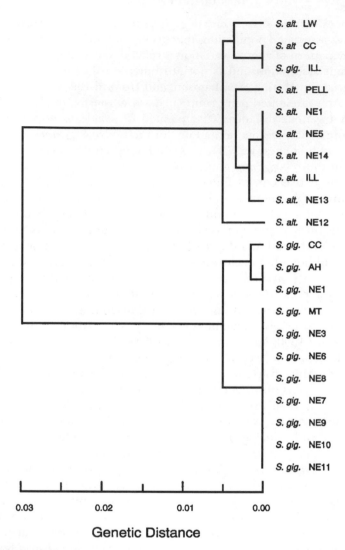

FIGURE 7.1. Estimation of a phylogenetic tree for twenty-one populations of *Eurosta solidaginis* collected from *Solidago altissima* and *S. gigantea* based on UPGMA (Sneath and Sokal 1973) clustering of the Nei's (1978) genetic distance matrix. (Redrawn from Waring, Abrahamson, and Howard 1990, "Genetic differentiation among host-associated populations of the gall-maker *E. solidaginis* [Diptera: Tephritidae]," *Evolution* 44:1648–1655, with kind permission from The Society for the Study of Evolution)

were W-clade haplotypes. Southern Michigan had populations composed of individuals from both clade haplotypes. Our findings are consistent with the suggestion that all *S. gigantea*-associated *Eurosta* populations studied to date have been derived from a single shift from *S. altissima* to *S. gigantea* (Waring, Abrahamson, and Howard 1990). Furthermore, DNA-sequence variation indicates that the shift occurred in the northeastern U.S. (fig. 7.2; Brown, Abrahamson, and Way 1996).

Because there have been so few studies of host-specific genetic differentiation in natural populations (but see Feder and Bush 1989a,b, 1991; Feder, Chilcote, and Bush 1988, 1989a,b, 1990a,b), our series of studies with *E. solidaginis* provides important evidence for the potential role that host shifts can play in the genetic differentiation of herbivore populations and their subsequent speciation (Waring, Abrahamson, and Howard 1990; Craig et al. 1993, 1994; Abrahamson et al. 1994; Brown et al. 1995; Brown, Abrahamson, and Way 1996). As we will see, the most parsimonious explanation for the genetic differences between these host-associated populations is limited gene flow based on assortative mating and strong oviposition preferences within each host-associated population coupled with strong divergent selection for physiological adaptation to their respective host plants.

7.6 BEHAVIORAL AND ECOLOGICAL EVIDENCE FOR HOST RACES

Five questions were framed to distinguish whether *E. solidaginis* host-associated populations are one undifferentiated species, host races, or two sibling species (Craig et al. 1993): (1) Are the host-associated populations truly sympatric? (2) Do emergence periods of the flies on the two host plants differ, creating allochronic reproductive isolation? (3) Do flies reared from different host plants have oviposition preferences for their own host plant? (4) Is there assortative mating independent of host-plant preference? (5) Does assortative mating occur due to host-plant preference? Recent work by our collaborators Timothy Craig, Joanne Itami, and John Horner (unpub.) has sought to answer two additional questions: (6) Are the two populations

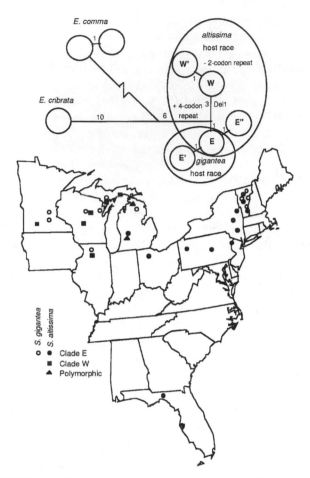

FIGURE 7.2. *Top:* Maximum parsimony network of relationships (fifty steps, C.I. = 1.00) among *E. solidaginis* and outgroup species mtDNA haplotypes. Branch lengths are proportional to the number of unambiguous nucleotide substitutions. Insertion and deletion events are inferred on the tree, but were not used to determine it. Del1 is at the 3' end of the cytochrome oxidase I gene, and the 4-codon repeat is near the 5' end of the sampled cytochrome oxidase II sequence. *Bottom:* Geographical distribution of mitochondrial DNA haplotypes from *S. altissima* and *S. gigantea* fly populations. (Redrawn from Brown, Abrahamson, and Way 1996, "Mitochondrial DNA phylogeography of host races of the goldenrod ball gallmaker, *Eurosta solidaginis* [Diptera: Tephritidae]," *Evolution* 50 : 777–786, with kind permission from The Society for the Study of Evolution)

under disruptive selection for host-plant use? (7) How is the ability to survive on each host plant inherited? The answers to these questions indicate that the host-associated, genetically divergent populations of *E. solidaginis* are host races (Craig et al. 1993, 1994; Brown et al. 1995; Hess, Abrahamson, and Brown 1996; Abrahamson, Perot, and Brown, unpub. data; Craig, Itami, and Horner, unpub. data). The following paragraphs will outline the evidence that provides answers to each of these questions.

Sympatry of Host Plants

We have found numerous sites with truly sympatric populations of *S. altissima* and *S. gigantea* across the northern tier of eastern and central states within the U.S. and in portions of southeastern Canada (e.g., Waring, Abrahamson, and Howard 1990; Craig et al. 1993, 1994; Abrahamson et al. 1994; Brown et al. 1995; Sumerford and Abrahamson 1995; Brown, Abrahamson, and Way 1996). Although *S. altissima* is more tolerant of edaphic conditions than *S. gigantea,* these two species sympatrically occur at a considerable number of sites (Givens and Abrahamson, unpub. data; Abrahamson, Ball, and Houseknecht, unpub. data). Perhaps the best measure of the degree of sympatry typical of these host plants is that clones of both host plants were interdigitated in seventeen of twenty Minnesota fields containing sympatric populations of *Eurosta* host-associated populations (Craig et al. 1993). We can answer our question about host-plant sympatry by stating that over much of *E. solidaginis'* range, *S. altissima* and *S. gigantea* occur sympatrically and interdigitated.

Emergence Times

We addressed the second question about whether the emergence periods of the flies on the two host-plant species were different by rearing adult flies in outdoor, screened cages from galls of both host-plant species collected at twenty sympatric sites near Minneapolis, Minnesota. Emergence times differed for the two host-associated populations during two separate growing seasons (spring of 1989 and 1990; Craig et al. 1993). In both years the *S. gigantea*-associated populations emerged over a week

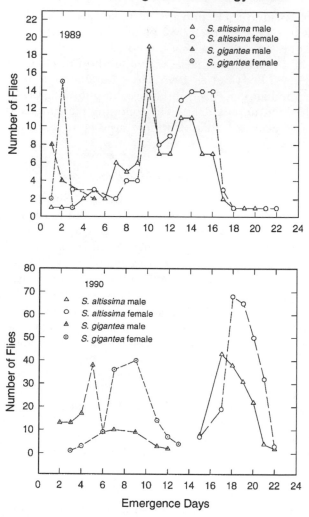

FIGURE 7.3. Emergence phenology of *Eurosta solidaginis* expressed as the number of days after the first fly emerged. *Top:* 1989, galls collected from a sympatric population located in Washington County, Minnesota; day 1 is May 17. *Bottom:* 1990, galls pooled from three sympatric fields located in Anoka County, Minnesota; day 1 is May 21. (Redrawn from Craig et al. 1993, "Behavioral evidence for host-race formation in *Eurosta solidaginis*," *Evolution* 47: 1696–1710, with kind permission from The Society for the Study of Evolution)

earlier than *S. altissima*-associated populations. The mean difference in emergence time was nearly 10 days in 1989 while in 1990 the difference was approximately 13 days (fig. 7.3).

We interpreted the less distinct emergence times in 1989 compared to 1990 as an effect of differences in spring temperatures. Galls in the spring of 1989 experienced a rapid change from cold temperatures to relatively summerlike temperatures during larval pupation. In 1990, temperatures during the *Eurosta* pupation period were more springlike without the heat wave that occurred in the spring of 1989. If the development rate of the *S. altissima* host-associated population increases more for a given increase in temperature than the *S. gigantea* host-associated population, we would see the difference in emergence times diminish in warm springs compared to cool springs.

We examined the effect of incubation temperature on emergence time for both host-associated populations using approximately 2000 galls of each host-associated population collected in late November 1993 from four sites in New England. After collection, galls were stored at $-8°C$ until February 1994. Beginning in early February, 333 galls of each host-associated population were randomly assigned to one of six constant-temperature regimes in environmental growth chambers at 10.1, 13.4, 16.8, 20.2, 23.5, and 26.9°C. Relative humidity was held constant at 80% for all temperature treatments, and galls received a photoperiod of 15:9 hr (light:dark).

No flies emerged from galls treated with the 10.1°C regime. Dissection of these galls after more than 120 days showed that *Eurosta* larvae had died at an earlier stage of incubation. At all higher temperatures, males from *S. gigantea* galls emerged in the shortest period of time followed by females from galls on *S. gigantea,* then males from *S. altissima* galls; females from *S. altissima* galls emerged last (fig. 7.4; Abrahamson, Perot, and Brown, unpub. data). This order of emergence was identical to that determined for galls incubated under outdoor conditions in Minnesota (Craig et al. 1993). A three-way analysis of variance showed all three main effects, host-associated population ($F_{1595} = 1010.2$, $P < 0.001$), fly sex ($F_{1595} = 47.4$, $P < 0.001$), and incubation temperature ($F_{4595} = 3030.9$, $P < 0.001$) were highly significant. One two-way interaction was signifi-

Host-Race Emergence by Temperature

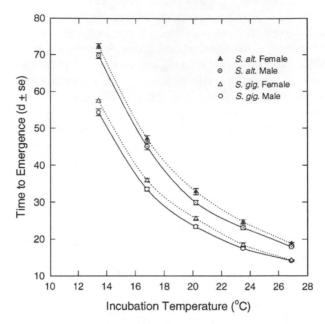

FIGURE 7.4. Time to emergence (days ± standard error) for male and female flies of both the *S. altissima* and *S. gigantea* host races collected from four sites in New England. Three hundred thirty-three galls of each host race were randomly assigned to one of six constant-temperature regimes in environmental growth chambers at 10.1, 13.4, 16.8, 20.2, 23.5, and 26.9°C. Relative humidity was held constant at 80% for all temperature treatments, and galls received 15 hours of light and 9 hours of dark. The second-order regression equation for males derived from *S. gigantea* was $Y = -12.7X + 0.2X^2 + 179.8$, $r^2 = 0.994$; for females from *S. gigantea*, $Y = -12.6X + 0.2X^2 + 182.5$, $r^2 = 0.994$; for males from *S. altissima*, $Y = -15.6X + 0.3X^2 + 225.6$, $r^2 = 0.996$; and for females from *S. altissima*, $Y = -15.1X + 0.3X^2 + 223.4$, $r^2 = 0.996$. (From Abrahamson et al., unpub. data)

cant—host-associated population × incubation temperature ($F_{4595} = 46.3$, $P < 0.0001$)—indicating that the two host-associated populations responded differentially to incubation temperatures. The interaction of fly sex and incubation temperature approached significance ($F_{4595} = 2.3$, $P = 0.058$) due to a reduction of emergence-time differences between the sexes at warmer temperatures. Second-order regressions of time to emergence on incubation temperature explained at least 99.4% of the variation in emergence time for each of the four populations. These regressions indicate that the rate of development of flies from *S. altissima* galls increased more rapidly for a given increase in incubation temperature than did development rates for *S. gigantea* flies (fig. 7.4). The consequence is that emergence times are more similar during springs with warm temperatures and are more different in springs with cooler temperatures.

We have also confirmed this pattern of allochronic emergence of host-associated populations at sympatric and allopatric sites along a north-south transect in New England. A survey of twenty-three *Eurosta* populations along a transect stretching from Greenfield, Massachusetts, north along the Connecticut River Valley to northern New Hampshire and Vermont found three geographical zones based on the attack of *E. solidaginis:* (1) a zone in which *E. solidaginis* infested only *S. altissima* (northern Massachusetts), (2) a zone in which only *S. gigantea* were attacked (northern New Hampshire and Vermont), and (3) an intermediate zone in which both host plants were infested sympatrically (Brown et al. 1995). The times from overwintering diapause to adult emergence for these New England populations varied significantly according to sex of fly, host-plant association, and treatment. Male flies of both host-associated populations emerged significantly earlier than female flies (22.2 versus 23.0 days) and flies associated with *S. gigantea* emerged significantly earlier than those allied with *S. altissima* (20.6 versus 24.6 days) (fig. 7.5; Abrahamson et al. 1994).

Such allochronic emergence of host-associated populations should substantially reduce the amount of gene flow between them. It does not mean, however, that sympatric host-associated populations are completely reproductively isolated. Flies frequently live 10–14 days under experimental conditions (Craig et al. 1993; Hess, Abrahamson, and Brown 1996), and they can live for several weeks in cool, moist environmental chambers (Craig et al. 1993). Thus, in spite of marked differences in emergence times, the life spans of the host-associated populations overlap, making mating between populations possible (Craig

Emergence Times of Host Races

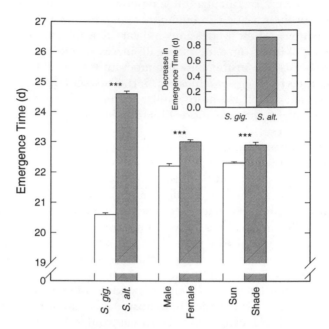

FIGURE 7.5. Emergence phenology of *Eurosta solidaginis* based on over three thousand galls of each host race from twenty-three New England populations. We incubated these galls out-of-doors in either a shade (galls received no direct solar radiation) or a sun (galls received up to 7 hours of direct solar radiation) treatment for 2 weeks after their removal from cold storage. Galls were then enclosed in individual emergence cups and monitored for emergence in an air-conditioned laboratory held at approximately 21°C. (Data from Abrahamson et al. 1994)

et al. 1993). Operating to reduce interpopulational mating, however, is the greater mating success of younger *E. solidaginis* adults relative to older individuals. We have repeatedly observed a loss of mating vigor among adult flies after the first 2–3 days following emergence (Abrahamson, pers. obs.).

We conducted an experiment to assess the impact of fly age on assortative mating since fly age might influence mating choice (Craig et al., unpub. data). A constant number of forty-five virgin males and females from each of three populations

Fly Age Affects Mating

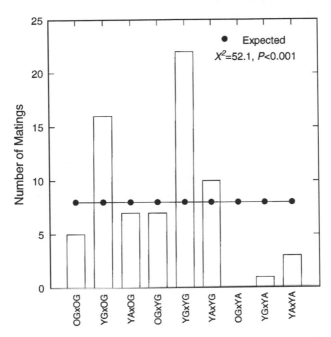

FIGURE 7.6. The impact of fly age on assortative mating between caged host races with no plants present. The fly population was maintained at forty-five virgin males and females from each of three populations' 7-day-old *S. gigantea* flies (OG), 1-day-old *S. gigantea* flies (YG), and 1-day-old *S. altissima* flies (YA). *X*-axis labels are given as male × female. The filled circles represent the expected mating frequency if matings occurred at random and the bars the observed mating frequency. (From Craig et al., unpub. data)

(7-day-old flies from *S. gigantea*, 1-day-old flies from *S. gigantea*, and 1-day-old flies from *S. altissima*) were allowed to mate in a cage with no plants. The seventy-one recorded matings occurred non-randomly. Mating frequency was influenced by which host-associated population a fly came from, age of the fly, and sex (fig. 7.6). Males and females derived from *S. gigantea* mated more frequently than either sex from *S. altissima*. Males from *S. altissima* mated more frequently than females from *S. altissima* and among the flies from *S. gigantea*, young flies mated more frequently than old flies. The matings between old females from *S. gigantea* and young males from *S. altissima* demonstrate the possibility of gene flow in natural populations given the emergence phenology of these two host-associated populations (*S. gigantea* flies emerge, on average, about 12 days earlier than *S. altissima* flies).

Oviposition Preferences

Females of both host-associated populations strongly preferred to oviposit into the buds of their own host-plant species (Craig et al. 1993). Acceptance of the two host plants for oviposition was measured by counting oviposition scars in host-plant buds under both no-choice (using cages placed over individual *Solidago* ramets) and choice conditions (using garden-grown replicates of genotypes of both host plants). Under no-choice conditions, a total of thirty-nine out of sixty-nine flies tested on their own host plant oviposited into that plant, while only one of seventy flies tested on the alternate host oviposited after 24 hours. The results of our host-plant choice experiment conclusively showed that the host-associated populations have very strong oviposition preferences for their own host plant (fig. 7.7). Only one female out of the thirty-three tested from *S. altissima* and one female out of thirty-two tested from *S. gigantea* oviposited on the alternate host plant.

Assortative Mating

Each host-associated population demonstrated strong assortative mating based on host-plant preference but only very weak assortative mating without host plants (Craig et al. 1993). We conducted a series of experiments to examine whether (1) there was assortative mating, and (2) if mating preference was affected by host-plant preference. Adult flies from both host plants were mated in outdoor screen cages without host plants, first with one sex from each host-associated population and subsequently with both sexes (marked with population-specific small color dots on the upper thorax or middle leg) from both host-associated

216

Host-Race Oviposition Choice

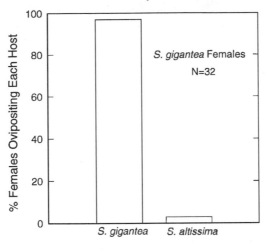

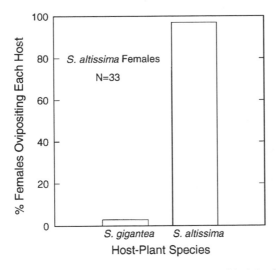

FIGURE 7.7. Oviposition choices by female *Eurosta solidaginis* when presented with both host-plant species. *Top: Solidago gigantea* population. *Bottom: Solidago altissima* population. (Redrawn from Craig et al. 1993, "Behavioral evidence for host-race formation in *Eurosta solidaginis,*" *Evolution* 47:1696–1710, with kind permission from The Society for the Study of Evolution)

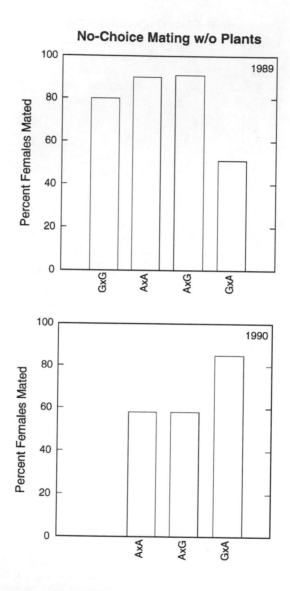

FIGURE 7.8. Percentage of *Eurosta* females that mated with males from their own host race or from the alternate host race in cages without host plants and with mates from only one host race present in (*top*) 1989 and (*bottom*)

populations. Flies of both sexes from one host-associated population were also mated in the presence of both host plants in both a garden experiment and a field experiment.

We found that mating readily occurred between *S. altissima*- and *S. gigantea*-associated flies in the absence of (1) mates from their own host-associated population, and (2) host plants (figs. 7.8 and 7.9). When males and females from both host-associated populations were contained in cages without host plants, the frequency of matings between host-associated populations was high (approximately 38% of the matings that took place); however, this interspecific mating frequency was significantly lower than the frequency of matings within host-associated populations (fig. 7.10).

However, when host plants were included within the cages, both host-associated populations mated primarily on their own host plant and rarely mated interspecifically (fig. 7.11). In both garden and field situations, male flies of both host-associated populations rested on the buds of their own host plants and waited for an opportunity to mate. Unmated females flew to the buds of their respective host plants to mate or rest.

We found that a major reduction in interpopulational matings occurred when both sexes from both host-associated populations were tested in the presence of both host plants (fig. 7.12). As expected from the previous experiment, males of both host-associated populations waited on their respective host-plant species for females to fly in to mate. Of the twenty-four matings that took place on the host plants, twenty-three were between individuals of the same host-associated population. An additional six matings took place on the neutral surfaces of the cage but four of these were between members of different host-associated populations. We conclude that these host-associated

---

1990. Mating labels on *X*-axis are male × female; A = *S. altissima* and G = *S. gigantea*. Sample sizes 1989: G × G = 75, A × A = 197, A × G = 24, G × A = 25, 1990: A × A = 40, A × G = 40, G × A = 40. (Redrawn from Craig et al. 1993, "Behavioral evidence for host-race formation in *Eurosta solidaginis*," *Evolution* 47:1696–1710, with kind permission from The Society for the Study of Evolution)

Mating Choice w/o Plants

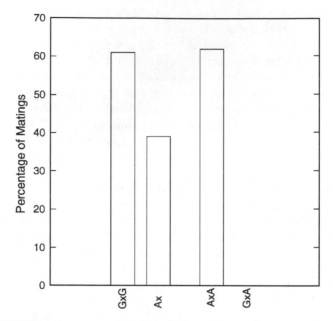

FIGURE 7.9. Percentage of matings within and between host races in cages without host plants. The gender of crosses on the X-axis is given as male × female; A = *S. altissima* and G = *S. gigantea*. (Redrawn from Craig et al. 1993, "Behavioral evidence for host-race formation in *Eurosta solidaginis*," *Evolution* 47:1696–1710, with kind permission from The Society for the Study of Evolution)

populations assortatively mate in relation to host-plant preference (Craig et al. 1993).

Effect of Host-Plant Abundances on Mating Fidelity

Both host races exhibit strong assortative mating and oviposition. However, because assortative mating is tied to host-plant preference, the relative abundance of the two host-plant species could alter the opportunities for between host-race matings. To examine this possibility, both host races were offered pure and

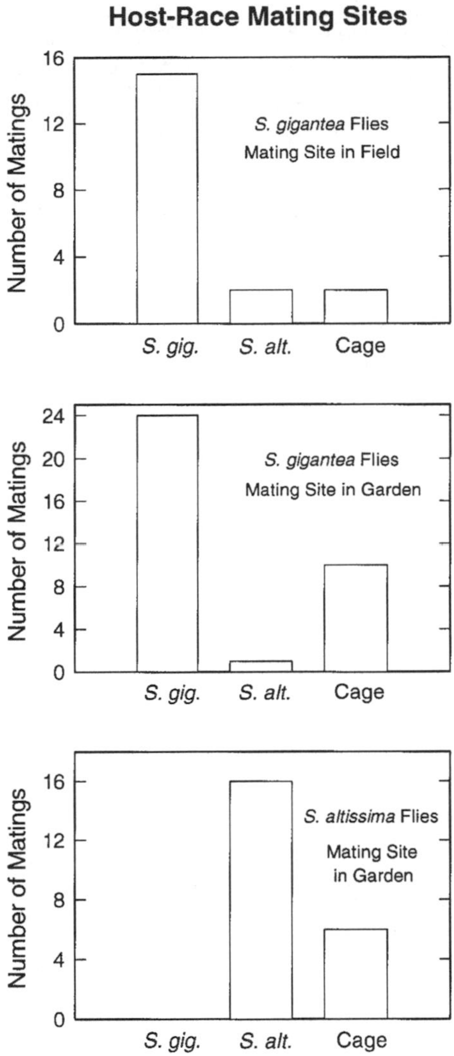

Host-Race Mating Sites

FIGURE 7.10. Location of matings (as indicated on the X-axis) in cages that contained males and females of only one host race and both species of host plants. *Top: Solidago gigantea* flies in the field. *Middle: Solidago gigantea* flies in the garden. *Bottom: Solidago altissima* flies in the garden. (Redrawn from Craig et al. 1993, "Behavioral evidence for host-race formation in *Eurosta solidaginis*," *Evolution* 47:1696–1710, with kind permission from The Society for the Study of Evolution)

Male Occurrence by Host Plant

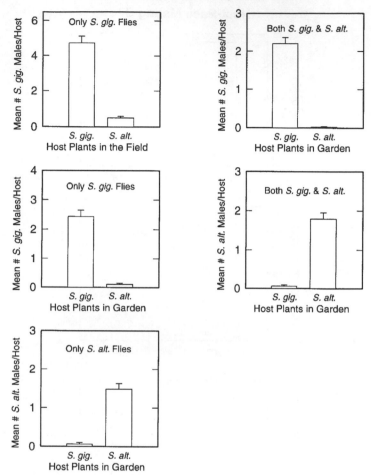

FIGURE 7.11. The mean number of males observed on each host plant at 15-minute intervals in experiments where males were given a choice of host plants. *Top left: Solidago gigantea* flies in the field, both sexes from one host race, both host plants present. *Middle left: Solidago gigantea* flies in the garden, both sexes from one host race, both host plants present. *Bottom left: Solidago altissima* flies in the garden, both sexes from one host race, both host plants present. *Top right: Solidago gigantea* flies in the garden, both sexes from both host races, both host plants present. *Middle right: Solidago altissima* flies in the garden, both sexes from both host races, both host plants present. (Redrawn from Craig et al. 1993, "Behavioral evidence for host-race formation in *Eurosta solidaginis*," *Evolution* 47:1696–1710, with kind permission from The Society for the Study of Evolution)

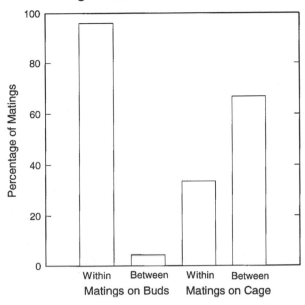

Mating w/ Both Host Races and Host Plants

FIGURE 7.12. Mating frequency within and between host races in a cage with both host races and both host plants. Matings are divided into two groups: those that occurred on the host plants, and those that occurred on the cage. (Redrawn from Craig et al. 1993, "Behavioral evidence for host-race formation in *Eurosta solidaginis*," *Evolution* 47:1696–1710, with kind permission from The Society for the Study of Evolution)

mixed arrays of both host plants (Craig, Itami, and Horner, unpub. data). Under all conditions, intra-host-race matings were most frequent because of strong host-plant association. There were, however, differences in the mating vigor and fidelity of the two host races. Males of the derived *S. gigantea* host race mated more frequently and less discriminately than those from the ancestral *S. altissima* host race. The consequence is that males of the *S. gigantea* host race can mate, at least under experimental conditions, with females of the *S. altissima* host race on *S. altissima* buds. Crosses in the reverse direction, however, were uncommon in this experiment due in part to the higher fidelity of *S. altissima* males. These results indicate that both the frequency and direction of mating between host races may be determined by the relative availability of host plants and differences in host-race behavior.

Offspring Performance and Adaptation to Host Plants

In a series of multigenerational experiments, males and females of both host races were crossed interspecifically and intraspecifically with subsequent oviposition on both species of host plants (Craig, Itami, and Horner, unpub. data). These experiments found highly significant differences in offspring survival from egg to adult among different crossing groups (fig. 7.13). Both host races survived poorly on the alternate host plant primarily due to a lack of gall induction and early larval death. The reduced survival of offspring (P1) from one host race on the alternate host plant as compared to survival on their own host plant indicates that different physiological adaptations may be necessary for *Eurosta* survival on each host-plant species. However, the requirements for survival on one host are not completely incompatible with the other since a few individual P1s did survive on the alternate host plant. These findings illustrate the adaptive nature of the strong preference each host race has for ovipositing on its own host plant, but also that a few alternate host-plant genotypes are suitable for survival.

Additional evidence for the adaptation of each host race to their respective host plant includes the finding that a lower proportion of the ramets ovipunctured by females carrying hybrid offspring of the two host races or backcross progeny formed galls relative to pure host races. While the primary source of mortality for hybrid and backcross progeny varied according to the direction of the cross, these progeny suffered either very low rates of gall induction (e.g., offspring of *S. altissima* females × *S. gigantea* males) or died at an early larval stage (e.g., offspring of *S. gigantea* females × *S. altissima* males; fig. 7.14). Such mortality would produce strong selection favoring females who oviposit on their own host plant. Furthermore, since assortative mating occurs on the host plant, selection against hybrids would result in selection for host fidelity in each host race. Thus for

Ovipunctured Ramets Forming Galls

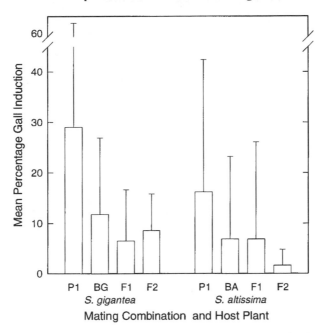

FIGURE 7.13. Mean (± standard deviation) percentage gall induction (as the percentage of eggs that formed galls) for mating combinations on the two host plants (1993 generation). Mating combinations on *Solidago altissima* were *S. altissima* female × *S. altissima* male (P1), the back-cross F1 male × *S. altissima* female (BA), and F1 male × F1 female (F2). Mating combinations on *S. gigantea* were *S. gigantea* female × *S. gigantea* male (P1), the back-cross F1 male × *S. gigantea* female (BG), and F1 male × F1 female (F2). (From Craig, Itami, and Horner, unpub. data)

Eurosta, selection acting on both mating-site choice and oviposition choice should cause divergence in host preference (Craig, Itami, and Horner, unpub.).

Modes of Inheritance of Survival and Preference

The experiments of Craig, Itami, and Horner (unpub. data) conclusively show that both host races possess the genetic variation necessary for survival on the alternate host plant. This find-

Mortality Sources for Hybrids

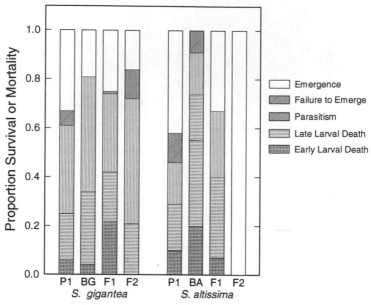

FIGURE 7.14. The proportion of larva that induced galls that survived or died during five stages of development during the 1993 generation. Mating combinations on *Solidago altissima* were *S. altissima* female × *S. altissima* male (P1), the back-cross F1 male × *S. altissima* female (BA), and F1 male × F1 female (F2). Mating combinations on *S. gigantea* were *S. gigantea* female × *S. gigantea* male (P1), the back-cross F1 male × *S. gigantea* female (BG), and F1 male × F1 female (F2). (From Craig, Itami, and Horner, unpub. data)

ing suggests that no mutations for survival would be necessary for a host shift and subsequent sympatric speciation.

The genes involved in survival likely are inherited at an autosomal locus or loci (Craig, Itami, and Horner, unpub. data). This conclusion is based on the fact that our colleagues found no evidence that survival ability was either maternally inherited or sex linked. Furthermore, the low hybrid survival they found indicates that there is underdominance in the interaction of survival genes on the two host-plant species.

Our colleagues have suggested that there are multiple alleles at many loci that optimize survival on the two host plants. This conclusion is based on the finding that hybrids and pure host races suffered low survival rates on the alternate host plant owing to mortality sources acting at all stages of their life history. Different genes are likely involved since it is improbable that mortality would occur at different life stages if only one gene were involved.

One Undifferentiated Species, Host Races, or Two Sibling Species?

These host-associated populations are partially reproductively isolated as a result of allochronic emergence, a strong preference for mating on their host plants, and reduced survivorship of hybrid F1, F2, and backcross offspring relative to the pure host-associated populations (Craig et al. 1993; Craig, Itami, and Horner, unpub. data). Furthermore, although they are very similar genetically, these host-associated populations are distinct at several loci coding for proteins (Waring, Abrahamson, and Howard 1990) and exhibit genetic differentiation related to both host-association and geography within a portion of the cytochrome oxidase gene in the mitochrondrial genome (Brown, Abrahamson, and Way 1996).

Nonetheless, gene flow is possible between these host-associated populations. In Craig, Itami, and Horner's (unpub. data) hybridization experiments, hybrid F1s survived well on some host-plant genotypes, and these hybrids exhibited considerable fertility in the production of F2 and backcross offspring. There is little difference in the viability of P1, F1, F2, and backcross progeny as measured by mating and oviposition rates.

Efforts to measure gene flow in the field have not identified any allozyme markers (Itami et al., unpub. data) so we cannot be certain that hybrids occur under field conditions. A combination of prezygotic and postzygotic isolating mechanisms may completely isolate host-associated populations. Alternatively, the absence of allozyme markers could be a consequence of sufficient gene flow between the populations that alleles do not become fixed. Yet behavioral data suggest that gene flow does occur in the field. *Eurosta* adults have a strong preference for the host-plant species from which they emerged (Craig et al. 1993, 1994), and the F1 hybrids in Craig, Itami, and Horner's (unpub. data) experiments exhibited a different host preference than pure host-race individuals. Their experiment comparing oviposition preference found that 0% of the pure host-race females and 18% of F1 hybrid females oviposited on both host plants. Furthermore, most F1 females exclusively oviposited on *S. gigantea*. Consequently, Craig, Itami, and Horner (unpub.) suggest that intermediate host preference is an indication of hybrid status. These workers consistently have found that a small proportion (2–3%) of presumed pure host-race individuals collected from the field exhibit an intermediate host preference. Such preference suggests that there are naturally occurring hybrids in the field. The existence of hybrids would account for the lack of Hardy-Weinberg equilibrium at the HBDH locus (Waring, Abrahamson, and Howard 1990) in sympatric populations. It is likely that the amount of gene flow varies among fields owing to differences in host-race abundance, the degree of sympatry of host-plant species, and the presence of "benign" host-plant genotypes that enable the survival of F1s (Itami et al., unpub.).

A host race has been defined as "a population of a species that is partially reproductively isolated from other conspecific populations as a direct consequence of adaptation to a specific host" (Diehl and Bush 1984). According to this definition, *E. solidaginis* host-associated populations are host races. They exhibit all the reproductive, ecological, and evolutionary characteristics expected of such-defined host races. However, because Jaenike (1981) defined host races as populations solely isolated by plant-preference differences, *E. solidaginis* may not meet Jaenike's definition given that allochronic emergence of host-associated populations may be important to their reproductive isolation (Craig et al. 1993).

7.7 FACTORS FACILITATING OR MAINTAINING HOST SHIFTS

There are a variety of ecological and behavioral factors that may operate to facilitate a host-plant shift and subsequently to promote speciation. These likely include phenological traits of both the host plants and herbivores that enable a host shift in the first place. Other factors include preadaptation of the herbivore to a novel host; larval or adult conditioning that results in altered host preference; intense versus weak competition on the ancestral and derived hosts, respectively; disruptive selection acting on host races for adaptation to the two host plants; and escape to enemy-reduced space on the derived host. For gallmakers we must also examine the role of any host-plant traits that affect gall morphology, color, or physiology.

Effects of Phenology

Tom Wood and his associates have demonstrated that reproductive isolation on novel host plants can be rapid and selection for adaptations to a new host can be strong when herbivore emergence is keyed to host-plant phenology (e.g., Wood 1980, 1986; Wood, Olmstead, and Guttman 1990). Because of the necessity of infesting hosts during their vulnerable periods, the emergence times of most herbivores are strongly correlated to the phenology of their host plant. The consequence of this correlation is that a shift to an alternate host plant should result in strong selection pressures for altered emergence times. If reproductive isolation is established on a novel host, strong selection should follow for adaptations that promote survival on this host.

The two host races of *E. solidaginis* are partially reproductively isolated by a difference of approximately 12 days in emergence. To gain insight into whether the differences in adult emergence times could be the consequence of differential host-plant phenology, we examined the growth phenologies of the host plants for both *Eurosta* host races. We compared the growth rates of host-plant populations in Minnesota, where both goldenrod species are infested, with populations of the same *Solidago* species in Pennsylvania, where only *S. altissima* is attacked but where both host plants are present (How, Abrahamson, and Craig 1993; Abrahamson et al. 1994). Daily host-plant growth rates varied widely in Minnesota and Pennsylvania for both host plants in response to factors such as temperature and cloud cover. However, at the Pennsylvania site, ramets of *S. altissima* were taller, more uniform in size, and grew faster than *S. gigantea* ra-

Host-Plant Height at Oviposition

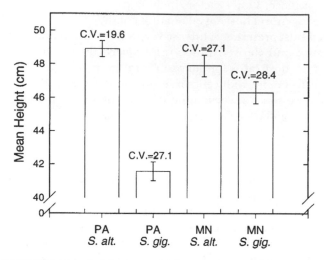

FIGURE 7.15. Mean height (± standard error) and coefficient of variance (C.V.) for goldenrod populations of the two host-plant species in Pennsylvania and Minnesota fields during their respective *Eurosta solidaginis* oviposition period. *S. alt.* = *Solidago altissima*, *S. gig.* = *S. gigantea*. (Redrawn from How, Abrahamson, and Craig 1993, "Role of host plant phenology in host use by *Eurosta solidaginis* [Diptera: Tephritidae] on *Solidago* [Compositae]," *Environmental Entomology* 22:388–396, with kind permission of The Entomological Society of America; Abrahamson et al. 1994)

mets, particularly during *Eurosta*'s oviposition period (fig. 7.15). Yet there were no significant differences between host plants in measured growth characteristics in Minnesota where both host plants are infested. Thus, the growth traits of *S. altissima* and *S. gigantea* converge in Minnesota (and although untested, convergence may be the rule in other regions where both host plants are infested; e.g., New England). Such convergence of host-plant phenology makes a host shift more likely. At the same time, the divergence of growth traits in Pennsylvania would greatly reduce the likelihood of a host shift (How, Abrahamson, and Craig 1993; Abrahamson et al. 1994).

We also established replicated clones in a Pennsylvania common garden to monitor the growth of the two host-plant species from both locations. *Solidago altissima* from both locations had a longer growth period and consequently a broader time span of reactivity to *Eurosta* than *S. gigantea*. Growth measurements showed that *S. gigantea* ramets peaked earlier and declined more quickly than those of *S. altissima* (How, Abrahamson, and Craig 1993). Given these host phenologies, the earlier emergence of the *S. gigantea* host race compared to the *S. altissima* host race appears adaptive.

In a greenhouse experiment, we examined the oviposition preference of adult *S. gigantea*-infesting flies from Minnesota as well as the galling susceptibilities of *S. gigantea* plants from both Pennsylvania and Minnesota (How, Abrahamson, and Craig 1993). These flies showed a strong oviposition preference for Minnesota *S. gigantea* ramets when simultaneously offered *S. gigantea* ramets from Minnesota and Pennsylvania. This preference was likely a consequence of the Minnesota ramets being both taller and faster growing at the time of oviposition than the Pennsylvania ramets (table 7.3). We have seen such a relation-

TABLE 7.3. Galling susceptibility in *Solidago gigantea* of Minnesota (colonized) and Pennsylvania (noncolonized) origin.

Parameter	Minnesota	Pennsylvania
Cumulative growth (cm)		
Ovipunctured	13.7 ± 0.5	9.5 ± 0.5
	($n = 206$)	($n = 113$)
Not ovipunctured	7.7 ± 0.6	7.8 ± 0.5
	($n = 88$)	($n = 181$)
Height at first oviposition (cm)		
Ovipunctured	12.9 ± 0.2	10.4 ± 0.3
	($n = 206$)	($n = 113$)
Not ovipunctured	7.8 ± 0.4	8.8 ± 0.2
	($n = 88$)	($n = 181$)
Growth rate at oviposition	7.2 ± 0.2	5.1 ± 0.2
	($n = 293$)	($n = 293$)

Source: From How, Abrahamson, and Craig 1993.
Notes: Cumulative growth and height are expressed as means ± standard errors. Cumulative growth represents change in height throughout the 1-month observation period. Growth rate at oviposition represents change in height during the 2 weeks following first introduction of *Eurosta solidaginis*.

ship between host-plant size and vigor, and *Eurosta* oviposition preference during other experiments with the *S. altissima* host race (Anderson et al. 1989; Horner and Abrahamson 1992 and submitted).

When Minnesota *S. gigantea*-attacking flies oviposited into Pennsylvania ramets of *S. gigantea,* these ramets frequently supported gall formation under greenhouse conditions (How, Abrahamson, and Craig 1993). Furthermore, the Pennsylvania *S. gigantea* galls typically contained a living *Eurosta* larva at the time of ramet senescence. These findings indicate that the absence of galls on *S. gigantea* ramets in Pennsylvania is not due to the inability of the gallmaker to stimulate gall formation or of these ramets to react to the gallmaker. Yet as expected, Minnesota *S. gigantea* ramets produced galls with larger gall diameters, greater gall fresh and dry masses, and higher peak larval dry mass than Pennsylvania galls (table 7.4). While these findings could arise from the greater vigor of the Minnesota *S. gigantea* host plants, they may suggest that *E. solidaginis* host races are adapted to geographical variation among its hosts.

A critical question is whether such differences in host-plant phenology affect larval survival. In separate experiments con-

TABLE 7.4. Mean ± standard error for gall and larval characteristics by host origin.

Parameter	Host Origin	
	Minnesota	Pennsylvania
Gall diameter (mm)	20.64 ± 0.77	14.76 ± 0.98
	($n = 37$)	($n = 11$)
Gall fresh mass (g)	3.54 ± 0.29	1.33 ± 0.20
	($n = 37$)	($n = 11$)
Gall dry mass (g)	1.23 ± 0.08	0.59 ± 0.07
	($n = 37$)	($n = 11$)
Eurosta fresh mass (g)	31.5 ± 2.06	24.3 ± 3.25
	($n = 20$)	($n = 11$)
Eurosta dry mass (g)	12.4 ± 0.78	9.2 ± 1.48
	($n = 19$)	($n = 11$)

Source: From How, Abrahamson, and Craig 1993.
Note: Means for Minnesota and Pennsylvania hosts are statistically significant at $P < 0.05$ for all parameters according to analysis of covariance.

ducted in Minnesota during two growing seasons, Horner et al. (unpub. data) investigated if emergence times of *Eurosta* host races are keyed to the time periods of greatest vulnerability of each host plant. This was done by removing galls from cold storage at different times. Flies emerged during four weekly intervals (year 1 study) that ranged from 1 week before normal attack by the *S. gigantea* host race to 1 week after normal attack by the *S. altissima* host race (remember that the *S. gigantea* host race emerges, on average, about 12 days earlier than the *S. altissima* host race); or three intervals (year 2 study) that corresponded to (1) the normal period of attack for the host race on *S. gigantea*, (2) the normal period of attack for the *S. altissima* host race, and (3) the period between these two. Ramets of both host plants were exposed during each of the intervals of gallmaker attack in both years.

In the first-year study, no survival differences were observed for the two host races during the various periods of attack. However, in the second-year study, larval survival was higher during the earliest time interval for the *S. gigantea* host race than in the other time intervals. There was no survival difference among the various periods of attack for the *S. altissima* host race. This latter result was probably due to the inordinately low survival of *Eurosta* on *S. altissima* (see fig. 7.13). Taken together, our results indicate that (1) the derived *S. gigantea* host race may be more sensitive to host-plant phenology than the ancestral host race, and (2) environmental conditions from growing season to growing season may affect host-plant phenology and its influence on the survival of the two host races (How, Abrahamson, and Craig 1993; Horner et al., unpub. data).

An earlier greenhouse experiment using only *S. altissima* ramets of varying age had similar results; however, we also found that ramet age at the time of attack significantly affected gall size (Weis and Abrahamson 1985). This suggests that at least the phenology of the ancestral host race is keyed to the period of suitability for gall growth. To further examine this, we compared the rates and durations of gall growth of individual galls for one long-established, garden-grown cohort of each host race at Cedar Creek, Minnesota, during the 1993 growing season (Brown and Abrahamson, unpub. data). The gall diameters of the *S. al-*

tissima host race were smaller than those of the *S. gigantea* host race throughout the period of gall growth. *Solidago altissima* gall diameters ranged from 6.3 ± 1.4 (SD) mm at the first measurement to 17.3 ± 2.4 mm 33 days later (and after galls had reached their maximum diameter). During the same period, *S. gigantea* gall diameters increased from 12.5 ± 4.4 mm at the beginning of our census to 19.9 ± 4.5 mm 33 days later. Gall growth rates (mm/day) varied widely for both host races. Initial rates were as high as 1.0 mm/day for the *S. gigantea* host race, but such rates quickly fell (within 15–22 days) to lower levels. However, the mean rates of diameter increase for galls on *S. altissima* did not fall as quickly as those on *S. gigantea*. These results coupled with our measurements of host-plant ramets (How, Abrahamson, and Craig 1993) indicate that *S. gigantea* host plants have an earlier and narrower "window of reactivity" to the *Eurosta* gallmakers than do *S. altissima* hosts. Such reactivity differences should create strong selection pressures among gallmakers that favor emergence times keyed to the windows of greatest reactivity of each host plant. The striking difference in emergence times of the two host races is a likely consequence of such selection pressures. Such differences in plant phenology and vigor among potential host plants may create obstacles to the adoption of new hosts. Such obstacles would make host shifts very rare events that occur only under particular conditions (How, Abrahamson, and Craig 1993; Abrahamson et al. 1994).

Effects of Host-Plant Traits (Stem Color)

Even though a gall is plant tissue, its development is controlled by both the gallmaker's and the host-plant's genotypes (Weis and Abrahamson 1986). While the gallmaker exerts substantial control over gall development (hence our ability to identify gallmakers by the morphology of their gall; Felt 1917, 1940), the host plant also has considerable influence over the gall's phenotype (Weis and Abrahamson 1986). The consequence of this genetic interaction in the gall's phenotype is that a gallmaker can acquire immediately a new gall phenotype if it shifts to a novel plant species. If this acquired phenotype varies in characters that influence adult emergence time or other traits, reproductive isolation of the derived population from the ancestral population could occur quickly and without genetic modification (Abrahamson et al. 1994).

The emergence times of adult gallmakers that overwinter within their gall as diapausing larvae or puparia may be influenced by traits including the gall's reflectance value. Since reflectance will influence a gall's absorbance of solar radiation, the internal temperatures of a gall may vary according to the state of this trait. Variations within and between alternate host-plant species for stem color and reflectance could be crucial to enhance the reproductive isolation of ancestral and derived host races of spring-emerging, temperate-zone gallmakers. Such variation could potentially alter development rates in gallmakers like *E. solidaginis* (that overwinter in their gall). Modest differences in gall reflectance on the sunny days of late winter and spring may cause variations of several degrees within the larval chamber (Abrahamson et al. 1994).

We became interested in this possibility for the reproductive isolation of *E. solidaginis* host races because of the degree of variation in host-plant stem colors of *S. altissima* and *S. gigantea*. While individual genotypes of both host plants can have stem colors that vary from green to purple, *S. altissima* stems and hence galls are generally on the green end of the continuum during the growing season, while *S. gigantea* stems and consequently galls are on the purple end. Stems of *S. gigantea* ramets typically also have a white, waxy glaucous layer that often covers its galls. Overwintering galls of both host races lose their green coloration and hence range in color from off-white to deep purple.

We examined the relationships among gall color, gall diameter, gall heating, and emergence times for both host races of *E. solidaginis* during the diapausing larva's (March) or pupa's (May) development period (Abrahamson et al. 1994). Two sets of galls were used: one from Cedar Creek Natural History Area near Bethel, Minnesota, and the other from New England sites along the Connecticut River Valley (from northern Massachusetts to northern New Hampshire and Vermont). Minnesota galls were collected during mid-autumn and stored outdoors in Minnesota within bird-predator-excluding screen cages until their shipment to Lewisburg, Pennsylvania, for analysis in either March or May. The New England galls were collected from natural field populations in late March and subsequently stored in a freezer until May. The samples of both host races from Minnesota were used to assess gall reflectance and heating characteristics, while the New England galls were used to evaluate differences in host-race emergence time and whether environmental effects would differentially alter the emergence times of each host race.

As we have found in other studies (e.g., Lichter, Weis, and Dimmick 1990; Craig et al. 1993; Brown et al. 1995; Sumerford and Abrahamson 1995), the mean diameter of *S. gigantea* galls was consistently larger than that of *S. altissima* galls (table 7.5). An analysis of covariance using gall diameter as the covariate showed that *S. gigantea* galls reflected significantly more red light (from a 0.63 mm diameter beam from a helium-neon laser @ 633 nm) than *S. altissima* galls in March (covariate: $F_{1185} = 54.8$, $P < 0.001$; species effect: $F_{1185} = 11.6$, $P < 0.001$). Near emergence time (May), an analysis of variance indicated that *S. gigantea* galls continued to reflect more red light than *S. altissima* galls ($F_{1210} = 9.5$, $P < 0.002$); however, the difference in reflectance by host disappeared when diameter was added as a covariate (Abrahamson et al. 1994). Notwithstanding, reflectance from galls is not simply a diameter effect. A regression of reflectance on gall diameter within each species resulted in nonsignificant relationships (*S. altissima* $r^2 = 0.03$, NS, $n = 112$; *S. gigantea* $r^2 < 0.01$, NS, $n = 100$). The enhanced red reflectance of *S. gigantea* galls is due to their more reddish color and the waxy covering of their gall's epidermis.

Our gall-heating studies found that larval-chamber temperatures were strongly related to solar radiation in March ($r = 0.65$, $P < 0.001$, $n = 197$) and to block (a measure of time of day) and solar radiation (multiple $r = 0.36$, $P < 0.001$, $n = 212$) near emergence time in May (Abrahamson et al. 1994). Unexpectedly, there was no significant variation between host races for either larval-chamber temperature or the difference between larval-chamber temperature and ambient temperature.

The nonsignificant difference of larval-chamber temperatures for the two host races leads us to predict that gall color would

TABLE 7.5. Mean gall diameter (± standard deviation) for outdoors over-wintered galls from both host races collected at a sympatric site on the Cedar Creek Natural History Area near Bethel, Minnesota.

Season	Gall Diameter (mm)	
	S. altissima	*S. gigantea*
March	20.2 ± 2.4	25.2 ± 2.8
	F = 179, df = 1,195, P < 0.001	
May	16.9 ± 3.1	22.9 ± 2.8
	F = 196, df = 1,186, P < 0.001	

Source: From Abrahamson et al. 1994.

not alter *Eurosta* host-race emergence times. However, we tested this possibility using over three thousand galls of each host race from twenty-three New England populations (Abrahamson et al. 1994). We incubated these galls outdoors in either a shade or a sun treatment for 2 weeks after their removal from cold storage (the time from overwintering larval diapause to adult emergence for *Eurosta* attacking *S. altissima* is approximately 21 days when held in environmental chambers at a constant 20°C; Uhler 1951). Galls were then enclosed in individual emergence cups and monitored for emergence in an air-conditioned laboratory held at approximately 21°C. Host species (*S. gigantea* host race emerged earlier), sex (males of both host races emerged before females), and treatment significantly affected the day of emergence. Most relevant in the context here was the finding that flies from sun-treated galls emerged before those from shaded galls (22.3 versus 22.9 days; fig. 7.5). The interaction of host race and treatment was also significant because of the more rapid emergence of the *S. altissima* host race (0.9 days faster) relative to the *S. gigantea* host race (0.4 days faster) under the sun treatment as compared to the shade treatment (Abrahamson et al. 1994). So while the gall's environment did influence the host races' development rates and hence their emergence times, the pattern of emergence-time shifts was opposite that necessary for us to conclude that host-plant traits may have facilitated the observed differences in the emergence times of these host races. We conclude that the earlier emergence time of the *S. gigantea*

237

host race is not due to their redder galls. If the gall-color differences were responsible for the distinctive emergence times and partial reproductive isolation of the host races, we should have found that sun-treated *S. gigantea* galls emerged more rapidly than sun-treated *S. altissima* galls. While the acquisition of gall color from host-plant stem color traits does not appear to facilitate reproductive isolation for *E. solidaginis* host races, this type of appropriation of host characteristics may be crucial in other cases of host-race formation.

Role of Preadaptation for Novel Host Plants

Adaptation to one host plant may preadapt an herbivore to recognize and/or survive on a closely related, but novel, host. Because of shared phylogeny, some congeneric host plants may differ only slightly in the "hurdles" they offer through requirements for herbivore survival. Such preadaptation for host-plant utilization relaxes the need for immediate genetic changes in the shifted herbivore population. We know, for example, that many herbivores can survive on host plants that they do not recognize as hosts (e.g., Rausher 1984; Via 1984; Butlin 1987; Futuyma and Phillipi 1987). This implies that some herbivores can establish populations on novel hosts through a change in host recognition alone.

Our studies of host preference and recognition described above (Abrahamson, McCrea, and Anderson 1989) showed that Pennsylvania *E. solidaginis* have strong preference for *S. altissima* over several sympatrically occurring *Solidago* species. However, this study also established that ovipositing females will occasionally ovipuncture alternate *Solidago* species (e.g., *S. gigantea, S. canadensis*) under no-choice conditions. Furthermore, our findings also document that each host race of *E. solidaginis* can survive, albeit poorly, on the alternate host plant (Craig, Itami, and Horner, unpub. data). These results establish the possibility that a host shift through an oviposition mistake or other means is at least possible even though its probability is low (see chapter 5; Larsson and Ekbom 1995).

Congeneric host plants often overlap in traits (e.g., chemical cues, phenology) important to host choice and/or herbivore survival (as we saw for Minnesota populations for *S. altissima* and *S. gigantea;* How, Abrahamson, and Craig 1993). Furthermore, herbivore adaptation to variation among the genotypes within the ancestral host species could yield a population of herbivores with differing character states that might facilitate a host shift to and subsequent survival on a novel host species (e.g., variation in emergence times and physiology). Herbivores adapted to use the extreme host genotypes on the ancestral host plant may possess the necessary preadaptations for utilization of overlapping genotypes of a novel host-plant species (Craig, Itami, and Horner, unpub. data).

An important question relative to the establishment of a derived host race is whether there is sufficient genetic variability within the populations of an herbivore to enable survival on extreme genotypes. If such variation exists, then particular combinations of host genotype and insect genotype could form an effective "bridge" between the alternate host plants. Several studies have documented that genetic variability exists within the host plants for survival of *Eurosta* (e.g., Maddox and Root 1987; McCrea and Abrahamson 1987; Horner and Abrahamson 1992 and submitted; Craig, Itami, and Horner, unpub. data). However, we did not know if genetic variability for performance existed within host races or whether there were interactions between host-plant genotype and insect genotype that were related to performance. To examine these possibilities, replicate *S. gigantea* fragments from four genotypes were exposed to attack by individual mated females (Horner et al., unpub. data). Gall formation (and hence offspring survival) varied significantly by sibship ($P < 0.001$). Furthermore, there was a significant fly sibship × plant genotype interaction ($P < 0.03$), suggesting that the survivorship of particular fly genotypes varies with plant genotype.

Craig, Itami, and Horner's (unpub. data) exploration of whether plant genotype influenced the survivorship of the hybrid offspring found that there was a highly significant difference in the rate of gall formation among plant genotypes on *S. gigantea* ($F_{5,92} = 13.9$, $P < 0.001$), while on *S. altissima* there was no significant difference in spite of large variations in gall numbers formed per host genotype. These data suggest that some

plant genotypes may represent particularly benign environments where a larva with an unusual genotype may survive. Such
host genotypes could potentially function as "bridges" for gene
flow between the host races. An occasional oviposition mistake
on the alternate host plant could survive on some host genotypes and subsequently this offspring could mate with flies of the
alternate host race. While there are other potential causes of
gene flow between these host races, these results do suggest a
possible mechanism for genetic interactions between host races.

The Role of Host-Plant Chemical Cues

Host-plant chemical cues allow *E. solidaginis* to recognize potential hosts (Abrahamson et al. 1994). The significance of such
cues was illustrated when we presented *Eurosta* females with an
artificial bud (a strip of sponge $\approx 0.5 \times 1 \times 6$ cm wrapped with
polyester/rayon gauze) containing plant extracts (Abrahamson
et al. 1994; see chapter 5.2). By using either stems or extracts
(applied to the artificial buds) from greenhouse-grown *S. altissima* and *S. gigantea,* we learned that *E. solidaginis* females can
discriminate between real and untreated artificial buds. Furthermore, *E. solidaginis* reared from *S. altissima* attempted to oviposit
on artificial buds containing a crude methanolic extract of *S.
altissima* (containing such potential chemical cues as flavonoids,
terpenoids, and simple phenolics) even though the physical
cues of a real bud were absent. Our study also found that the
relative concentrations of chemical cues may be critical. A
smaller proportion of *S. altissima* flies accepted artificial buds
when painted with low concentrations of *S. altissima* extract than
when treated with standard concentrations (table 5.4).

Important to the potential for a host shift, all *S. altissima* flies
tested rejected artificial buds painted with standard amounts of
S. gigantea extracts. The rejection rate did decrease, although
not significantly, when the concentration of *S. gigantea* was lowered. We also tested whether the addition of extracts of *S. gigantea* might function as a deterrent to oviposition by Pennsylvania
E. solidaginis that normally attack only *S. altissima.* Flies that originally accepted an artificial bud treated with *S. altissima* extract
would not accept a similar artificial bud once a standard concen-

tration of *S. gigantea* was added (table 5.4). Although not significant, this rejection rate did decrease if a reduced concentration of *S. gigantea* extract was used. Conversely, we found that adding standard concentrations of *S. altissima* extracts to artificial *S. gigantea* buds with standard extract levels lowered their rejection rate. To date, we have not tested the role of chemical cues in *Eurosta* from *S. gigantea* galls.

Host choice by ovipositing *Eurosta* appears to be based on a complex interaction of chemical cues (Abrahamson et al. 1994). Differential acceptance of plant species by the host races could result from variations in the amount or ratio of secondary chemicals. Such variation may encourage a host shift, particularly if the extremes of the amounts or ratios of chemicals within similar hosts nearly overlap and if host phenologies are similar. A host shift becomes more probable if individual *Eurosta* react positively to host-specific compounds or ratios of compounds in a new host. As we saw above, the phenologies of the two host plants sufficiently overlap in some regions that an alteration of reactivity by *Eurosta* to plant cues has the potential to facilitate a host shift (How, Abrahamson, and Craig 1993; Abrahamson et al. 1994). However, our genetic and behavioral findings (Waring, Abrahamson, and Howard 1990; Craig et al. 1993; Brown, Abrahamson, and Way 1996) indicate that only one host shift has occurred from *S. altissima* to *S. gigantea*. The implication is that the opportunity for *Eurosta* to alter sensitivity to plant chemical cues may be limited.

Effect of Female Age on Oviposition

It is conceivable that young, mated females are relatively more choosy about host-genotype choice and accordingly are slower to oviposit than females nearing the end of their life. Older females that more quickly inject eggs in any reasonably acceptable host may gain fitness over females that infrequently oviposit and subsequently die while still containing many eggs. We explored the relationship between oviposition choice and female age by offering 1-day-old and 7-day-old females eight stems with one each from four clones of *S. altissima* and four clones of *S. gigantea* (Brown and Abrahamson, unpub. data). Cut stems were ran-

CHAPTER 7

TABLE 7.6.   The numbers of 1-day-old (young) and 7-day-old (old) central Pennsylvania *Eurosta solidaginis* females emerged from *Solidago altissima* that oviposited, ovipunctured without egg injection, or did not ovipuncture.

Age	Total	No Punctures	Punctured S. altissima	Punctured S. gigantea	Oviposited S. altissima	Oviposited S. gigantea
Young	51	42	9	0	3	0
Old	91	69	22	3	10	0

Fisher's Exact Test, $P < 0.05$

Source: Brown and Abrahamson, unpub. data.

Note: Females were tested for 1 hour in cages containing eight randomly chosen ramets from four clones each of *S. altissima* and *S. gigantea*. Rhizomes of host plants were collected along the Connecticut River Valley in New Hampshire and Vermont and transplanted to a greenhouse at Bucknell University. Oviposition trials were conducted in mid-July.

domly assigned positions in Ehrlenmeyer flasks containing water within a 40 × 40 × 55 cm cage. Flies were placed on a leaf of the stem in the middle-right position and were left undisturbed for 1 hour. Six flies were run simultaneously at temperatures of 25–30°C and light levels of $400-500\mu Em^{-2}sec^{-1}$. The age of *S. altissima* females did not affect their tendency to ovipuncture or inject eggs into *S. gigantea* (table 7.6). The three 7-day-old females that did puncture *S. gigantea* buds also punctured *S. altissima* buds.

Conditioning Effects

It is possible that choice of either mating and oviposition site could be altered by adult or larval conditioning. If such conditioning occurs, it relaxes the need for genetic changes in host preference while increasing the likelihood of reproductive isolation (Maynard Smith 1966; Bush 1975a,b; Diehl and Bush 1989). For example, adult conditioning among individuals of *Rhagoletis pomonella* alters subsequent oviposition choices (Prokopy et al. 1982; Papaj and Prokopy 1988). If adult *Eurosta* prefer to mate and oviposit on the plants from which they emerged, then oviposition "mistakes" have the potential to result in offspring with mating-site and oviposition-site preferences that differ from their parents. Such oviposition "mistakes" could facili-

242

tate an initial shift to a new host plant without any genetic change or could lead to limited gene flow between existing host races.

One scenario in which adult experience might influence oviposition preference is if gravid females experience desperation to decrease their egg loads at sites containing large stands of the alternate host. Under such conditions, they may occasionally accept the alternate host for oviposition. This could lead to increased gene flow between host races. To test this possibility, flies derived from *S. gigantea* were reared out of puparia that were dissected from galls (to preclude exposure to potential host-plant cues during emergence). Adult flies were then mated without host plants and the mated females were initially exposed for at least two hours to *S. altissima* or to *S. gigantea*. After initialization, females were offered a choice of ten circularly arranged host-plant stems, five stems of *S. altissima* and five of *S. gigantea*. While all twenty females with an initial exposure to *S. gigantea* chose stems of *S. gigantea* for oviposition, nineteen out of twenty females (95%) initially exposed to *S. altissima* selected *S. gigantea* for oviposition. Only one female oviposited into *S. altissima* (Horner et al., unpub. data). These results with the *S. gigantea* host race lead us to conclude that adult conditioning has only a very minute effect, if any, on oviposition preference.

Competition Effects

The establishment of a herbivore population on a new host-plant species could also be encouraged by frequency-dependent selection operating through competition for host resources. Individuals of a derived host-associated population would initially experience lower fitness than members of the ancestral population due to poor adaptation to the new host plant. However, frequency-dependent selection could preserve the variation for host choice of the novel host if competition for host resources on the ancestral host plant is strong. The consequence would be that individuals that are poorly adapted to an abundant, but underexploited, novel host plant (where survival and fecundity are low) could have relative fitness levels that rival those of intensely competing individuals on the ancestral host (Rosenzweig and Shaffer 1978; Rausher 1984; Wilson and Turelli 1986; Feder et al. 1995).

The potential for strong intraspecific competition among *Eurosta* larvae on its ancestral host plant was suggested by the extreme attack (up to 100% of the ramets with up to forty-two punctures per bud in naturally occurring field clones) of some highly acceptable host genotypes by ovipositing *Eurosta* (Hess, Abrahamson, and Brown 1996). However, mature galls containing more than one larva are extremely rare (pers. obs.). Furthermore, *E. solidaginis* females are not deterred from ovipositing eggs into the buds of ramets that already contain multiple punctures or eggs (Craig et al., in prep.). As a result, *E. solidaginis*'s survival might be as high or higher on a novel, closely related host where larval competition is nonexistent or low in spite of poor physiological adaptation.

As we described in chapter 5.1, we manipulated *Eurosta* larval densities in a greenhouse study to explore larval competition levels on the ancestral host (Abrahamson et al. 1994; Hess, Abrahamson, and Brown 1996). While the oviposition of eggs increased with the number of ovipunctures over a wide range of attack levels, all ramets had similar numbers of larvae per bud just before gall appearance (fig. 5.6). This result suggests that rapid, intense larval competition and larval death accompany higher rates of ovipuncturing.

We also observed that the occurrence of necrotic, hypersensitive plant responses (causing early larval death) increased with ovipuncture rate (chapter 3). While such necrosis commonly occurs in resistant genotypes of *S. altissima,* it is uncommon in susceptible genotypes (Anderson et al. 1989). The triggering of this hypersensitive response at higher oviposition levels in the susceptible genotype used in the larval-competition experiment suggests that there may be a selective advantage to *Eurosta* that oviposit eggs into buds with fewer punctures regardless of the host genotype (Abrahamson et al. 1994).

Intense competition among larvae in the buds of high-ranked ancestral host genotypes would reduce the availability of high-quality buds for ovipositing *Eurosta.* Such a pattern of host usage could cause the breakdown of acceptance thresholds for preferred, high-ranked hosts (Jaenike 1990) and could initiate oviposition into low-ranked genotypes or novel host species. Although this possibility remains to be experimentally tested for either host race, oviposition data for high-ranked and low-ranked, naturally occurring *S. altissima* clones showed that the rate of ovipuncturing was initially higher on high-ranked clones. However, as high-ranked clones were exploited, the relative rate of ovipuncturing of low-ranked clones increased (Anderson et al. 1989).

Enemy-Free Space Effects

Differential levels of attack by natural enemies on herbivore populations using ancestral versus novel hosts could alter their relative fitness. If colonization of a novel host plant markedly reduces mortality because of escape from natural enemies, the survivorship on the new host may be equal to or exceed that on the ancestral host in spite of poor adaptation to the novel host (Price et al. 1980; Abrahamson et al. 1994; Brown et al. 1995; Feder 1995; Sumerford and Abrahamson 1995).

The five primary natural enemies of *E. solidaginis,* the parasitoid wasps *Eurytoma obtusiventris* and *E. gigantea,* the inquiline beetle *Mordellistena unicolor,* as well as the black-capped chickadee and downy woodpecker, differentially attack galls of the two host races (Abrahamson et al. 1994; Brown et al. 1995; Sumerford and Abrahamson 1995). A shift to a novel host plant could alter the cues available to natural enemies since the resultant gall could differ in gall size or morphology. Consequently, the acquisition of a new gall phenotype through a host shift has the potential to reduce the susceptibility of the gallmaker to its natural enemies, and hence increase its fitness.

Gall phenotype (i.e., gall size) is critical to *Eurosta*'s survivorship as a result of size-dependent attack by two natural enemies. The parasitoid wasp *E. gigantea* preferentially attacks smaller galls while birds strike larger galls (Weis and Abrahamson 1985; Abrahamson, McCrea, and Anderson 1989; Weis, Abrahamson, and Andersen 1992). Furthermore, the *S. gigantea* host race develops significantly larger galls than the *S. altissima* host race (Lichter, Weis, and Dimmick 1990; Abrahamson et al. 1994;

245

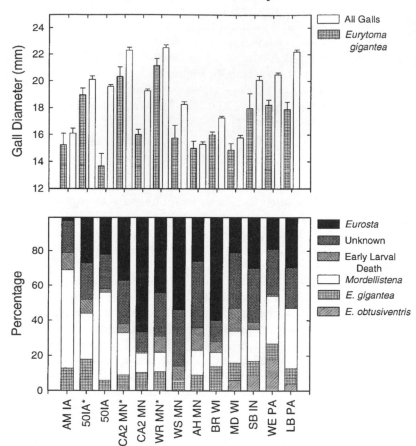

FIGURE 7.16. *Top:* Mean gall diameter (mm ± standard error) for all galls and galls parasitized by *Eurytoma gigantea* at sites along an east-west transect from Lewisburg, Pennsylvania, to Ames, Iowa. *Bottom:* The percentage occurrence of *Eurosta* and its natural enemies by site. *Solidago gigantea* populations are denoted by asterisk (*) next to site names (the last two letters of site names indicate state; see fig. 7.18 for site acronyms). (Redrawn from Abrahamson et al. 1994)

Brown et al. 1995; Sumerford and Abrahamson 1995). To determine if *Eurosta*'s host shift has been encouraged by acquisition of enemy-free space, we compared the sources of gallmaker mortality for the ancestral and derived host races at both allopatric and sympatric sites over a wide geographic area (Abrahamson et al. 1994; Brown et al. 1995; Sumerford and Abrahamson 1995).

Our expectation was that the *S. gigantea* host race would escape at least some attack from *E. gigantea* given its larger mean gall size and the restriction of the parasitoid *E. gigantea* to smaller galls due to limitations in ovipositor length (*E. gigantea* attacks *Eurosta* galls after maximum gall size is reached; Weis and Abrahamson 1985; Weis, Abrahamson, and McCrea 1985). To the contrary, we found that the *E. gigantea* that parasitized the *S. gigantea* host race attacked significantly larger galls than when this parasitoid infested the *S. altissima* host race (fig. 7.16). Consequently, there was no escape from *E. gigantea* parasitism for the derived *S. gigantea* host race.

Eurytoma gigantea was able to attack the larger *S. gigantea* galls because of the significantly longer ovipositors of wasps emerging from the *S. gigantea* host race compared to wasps from the *S. altissima* host race (Brown et al., unpub. data). Since there was a significant relationship between this parasitoid's ovipositor length and the size of the gall from which it emerged (r = 0.40, $P < 0.01$), *E. gigantea* that shifted to attack the *S. gigantea* host race may have experienced an immediate increase in mean ovipositor length. This increase in ovipositor length could be a consequence of phenotypic plasticity in body size based on natal gall size (Weis, Wolfe, and Gorman 1989). The shifted parasitoid population could have consisted, within a generation or two, of females with long enough ovipositors to inflict similar parasitism rates on *S. gigantea* host race as were inflicted on the *S. altissima* host race (Sumerford and Abrahamson 1995).

Eurytoma obtusiventris parasitizes *S. altissima* galls of all sizes because it attacks *Eurosta* larvae before gall formation (Weis and Abrahamson 1985). Observations of this parasitoid's searching behavior showed that it preferentially searches the ancestral host for fly larvae (fig. 7.17; Brown et al. 1995). Because *E. obtusiventris* uses host-plant cues to seek its host (itself a host-plant

E. obtusiventris Uses Host Cues

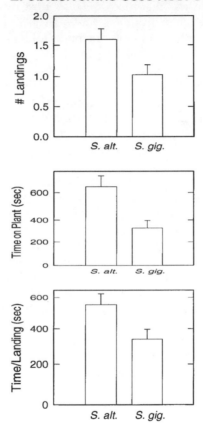

FIGURE 7.17. Means ($\pm$ standard error, $n = 57$) for number of landings (*top*), time spent on plant (*middle*), and time spent per landing (*bottom*) by *Eurytoma obtusiventris* observed in a 40 × 40 × 25 cm arena. Wasps were allowed to search among four randomly placed ramets each of *S. altissima* and *S. gigantea* for 30 minutes. (Redrawn from Brown et al. 1995)

specialist), this parasitoid could provide strong facilitation for a shift in host use by *E. solidaginis.*

Eurytoma obtusiventris is abundant in galls on eastern U.S. populations of *S. altissima,* occurring in as high as 40% of the galls (fig. 7.18). However, it is rare in the galls of north-central U.S. populations of *E. solidaginis* (Lichter, Weis, and Dimmick 1990; Abrahamson et al. 1994; Brown et al. 1995; Sumerford and Abrahamson 1995). Where this parasitoid does appear in appreciable numbers, it attacks the ancestral *S. altissima* host race and only very rarely infests the *S. gigantea* host race. The net effect is that the derived *S. gigantea* host race can survive better than the *S. altissima* host race in spite of the typically high larval mortality rates of the *S. gigantea* host race (e.g., Menominee, Michigan). In the north-central U.S. where this parasitoid is uncommon, the *S. altissima* host race can survive at significantly higher rates (e.g., Carlos Avery 1 and Carlos Avery 2, Minnesota).

Because of the abundance of *E. obtusiventris* in the northeastern U.S., we explored its host-plant distribution within this region by sampling twenty-three New England populations along a transect from northern Massachusetts to northern New Hampshire and Vermont. The transect was located so that it stretched across the overlap zone of the *S. altissima* and *S. gigantea* host races. Our analyses showed that the derived *S. gigantea* host race experienced nearly a 2-fold survivorship advantage over that of flies on *S. altissima* (table 7.7; Brown et al. 1995). This survival advantage was the consequence of significantly higher bird predation (primarily downy woodpeckers) and parasitism by *E. obtusiventris* on the ancestral host race. The net result of higher survivorship of the derived host race was despite significantly higher levels of early larval death (likely due to poor adaptation) and mortality due to *M. unicolor* on the derived host race (Brown et al. 1995). When we examined the levels of parasitism by *E. obtusiventris* at all sympatric sites sampled on the New England transect and in other studies (Abrahamson et al. 1994; Sumerford and Abrahamson 1995), we found that the relative survivorship of the derived host race at sympatric sites was positively correlated ($r = 0.90$, $P < 0.01$) with the level of attack by *E. obtusiventris* on the ancestral host race. Furthermore, the survivorship ratio (derived host race/ancestral host race) was

Parasitism and Survival of *Eurosta*

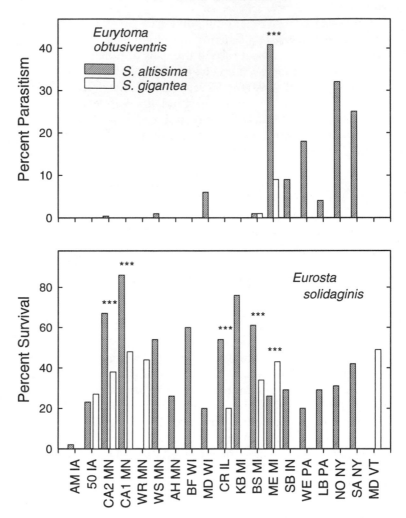

FIGURE 7.18. *Top:* Rates of parasitism on *Eurosta solidaginis* by *Eurytoma obtusiventris* and *bottom:* survival of *E. solidaginis* on *Solidago altissima* (filled bars) and *S. gigantea* (open bars) for nineteen sites from Vermont to Iowa. All sites harboring the *S. gigantea* host race (indicated in the bottom figure) do not necessarily have *Eurytoma obtusiventris* parasitism (top figure). Site acronyms are as follows: AM IA = Ames, IA (*n* = 141), 50 IA = Route 50

positively correlated with the percentage of mortality due to *E. obtusiventris* (r = 0.88, *P* < 0.01) (Brown et al. 1995).

The enhanced mortality due to bird attack on the *S. altissima* host race on the New England transect was a surprise given that previous studies (Weis and Abrahamson 1986; Weis, Abrahamson, and Andersen 1992; Weis and Kapelinski 1994) had established the strong relationship of larger gall size and bird attack within *S. altissima* galls. We suspect that the higher incidence of bird attack on many populations of the *S. altissima* host race may be due to the common occurrence of *S. altissima* near woodland habitats. Downy woodpeckers frequent woodlands and only forage old fields near woodlands (Weis and Abrahamson 1986). Likewise, the relatively lower rate of bird predation on the *S. gigantea* host race could be a consequence of the habitat affinities of its host plant. *Solidago gigantea* requires more mesic conditions than *S. altissima* and is typical of roadside ditches, floodplains, and wet meadows (pers. obs.).

When abundant, *M. unicolor* contributes substantially to the mortality of *Eurosta* host races. However, because its predation is variable from site to site among host-race populations (fig. 7.16) or is higher on the derived host race (table 7.7), *Mordellistena* is not likely a selective force that would promote a host shift (Abrahamson et al. 1994; Brown et al. 1995).

---

near Ames, IA (*n* = 107 *S. altissima, n* = 155 *S. gigantea*), CA2 MN = Carlos Avery 1989, MN (*n* = 460 *S. altissima, n* = 269 *S. gigantea*), CA1 MN = Carlos Avery 1991, MN (*n* = 1579 *S. altissima, n* = 542 *S. gigantea*), WR MN = Wright County, MN (*n* = 195), WS MN = Washington County, MN (*n* = 158), AH MN = Arden Hills, MN (*n* = 238), BF WI = Black River Falls, WI (*n* = 512), MD WI = Madison, WI (*n* = 296), CR IL = Castle Rock, IL (*n* = 319 *S. altissima, n* = 230 *S. gigantea*), KB MI = Kellogg Biological Station Hickory Corners, MI (*n* = 392), BS MI = University of Michigan Biological Station, Pellston, MI (*n* = 340 *S. altissima, n* = 375 *S. gigantea*), ME MI = Menominee, MI (*n* = 659 *S. altissima, n* = 489 *S. gigantea*), SB IN = South Bend, IN (*n* = 96), WE PA = Western PA (*n* = 374), LB PA = Lewisburg, PA (*n* = 363), NO NY = North Petersburg, NY (*n* = 260), SA NY = Salem, NY (*n* = 377), and MD VT = Mad River, VT (*n* = 280). An asterisk indicates a significant difference at least at the *P* < 0.05 level. (Redrawn from Abrahamson et al. 1994; data from Sumerford and Abrahamson 1995 and Brown et al. 1995)

TABLE 7.7. Numbers (and percentages) of surviving *Eurosta solidaginis* (ES) and numbers (and percentages) killed by early larval death (ELD), *Eurytoma obtusiventris* (OBT), *Eurytoma gigantea* (GIG), unknown causes (UNK), *Mordellistena unicolor* (MORD), and downy woodpeckers (BIRD).

	ES	ELD	OBT	GIG	UNK	MORD	BIRD
Solidago gigantea	1046	475	12	308	399	490	138
	(36.5)	(16.6)	(0.4)	(10.7)	(13.9)	(17.1)	(4.8)
Solidago altissima	751	93	1101	424	43	95	717
	(20.8)	(2.6)	(30.5)	(11.8)	(12.0)	(2.6)	(19.8)

Source: From Brown et al. 1995.
Note: Samples have been lumped across twenty-three populations in a transect from Greenfield, Mass., north along the Connecticut River Valley.

The patterns of survivorship and mortality for *Eurosta* host races from several studies suggest that the derived host race benefits from a selective advantage over the ancestral host race in some regions (i.e., parts of New England) despite its lack of adaptation to the novel host's physiology and/or chemistry (Brown et al. 1995). Our studies with *Eurosta* are not the only ones to illustrate the importance of natural enemies in host-race formation. Other examples where such escape from natural enemies has been important in host-race formation include the shift of bruchid beetles to novel hosts (Johnson and Siemens 1991a,b) and the shift of *Rhagoletis pomonella* from its ancestral hawthorn host to apples (Feder 1995). Escape from natural-enemy attack may be an important event in the success or failure of a host shift. Furthermore, such escape may be crucial to help maintain the derived populations on novel hosts and enable subsequent reproductive isolation and speciation.

7.8 THE FACILITATION OF A HOST SHIFT

Herbivores may be able to shift to a new host plant if a combination of some of the following factors occurs: (1) the novel host is phenologically similar to the ancestral host, (2) the adaptations to the ancestral host preadapt the herbivore to utilize the novel host, (3) traits of the novel host alter herbivore characters such as emergence times, (4) adults or offspring modify host preferences due to conditioning, (5) the intensity of herbivore competition is intense on the ancestral host, and/or (6) escape from natural enemies enhances survival on the derived host in spite of poor adaptation. The occurrence of several of these factors in combination would lower the barriers to the successful establishment of a novel host race, and as a result, would relax the need for a "genetic revolution" of adaptations for host-plant use (Bush 1975a,b, 1994; Craig et al. 1993). The probability of occurrence would be enhanced if the ancestral herbivore population displayed considerable variation in the range of host genotypes used. Herbivores adapted to extreme individuals of the ancestral host may perform reasonably well on some extreme individuals of the novel host. However, the phenologies of extreme individuals of the ancestral and a novel host must at least overlap first in order for a host shift to occur. Individual herbivores must recognize the novel host and attack it, and preadaptation must allow limited survival on the new host. Factors such as larval competition and natural-enemy attack could maintain the initial host shift if they cause sufficiently higher mortality to populations infesting the ancestral host plant to offset the mortality of populations on the novel host due to poor adaptation. Once the host shift has occurred, slight difference in host-plant phenologies between the ancestral and new host would be crucial to reproductively isolating the derived population (Butlin 1990). We know that reproductive isolation on new hosts can be rapid (with strong selection for adaptation) when herbivore emergence is cued by host-plant phenology (Wood 1980, 1986; Wood, Olmstead, and Guttman 1990; Keese and Wood 1991).

7.9 IS IT INCIPIENT SYMPATRIC SPECIATION?

The question of whether *Eurosta* host races were initiated in sympatry or allopatry cannot be determined definitively by the data available. However, we can assess whether the characteristics of today's populations are compatible with the conditions necessary for sympatric speciation. To do so, we must assume that extant populations have traits similar to those of ancestral populations. The evidence we have generated to date clearly

253

does not exclude host-race formation in sympatry. The behavioral, ecological, and genetic traits of these host races are currently such that geographic isolation would not be a prerequisite to their formation. However as we noted above, the concept of sympatric speciation by host-race formation remains highly controversial because of the restrictive conditions necessary for the required genetic changes (Tauber and Tauber 1989; but see Bush 1994 and Johnson et al. 1996). These restrictive conditions have led to arguments that sympatric speciation is highly unlikely (Futuyma 1983, 1986; Futuyma and Peterson 1985). However, for *E. solidaginis* some of these restrictions can be relaxed since, for example, mate choice and host choice are coupled.

We have shown that partial reproductive isolation results from strong host-plant association and that there is strong disruptive selection for host use (Craig et al. 1993; Craig, Itami, and Horner, unpub. data). Craig, Itami, and Horner's (unpub.) work illustrates that no mutations in genes related to survival are necessary for the shift from *S. altissima* to *S. gigantea* since a few larvae can emerge on the alternate host plant. Variation in the genes necessary for larval survival on different host-plant genotypes and phenotypes of *S. altissima* could have maintained the genetic variation that facilitated the host shift. Our studies repeatedly have documented that all *S. altissima* genotypes are not equally suitable for larval development (e.g., Weis and Abrahamson 1986; Anderson et al. 1989; Horner and Abrahamson 1992).

Furthermore, the two host-plant species utilized by *E. solidaginis* appreciably overlap in phenology in some regions of *Eurosta's* range, making extreme genotypes of these host plants similar to one another (How, Abrahamson, and Craig 1993). Finally, our studies have documented that the shift from *S. altissima* to *S. gigantea* has resulted in enemy-reduced space that offsets the higher mortality within the derived host race due to poor adaptation (Abrahamson et al. 1994; Brown et al. 1995; Sumerford and Abrahamson 1995). Such enemy-reduced space could maintain a host shift in sympatry. While we cannot prove absolutely that host-race formation in *E. solidaginis* came about in sympatry, the available data are certainly parsimonious with this explanation. We do not need to invoke geographic reproductive isolation to explain the present-day patterns.

In the absence of a time machine, we can never be certain that contemporary ecological conditions match past circumstances nor that present population-level situations can be extended to explain larger-scale, phylogenetic patterns (Rausher 1988a). Since speciation is a historical process, studies like ours of present-day behavioral, ecological, and genetic characteristics of taxa can only suggest the most parsimonious explanation of how speciation has proceeded (Wood and Keese 1990). Consequently, we will never know all the details of *Eurosta*'s speciation events. However, our studies do provide crucial insights into the processes of phytophagous insect speciation. Many of the characteristics that caused host-race formation in *Eurosta* likely still exist. Our studies of *E. solidaginis* suggest that (1) host-race formation can occur, (2) host shifts may enable host-race formation, (3) host shifts can occur with minimal or no genetic change, (4) establishment and maintenance of host races may not necessarily require geographic reproductive isolation, and (5) host races may serve as an intermediate step in speciation. Examples of host-race formation are limited (Bush and Howard 1986; Tauber and Tauber 1989; Bush 1994; Johnson et al. 1996); thus our group's series of studies on *E. solidaginis* provide an important example of host-race formation through a host shift. Furthermore, these studies document both its occurrence and the ecological, behavioral, and genetic conditions necessary for host-race formation.

The Third Trophic Level as an Agent of Selection

8.1 NATURAL ENEMIES AS SELECTIVE AGENTS ON HERBIVORE-PLANT INTERACTIONS

The evolution of the interaction between an insect herbivore and its host plant occurs in the context of a larger ecological community. Many environmental factors such as climate, nutrient availability, availability of plant mutualists, and abundance of competing herbivores influence the insect-plant relationship through effects on the availability and quality of host tissue. However in addition, plant use by an insect herbivore will evolve in a context that includes its own natural enemies. In a growing number of instances, it has been observed that plant traits can influence the mortality levels inflicted by parasites and predators on insect herbivores (Price et al. 1980; Price and Clancy 1986; Gross and Price 1988; Heinrich and Collins 1985; Keating, Hunter, and Schultz 1990; Barbosa, Gross, and Kemper 1991; Hare 1992; Hunter and Schultz 1993). The co-opting of plant defensive chemicals for predator deterrence by Monarch butterflies stands as a well-known example (Brower and Brower 1964). Sequestered phytochemicals can cause problems for parasitoids as well (Barbosa, Gross, and Kemper 1991). Such instances suggest that herbivore traits involved in host-plant choice and usage may evolve in response to selection exerted by parasitoids and predators.

Switching to the plant's perspective, natural enemies are a potential component of the defensive repertoire (Price et al. 1980; Vinson 1984; Hare 1992). This can make plant defensive chemistry a two-edged sword for the specialist herbivore that evolves to overcome its deterrent or toxic effects. The phytochemicals that many herbivores find repellent may cue searching parasitoids toward the adapted specialist (Baehrecke, Vinson, and Williams 1990; Geervliet, Vet, and Dicke 1994; Kester and Barbosa 1994). Plant traits such as these that facilitate attack can evolve through natural selection if two key requirements are met. First there must be a genetic variation in the plant trait that renders the herbivore more vulnerable to attack. Second, enemy attack must curtail damage to the plant. In other words, it is not enough that a facilitating trait causes enemies to lower herbivore fitness,[1] but by so doing it must also salvage plant fitness (Price et al. 1980; Moran and Hamilton 1981). And so, plants are more likely to encourage natural enemies that kill the herbivore early in its life history (e.g., egg parasitoids) than those that kill late (e.g., pupal parasitoids).

Evaluating the potential selection pressures exerted by natural enemies has been a key focus in our study of the *Solidago-Eurosta* interaction and underlies this and the next two chapters. Attack by the parasitoids *Eurytoma obtusiventris* and *E. gigantea,* the inquiline *Mordellistena,* and by the birds, downy woodpeckers and black-capped chickadees, could act on either plant or gallmaker traits, or both. Attack on the gallmaker could be an appreciable selection agent on facilitating plant characters if attack somehow prevented gall initiation or curtailed gall growth and development, such that the total damage to the plant is lessened. Detailing the timing of attack relative to gall development was the key to evaluating the potential for selection on plants. On the other hand, the gallmaker's plant-use characters could be under selection if those characters affect vulnerability to

[1] The presence of a facilitating plant in the population may reduce the overall density of herbivores in the subsequent generation, and thus reduce damage levels. It might be supposed that this delayed reduction in damage will lead to a selective advantage to facilitation, but under typical circumstances it will not. C. E. Bouton (pers. comm.) has pointed out that if insect herbivores freely disperse between plants at some point in their life cycle, then the benefit of increased herbivore mortality on facilitating plants in the current insect generation will be shared by all plants in the next generation. Herbivore dispersal will thus tend to reduce variance in plant fitness. A delayed benefit to facilitation would occur only if the insect herbivore population is subdivided into demes associated with individual, long-lived plants.

257

enemy attack. Entomologists and ecologists have noted in the past that *E. gigantea* is more frequently recovered from small galls (Ping 1915; Uhler 1951; Cane and Kurczewski 1976), while bird attack is concentrated on large galls (Uhler 1951; Cane and Kurczewski 1976; Schlichter 1978; Confer and Paicos 1985; Mecum 1994). Thus, we focused on gall size as an insect trait (see chapter 6) that could evolve under selection imposed by enemies (Weis and Abrahamson 1985, 1986; Weis, Abrahamson, and Andersen 1992).

8.2 PHENOLOGY OF ATTACK, GALL GROWTH, AND SELECTION FOR NATURAL ENEMY FACILITATION IN *SOLIDAGO*

In the *Solidago-Eurosta* system, it appears that enemy attack does not reduce the damage done by the gallmaker. For instance, bird attack occurs during the winter months (Cane and Kurczewski 1976; Schlichter 1978). By this time, the galled ramet has already produced its seeds and rhizomes, and so death of the gallmaker at this point can have no influence on plant fitness. However, we undertook a study to see if any of the parasitoids could arrest damage, and in turn if plant traits could facilitate their attack.

Published observations showed that *E. gigantea*-attacked galls are smaller than unattacked ones (Ping 1915; Uhler 1951; Cane and Kurczewski 1976). We hypothesized that this parasitoid attacks the gallmaker early in the season, killing the larvae before it has completed stimulating gall growth, thus taking advantage of the gallmakers "window of vulnerability" (Craig, Itami, and Price 1990). By so doing, this parasitoid would then reduce the plant's allocation to gall growth and thereby mitigate the decrement to reproduction caused by the gall (Hartnett and Abrahamson 1979; Stinner and Abrahamson 1979; Abrahamson and McCrea 1986b). This was tested in the field with a serial exposure/exclusion experiment. During early summer of 1983, over seven hundred newly formed galls were protected with parasitoid exclusion bags. Successive cohorts of bagged galls were exposed to attack for one-week intervals, rebagged, and then allowed to mature, after which they were dissected to discover their content. This experiment also allowed us to determine the attack sequence for *E. obtusiventris* and *Mordellistena.* Simultaneously we took frequent diameter measurements of marked galls to document their growth phenology.

The parasitoid *E. obtusiventris* is the first of the insect enemies to attack. It was found in all bagged cohorts (fig. 8.1), as would be expected, since it attacks the host during the egg stage, before gall initiation, and before the protective bags were applied. However, this wasp delays its larval development until the end of summer, after its host has reached full size (chapter 2). *Mordellistena* starts ovipositing in June and continues through July (fig. 8.1). However, it does not enter the gall's central chamber and kill the gallmaker until some months later (chapter 2).

Contrary to our initial expectation, *E. gigantea* does not oviposit early in the season. Instead, nearly all attack occurred after mid-July (fig. 8.1) when the gall has achieved full size (fig. 7.5). This showed that *E. gigantea,* like the other natural enemies, kills the gallmaker too late in its life cycle to benefit the plant by curtailing gall growth. Furthermore, it implied that *E. gigantea* is found in small galls because it is unsuccessful in penetrating large galls with its ovipositor. Is there then any way for *E. gigantea* to benefit goldenrod? A plant that reacts to the gallmaker by producing a small gall would facilitate parasitoid attack. It could be argued that plants that produce small galls will gain some fitness increment if *E. gigantea* attack reduces allocation to the gall during its maturation phase. However, reduced allocation at this later time is likely to be very small. Histological study has shown that the mitotic activity of the gall cells, and their cytoplasm content, begins to decline even before the gall reaches full size (Bross, Weis, and Hanzley 1992). The advantage of a small savings late in the season, caused by parasitoid attack, would probably pale compared to the direct benefit to the plant of producing a small gall in the first place. We conclude, therefore, that the selective pressure on the plant favoring parasitoid facilitation is at most a minor augmentation of whatever selection there might be on the plant for low reactivity already exerted by the gallmaker.

Parasitoid Attack Phenology

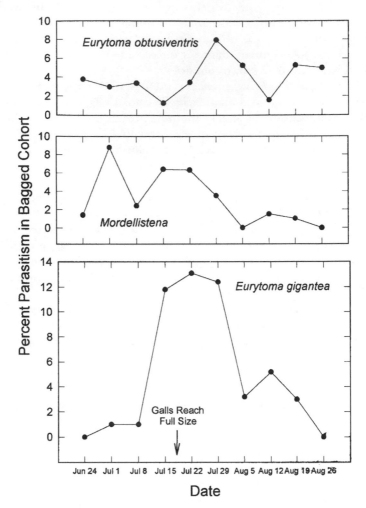

FIGURE 8.1. Attack phenology of the three parasitoid enemies of *Eurosta*. When galls first appeared, seven hundred galls were covered with mesh bags to exclude parasitoids. Each week a cohort of seventy galls was exposed to parasitoid attack and then retagged the following week. Each point represents the percentage of galls in a cohort to be parasitized by a parasitoid species. Points are arranged along the time axis at the midpoint of the exposure period. (From Weis and Abrahamson 1985, "Potential selective pressures by parasitoids on the evolution of a plant-herbivore interaction," *Ecology* 66:1261–1269; with permission)

8.3 GALL SIZE AND ENEMY ATTACK: IMPLICATIONS
FOR NATURAL SELECTION ON *EUROSTA*

The phenology of enemy attack indicates that small gall size is not caused by parasitoid attack, nor can bird attack be the cause of large gall size. This leaves open the reverse logical relationship—gall size casually influences *Eurosta* survival from attack. If so, natural enemies act as selective agents on the gallmaker, favoring larvae that produce galls of a protective size. In sections 8.4 and 8.5, we will present experimental results that establish this causal link. To set the stage for these investigations, however, we will present more information on patterns of gall size-dependent attack in natural populations. These data come from a long-term study of twenty *Eurosta* populations in central Pennsylvania (Abrahamson et al. 1989; Weis, Abrahamson, and Andersen 1992; Sumerford, Abrahamson, and Weis, in prep.). A wealth of information on selection was developed in this study, and we will return to it several times later in this and the two following chapters.

Our methods for this long-term study were very simple. Every April, we collected overwintered galls from each of twenty fields within a 20 km radius of Lewisburg, Pennsylvania. A 2 m × 2 m quadrat frame was dropped at fifty points in a regular grid in each field, and all galls with the quadrat were collected. If fewer than one-hundred galls were collected from the quadrats, additional galls were haphazardly collected from diagonal transects through the field (see Weis, Abrahamson, and Andersen 1992). The galls were then returned to the laboratory and measured to the nearest millimeter. Measurement was done with a drafting template that featured circular holes graduated in 1 mm increments; gall size was scored as the size of the smallest hole the gall would pass through. After measurement, each gall was dissected to determine the fate of the gallmaker. Each of the natural enemies leaves distinctive remains in the gall (see fig. 2.10), so a cause of death could be assigned when living *Eurosta* larvae were not found. In the following sections, galls exhibiting early larval death (see chapter 2) were excluded from analysis since it is probable that these gallmakers died before *E. gigantea* and birds attacked. Over the course of this

261

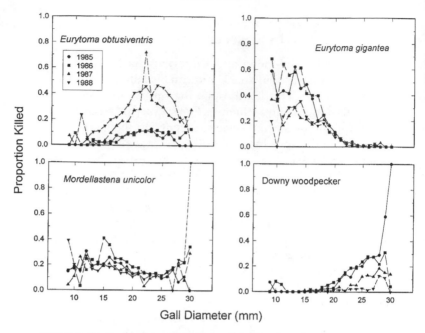

FIGURE 8.2. Gall size-dependent attack rates on *Eurosta* by the three parasitoids and by the downy woodpecker. (From Weis, Abrahamson, and Andersen 1992, "Variable selection on *Eurosta*'s gall size. I: The extent and nature of variation in phenotypic selection," *Evolution* 46:1674–1697; with permission)

study, we examined over 20,000 *Eurosta* individuals from one hundred population-generations (Abrahamson et al. 1989; Weis, Abrahamson, and Andersen 1992; Sumerford, Abrahamson, and Weis, in prep.).

This field data study confirms that two of the natural enemies have clear size-dependent attack rates (fig. 8.2). The parasitoid *E. gigantea* was found in 20–60% of all galls 18 mm or less in diameter, and in less than 1% of all galls 23 mm or greater in diameter. On the other hand, it was quite unusual for galls 18 mm or less to be opened by birds, but the probability of attack rose steadily with gall diameter above this point. The degree of size-dependent attack by the other two species is less clear. It appears that *E. obtusiventris* attack rises with gall size, perhaps peaking at some intermediate level. However, these data must be interpreted in light of this parasitoid's position within the attack sequence (chapter 2). First, this wasp oviposits before galls are initiated, and so gall-size cannot directly influence its foraging or oviposition behaviors. Size-independent attack is thus expected. However, this expectation is not reflected in end-of-the-year samples because many of the small galls that would have contained this larva were undoubtedly attacked later by *E. gigantea*, thus the apparent upward trend; *Mordellistena* attack may also distort the size-independent relationship. When corrected for this effect, *E. obtusiventris* attack appears even across gall sizes (Abrahamson et al. 1989). Size-dependent attack data for *Mordellistena* are untroubled by subsequent attack, and its parasitism rate seemed to drop slightly toward the middle of the size range. In interpreting each of the graphs in figure 8.2, one should keep in mind that because very small and very large galls are rare, the proportions killed by the various enemies in galls at the ends of the size distribution are based on small sample sizes ($n < 10$), even though total sample size exceeded three thousand in each of the four years depicted ($n > 200$ for intermediate gall sizes in each year). Thus, strong upward and downward swings at extreme gall sizes in figure 8.2 are undoubtedly due to sampling error. But the consistently low rate of attack by birds at the low end of the gall diameter distribution, and by *E. gigantea* at the high end, shows up very clearly despite small sample size, which testifies to the strength of the size-dependent effect on attack.

8.4 HOW GALL SIZE AFFECTS OVIPOSITION SUCCESS BY *EURYTOMA GIGANTEA*

Attack by *E. gigantea* occurs too late in host development to curtail gall growth, as we showed above (sec. 8.2). This leaves the possibility that small galls are more frequently attacked because they are vulnerable to penetration by the wasp's ovipositor. Several experiments established this to be the case.

TABLE 8.1.   Oviposition success by *Eurytoma gigantea* depends on gall size.

	Gall Diameter (mm)		
	< 12.5	$13.0–17.5$	> 18.0
Successes	13	9	1
Failures	2	3	15

Source: From Weis, Abrahamson, and McCrea 1985.
Note: $\chi^2 = 11.73$, $P = 0.0028$.

Gall Size and Ovipositor Length

A set of behavioral experiments (Weis, Abrahamson, and McCrea 1985) was carried out in the laboratory to test for differences in the vulnerability of *Eurosta* larvae in small and large galls. In the first experiment, we determined the accessibility of hosts in small, medium, and large galls (Weis, Abrahamson, and McCrea 1985). We offered galls of various sizes to 4–9-day-old wasps for oviposition. Galls were presented to wasps one at a time, in random order of size, inside small cages made from plastic cocktail glasses. Each wasp was watched for one hour, and oviposition attempts were noted. Nearly all galls were probed. Afterwards, probed galls were dissected in search of freshly laid eggs. Gall wall thickness was also measured at this time. Parasitoids were successful at ovipositing into galls in the smallest size category (<12.5 mm diameter) in about 80% of their tries (table 8.1). But only one of the sixteen attempts was successful on large (>19 mm) galls (and this was on the second smallest of that size category). Since we dissected only galls where oviposition attempts were made, all galls without wasp eggs can be considered true oviposition failures.

Considering the geometry of the gall and parasitoid, it seemed plausible that oviposition attempts would fail when wall thickness exceeded *E. gigantea*'s reach. To determine what that reach may be, we sacrificed the wasps from the accessibility experiment (and others) and measured their ovipositors, which would of necessity set the upper limit (Weis, Abrahamson, and McCrea 1985). Using the data on gall walls and on ovipositor lengths, we calculated the thickness-to-length ratio for all attempted attacks. Then we performed a logistic regression, with the ratio as the

Relative Sizes of Gall and *E. gigantea* Affect Success

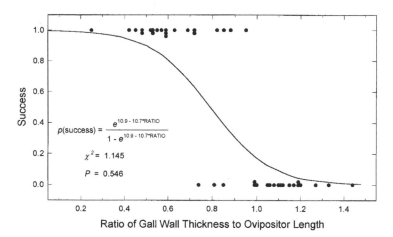

FIGURE 8.3. Success of *Eurytoma gigantea* on *Eurosta* depends on the ratio of the wasp's ovipositor length to the gall's wall thickness. When this ratio exceeded 0.95, all attempts failed. The low χ^2 value indicates that the data do not significantly differ from the predictions of the fitted logistic regression curve.

independent variable and success/failure as the binomial dependent variable (fig. 8.3). The data showed that all attempts to oviposit into galls where the ratio exceeded 0.95 were failures—in other words, a parasitoid with a 10 mm ovipositor could penetrate gall walls up to 9.5 mm thick.[2]

These findings established that gall size influences *E. gigantea*'s oviposition success. Larvae in large galls are outside the reach of wasps, which can successfully attack only if they deposit their eggs in the gall's central chamber. The fact that some *Eurosta* induce galls that are too large to penetrate means that insect genes contributing to large gall size also increase survival probability.

---

[2] The 0.95 ratio figure is based on the assumption that all galls are probed at their equator, where the wall thickness measurements were made. In fact, the wall thickness of a gall may be less at lattitudes above and below its equator, and wasps do probe these areas. Thus *E. gigantea*'s reach into a gall bay be somewhat less than 95% of its ovipositor length.

Does Wall Thickness and Ovipositor Length Explain It All?

In the experiments cited above, we established that the conditions for selection to act directly on gall diameter were met. However, the possibility that an unidentified background factor also could contribute to the correlation between gall diameter and parasitism rate was not eliminated. If gall size and *Eurosta* survival covary in part through their correlations to background environmental factors, then the intensity of selection imposed by *E. gigantea* will be less than it appears (see Falconer 1989; Lande 1979; Mitchell-Olds and Shaw 1987; Price, Kirkpatrick, and Arnold 1988; Rausher 1992).

Rather than conduct an exhaustive search for background factors (which had no guarantee of success), we tried an alternative approach. We reasoned that if all the factors behind the size-dependent recovery of *E. gigantea* had been identified, it should have been possible to construct a simulation model that reproduced the pattern of size-dependent parasitism seen in the field. Beyond gall dimensions and ovipositor lengths, we needed additional information on the parasitoids' behavior during attack to complete the simulation.

In the accessibility experiment it appeared that *E. gigantea* spent a lot of time probing galls that were too large to penetrate. Another experiment was performed to determine how long a large gall is handled, and how intensively it is probed, before it is rejected. At the same time, we wanted to compare this to the handling of small, penetrable galls. Eight females were offered one small (<15 mm) and one large (>21 mm) gall in random order in the plastic cup cages. We timed their encounters and counted the number of probes they made. It took wasps three times longer to reject an oversized gall than to successfully attack a small one, and likewise, it made three times as many probes before giving up (table 8.2).

The available information was then used to construct the simulation, which was constructed as a complex sampling scheme. In the scheme, a "gall" of given outer diameter was exposed to a "wasp" which was allowed to "probe" the gall P times. If the wall was less than 95% of the ovipositor's length, the attack was scored a success.

TABLE 8.2. Handling time and number of probes by *Eurytoma gigantea* on small (< 15 mm) and large (> 21 mm) galls.

Gall Size		Handling Time (s)	Number of Probes
Large	x	1830	8.00
($n = 8$)	x'	7.05	1.81
	SE	0.31	0.32
Small	x	590	2.63
($n = 8$)	x'	6.13	0.86
	SE	0.24	0.18
	$F_{1:11}$	5.39	7.05
	P	0.022	0.040

Source: From Weis, Abrahamson, and McCrea 1985.
Note: Mean (x) and log-transformed means (x') presented.

In the simulation results, the percent parasitism inflicted by *E. gigantea* fell in a sigmoidal fashion with gall size (fig. 8.4). Attack on galls larger than 23 mm is very rare and nonexistent on those above 25 mm. Individuals in galls smaller than 16 mm were at maximum vulnerability, with their probability of parasitism rising with the mean number of times galls are discovered. Field data for size-dependent recovery of *E. gigantea* (in this graph, corrected for subsequent attack by *Mordellistena*) plotted along with the simulation output show remarkable similarity between simulated and actual results. This agreement shows that any background factors that contribute to the correlation between gall diameter and *E. gigantea* parasitism rates are inconsequential.[3]

In summary, size-dependent attack by *E. gigantea* acts as a selective agent on *Eurosta* by exerting directional selection for increased gall size. At the end of this chapter and in the next we will return to *E. gigantea*'s foraging biology, focusing on some of the consequences of variation in gall size for foraging efficiency and reproductive success. Before that, we will consider birds as agents of selection.

[3] The field data most closely coincides with predicted parasitism rates when galls are discovered an average of three times. This strikes us as a high discovery rate, but there is the possibility that we overcorrected for multiple attacks by *Mordillistena*.

Simulated and Real Size-Dependent Attack

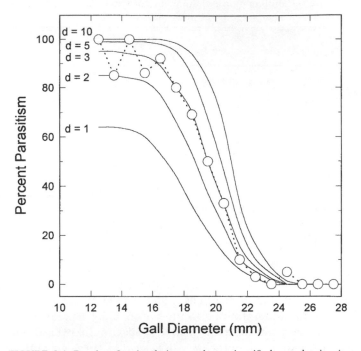

FIGURE 8.4. Results of a simulation to determine if observed ovipositor length, wall thickness, and number of probes adequately predicted size-dependent attack by *E. gigantea*. Simulation results appear as solid lines and actual field data as open circles and dashed lines. In each simulation run, a thousand simulated "galls" of each class were available for wasp discovery and probing. The "wall thickness" at each probe point was drawn from a normal distribution; the mean and variance of the distribution was based on measurements of field galls (Weis, Abrahamson, and McCrea 1985). The ovipositor length of the wasp was randomly drawn from another normal distribution, with mean and variance equal to those estimated from the slide-mounted ovipositors. The number of probes was drawn from a Poisson distribution with a mean of eight, which was the average number of probes made into oversized galls before rejection. The attack was scored a success if any of the probes were made at points where the gall wall thickness was less than 0.95 times the ovipositor length, and as a failure if it was not. We could not determine the number of times galls in the field are discovered by *E. gigantea*, and so made the assumption that discoveries follow a Poisson distribution. We set the mean number of discoveries to a series of values in order to account for the effect of variation in parasitoid population density on attack rates. (From Weis, Abrahamson, and McCrea 1985, "Host gall size and oviposition success by the parasitoid *Eurytoma gigantea*," *Ecological Entomology* 10:341–348, with permission)

8.5 GALL SIZE AND BIRD ATTACK RATES

Unlike the case of *E. gigantea,* it was obvious from the start that bird attack could not influence gall size. Woodpeckers do not bother with gallmakers during the spring and summer when galls are developing, and the gallmaker larvae are still small. Undoubtedly, other food sources are plentiful at this time. October is the earliest that these birds come out of their wooded habitat and into the open fields where goldenrod and its gallmakers are most often found (Confer and Paicos 1985). By this time, the galls have been at their mature size for 3 months. Although we could dismiss the possibility that attack influences size, experiments were needed to determine if large gall size causes increased attack rates directly, or, instead, if large gall size and high attack covary because both are correlated to background factors.

Several workers have noted that the intensity of downy woodpecker predation on *Eurosta* is strongest along field margins where woody plants provide cover (Schlichter 1978; Confer and Paicos 1985). If larger galls were concentrated there they would suffer disproportionate predation, even if woodpeckers had no size-dependent attack. Since small galls seemed to be haphazardly dispersed within fields, we thought this unlikely. Nonetheless, it was possible that size preferences by woodpeckers could change with distance from cover. For instance, close to the woody field margin, galls of any size would be attacked, but only large galls could lure the birds out into open habitat. This interaction could thus influence the correlation between gall size and attack probability. One of our undergraduate assistants, Thomas Richardson, performed a field experiment to determine if the apparent preference by woodpeckers for large galls could be altered by proximity to woody cover (T. E. Richardson et al., unpub. data). We set up artificial patches of galled goldenrod stems in an old field at varying distances to the wooded field edge. Each patch offered a mixture of large and small galls to foraging woodpeckers. The experiment was also designed to test the idea that the perching quality of the stem influenced attack. Stems that are leaning may give woodpeckers a better angle from which to peck, and so we set stems with large galls

269

TABLE 8.3.   Gall size and woodpecker attack in artificial plots.

	Gall Size and Orientation		
	Large Upright	Large Angled	Small Upright
Observed	38	38	2
Expected	27	27	27

Note: $\chi^2 = 29.4, P < 0.001$.

either at 45° angles or upright. Plots were checked three times a week, and when an attacked gall was found, it was replaced with another of the same size.

We found that size preference was not influenced by distance from the woody field margin. In fact, of the seventy-eight attacks scored, only two were on small galls. The remaining attacks were evenly divided between upright and angled stems (table 8.3). Although it did not affect size preference, proximity to cover had a very strong influence on overall attack rates (fig. 8.5). In our long-term selection study, we also found that *Eurosta* populations surrounded by large woodlots had higher predation rates than those that were not (Weis and Kapelinski 1994).

An objection could be raised that the small galls in Richardson's experiment were parasitized, and perhaps woodpeckers were able to distinguish them as such and reject them. If so, one would see size-dependent attack even though birds are not responding to size. Confer and Paicos (1985) suggested woodpeckers may have the ability to discriminate galls that have *Eurosta*. They found that in over 50% of attacks, the woodpecker made its extraction hole by enlarging the gallmaker's exit tunnel. This cannot be a random coincidence because the exit-tunnel entrance occupies only 1% of the gall surface area. Parasitized galls lack exit tunnels, and so birds searching for a tunnel may reject a gall occupied by one of the insect parasites.

We examined downy woodpecker attack on galls containing healthy *Eurosta* larvae compared to parasitized galls. The contents of over one thousand galled stems were determined by radiography (Mecum 1994; Mecum and Abrahamson, unpub. data), then we distributed them randomly into two field plots and monitored plots weekly for attempted (peck marks) and

Woodpecker Attack Drops in Open Fields

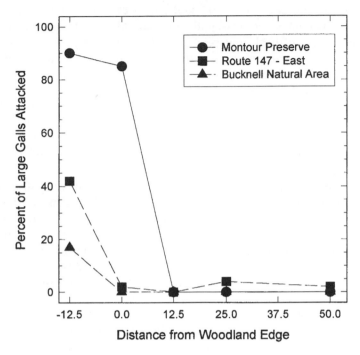

FIGURE 8.5. Woodpecker attack on *Eurosta* is most intense near the wooded field margins. In each of the three fields, artificial stands of galls were set up at five distances relative to the margin. Each plot was stocked with forty-five galled goldenrod stems, which were held in PVC pipes driven into the ground in a 3 × 15 grid (broad face parallel with the woodland margin) with 0.5 m spacing between pipes. Fifteen of the pipes held stems with small galls (<18 mm), another fifteen stems had large galls (>23 mm). The final fifteen pipes were set at a 45° angle, and stocked with stems bearing large galls. Stems were checked through February and March 1982, and when an attacked gall was found, it was replaced with another of the same size. (Richardson et al., unpub. data)

successful attack. The observed pattern of attack supports the hypothesis that exit tunnels cue the woodpecker to the presence of *Eurosta*. In a field study, Confer and Paicos (1985) examined naturally occurring galls for minor peck marks, which they too interpreted as a sign that the gall was examined and rejected.

The rejection rate of galls that contained healthy *Eurosta* larvae was only 12%, but 34% of parasitized galls were rejected. By this mechanism alone, bird-attacked galls would tend to be larger than average, since small galls are more likely to be parasitized. However, we think that rejection of parasitized galls is insufficient to explain the whole pattern of woodpecker attack. To determine if woodpeckers attack galls regardless of size but reject those that are parasitized or, alternatively, preferentially attack large galls, one needs to examine predation rates specifically on those galls that escaped parasitism. If birds cue in on the presence of *Eurosta* only, and not on gall size directly, the percent predation on unparasitized galls should be even across gall size. It is not. Figure 8.6 shows that smaller galls are less frequently preyed upon by birds even if parasitized galls are ignored. This indicates that even if small galls are more often rejected because they are more often parasitized, as Confer and Paicos (1985) suggested, small galls are less often targeted for probing in the first place. Because of this preferential attack, we can conclude that woodpeckers also act as a direct selective force on *Eurosta*'s gall size, but that it acts in the opposite direction to *E. gigantea*.

As a final question, if bird attack is more frequent on large galls, what is its behavioral basis? There is no answer at this time, but a few possibilities can be considered. The first is that small galls are below the level of visual acuity, and therefore go undetected. Although this could be true at far distances, it is unlikely for an insectivorous bird to be unable to detect an object about the size of its own head. Rather, the bird seems to actively discriminate against small galls. Birds seem to avoid fields where small galls abound just as much as they are attracted to fields with high densities of large galls (even after controlling for frequency of parasitized galls; see chapter 9). This avoidance could be the product of associative learning. After pecking and rejecting enough small parasitized galls, birds may narrow their foraging efforts to larger, more rewarding galls. This is an interesting topic for further research, but progress is hampered by the fact that bird attacks on galls are infrequent events from an ecologist's viewpoint. Although woodpeckers may be an important mortality source to *Eurosta*, *Eurosta* is only a minor compo-

Bird Attack on Galls Escaping Parasitism

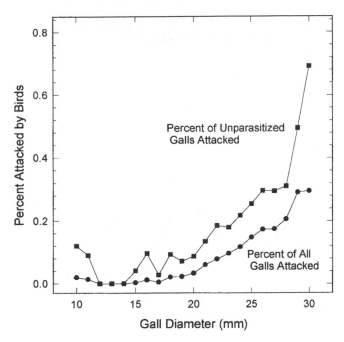

FIGURE 8.6. Gall size-dependent bird attack on *Eurosta*. The positive relationship between gall size and percent attack when only unparasitized galls are considered shows that birds respond to gall size per se, rather than to the presence of parasitoids. (Data from 1984 to 1988 pooled from long-term study) (Based on data from Weis, Abrahamson, and Andersen 1992)

nent in woodpecker diets. In over a hundred hours of observation, we and our associates have witnessed no more than five attacks.

8.6 PARASITOIDS, BIRDS, AND THE BALANCE OF SELECTION ON GALL SIZE

With respect to parasitoid attack, "fatter is fitter." However, the proclivity for downy woodpeckers and black-capped chickadees to attack the largest galls means the "fattest is not fittest"!

Gall Size Fitness Function

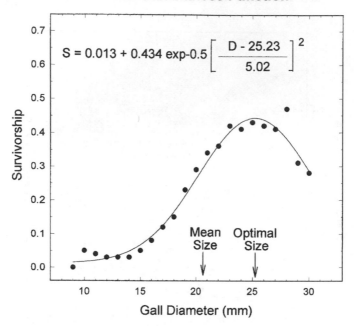

$$S = 0.013 + 0.434 \ \text{exp-}0.5 \left[\frac{D - 25.23}{5.02} \right]^2$$

FIGURE 8.7. Gall-size fitness function. (Data from 1984 to 1988 pooled from long-term study) (Based on data from Weis, Abrahamson, and Andersen 1992)

Preferential attack by birds can diminish, negate, or even reverse the upward selection pressure exerted by parasitoids. How do these opposing forces balance?

Overall, parasitoid and bird attack impose a selection regime on *Eurosta*'s gall size that has both stabilizing and directional components. This is illustrated in figure 8.7 which shows the fitness function for gall size, that is, the probability of a gallmaker surviving to the pupal stage as a function of gall diameter, from our long-term selection study (Weis, Abrahamson, and Andersen 1992). As the figure shows, larvae inducing the smallest galls (<10 mm) had only a 4% chance of surviving to the pupal stage. On the other hand, larvae in the largest galls (>30 mm) had a 27% chance. However, the highest survivorship, 42%, was seen among larvae in 25 mm galls. The higher survivorship of gall-makers in intermediate-sized galls demonstrates a stabilizing component of selection on gall size. This selection regime on *Eurosta* has been cited as one of the few documented cases of optimizing selection (Travis 1989), but variation in attack rates by parasitoids and birds cause stabilizing selection to be sporadic (see below). There is also a directional component to this selection regime. The "optimal" gall size of 25 mm is larger than the mean gall size of 20.9 mm, so that the net force of selection favors increased gall size. As we show in the next section, directional selection is very consistent across populations and generations.

8.7 VARIABLE SELECTION ON GALL SIZE

Although figure 8.7 illustrates the average selection regime that natural enemies impose on *Eurosta*'s gall size, the intensity and direction of selection are not constant. Spatial and temporal variation in the environment affects the ecological relationships among host plant, gallmaker, and natural enemies, and, as a result, induces variation in selection. The data from the extended study of natural-enemy attack afford the opportunity to explore the relationship between the population ecology of predator-prey and host-parasitoid interactions on the one hand, and the force of natural selection on the other.

Measuring the Strength of Selection on Gall Size

To determine how the wax and wane of enemy attack might influence selection, we must start with a way to measure selection's magnitude. Quantitative genetics offers an index that measures the intensity of directional selection on metric traits, such as gall diameter. This index can be easily understood if selection is viewed as a sampling process. Suppose a generation of a hypothetical species, with quantitative trait z, is born, matures, and then is exposed to a mortality factor that kills half the population. If an individual's value of z has no effect on the probability of death, then the mortality factor exerts no selection on z. The survivors will be a random sample from that generation, with

275

respect to trait z, and the frequency distribution of z (e.g., mean, variance) will be the same in the surviving "sample" as it was for the entire generation (fig. 8.8). If, however, vulnerability to the mortality factor changes with the value of z, selection occurs. When a high value of z reduces vulnerability, the surviving "sample" will not be randomly drawn for the generation, but instead will be biased toward higher values of z (fig. 8.8). The difference between the trait means for the whole generation and for the surviving "sample" is a measure of this bias, and as such is a measure of selection's magnitude. Formally, a statistic known as the directional selection differential can be defined as

$$S = \bar{z}^* - \bar{z}, \tag{8.1}$$

where z is the mean of trait z for the entire generation, $\bar{z}^*$ is the mean of the trait among the survivors. Positive values of S indicate selection for increased values of the trait, negative values indicate selection for a decrease, and a value of zero indicates no selection. When a trait is exposed to several episodes of selection, the total selection differential will be equal to the sum of the individual selection differentials at each episode (Arnold and Wade 1984). When the selection differential is divided by the phenotypic standard deviation, it is transformed into another statistic called the *selection intensity, i*. This converts the selection differential into standard deviation units, which then allows comparisons among populations with different mean phenotypes and among traits measured in different units. Because we will be making such comparisons, we will use the standardized index, i, for most purposes in this chapter.

Selection can also act on the variance in a trait. Disruptive selection causes an increase in variance, and stabilizing selection causes a decrease. A corresponding measure of sampling bias in variance can be used to quantify disruptive and stabilizing selection (fig. 8.8). Formally it is

$$j' = \frac{Var_{z^*} - Var_z + (\bar{z}^* - \bar{z})^2}{Var_z}, \tag{8.2}$$

where Var_z is the variance of trait z in the generation as a whole, Var_{z^*} is the variance among the survivors, and the factor $(\bar{z}^* - \bar{z})^2$ corrects for the variance reduction due to directional selection.

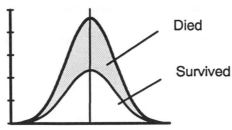

No Selection

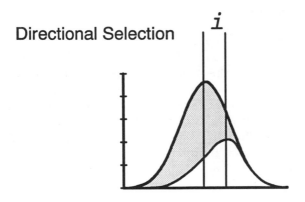

Directional Selection

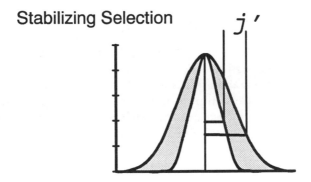

Stabilizing Selection

FIGURE 8.8. Meaning of the directional and stabilizing selection intensities. Each of these statistics is a measure of the sampling bias of survivor phenotype distributions caused by selection.

Division by Var_z standardizes j' as the proportional increase (disruptive) or decrease (stabilizing) in variance caused by selective factors.

These indices measure the magnitude of a selective force acting within a generation and not the amount of evolution it causes. However, S is an informative measure of the force of directional selection because its value equals the theoretically maximum evolutionary change in the phenotypic mean that can occur between the selected and the subsequent generation. This follows from the familiar quantitative genetic formula for predicting the response to selection

$$R = h^2S,$$

which states that the expected change in the phenotypic mean, R, is the product of S, the selection differential, and the heritability of the trait, h^2 (see Falconer 1989). As explained in chapter 6, the heritability of a trait is the proportion of its phenotypic variance that is caused by underlying additive genetic variance. If heritability were at its upper limit of 1.0, then evolution could occur at its maximum rate. As seen by the formula for predicted response, if h^2 were equal to 1.0, the population mean for the trait would be S units larger in the next generations. Remembering that the selection intensity is the standardized selection differential, i can be interpreted as the maximum evolutionary response to selection, expressed in standard deviation units.

The directional and variance selection intensities were measured in our long-term study of natural-enemy attack on the twenty central Pennsylvania *Eurosta* populations (Abrahamson et al. 1989; Weis, Abrahamson, and Andersen 1992; Sumerford, Abrahamson, and Weis, in prep.). Each spring, galls were systematically collected, measured, and dissected to determine the fate of each gallmaker and, if dead, the cause of mortality.

The Extent of Variation in Selection on Gall Size

As stated above, there is directional selection pressure on *Eurosta* to increase gall size. Figure 8.9 shows the directional selection intensity for sixteen populations across five generations (Abrahamson et al. 1989; Weis, Abrahamson, and Andersen

Net Selection Intensity on Gall Diameter

By Population and Generation

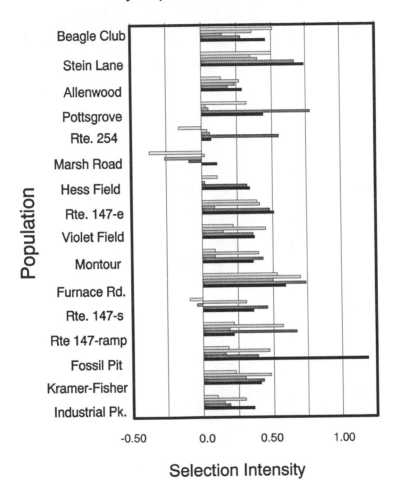

FIGURE 8.9. Directional selection intensities measured in the long-term selection study. The five bars for each population are (*top to bottom*) for the years 1983 to 1988. (Data from Abrahamson et al. 1989, and Weis, Abrahamson, and Andersen 1992)

Net Selection Intensity on Gall Diameter

By Population and Generation

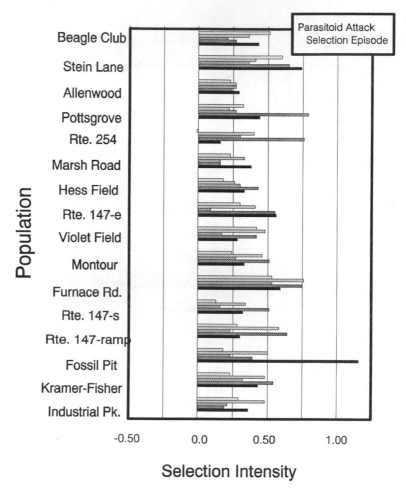

FIGURE 8.10. Directional selection intensities imposed by parasitoids only. As in figure 8.9.

1992). Clearly, in most populations and in most generations there is a positive selection intensity, with a mean of 0.32; in 63% of the eighty population-generations examined, selection intensity was significantly greater than zero (Weis, Abrahamson, and Andersen 1992). In only one case was it significantly less than zero. Though there is a strong tendency toward upward selection, the variability in selection's magnitude is reflected in the large coefficient of variation for i, which is 69%.

The variability in directional selection can be better understood by breaking selection down into its two distinct episodes— selection caused by parasitoid attack and selection caused by bird attack. Comparisons of figures 8.10 and 8.11 clearly show that the upward selection caused by parasitoids is far more consistent than downward selection by birds. The selection intensity for the parasitoid episode is significantly greater than zero in 88% of the population-generations (Weis, Abrahamson, and Andersen 1992), having a mean of 0.42 and a coefficient of variation of 44%. By contrast, for the bird episode of selection only 38% of the population-generations showed selection intensities different from zero (Weis, Abrahamson, and Andersen 1992), all in the negative direction. The mean selection intensity by birds was -0.07, with a coefficient of variation of 148%. Thus, the sporadic selection exerted by the generalist bird predators is seldom strong enough to negate or reverse selection by the specialist parasitoids, but it is a major source of the overall variation in directional selection.

The sporadic occurrence of bird attack also clouds the picture when considering the strength of stabilizing selection. When stabilizing selection is measured as j', nineteen of the sixty-four population-generations measured between 1984 and 1988 were significantly negative (i.e., selection is against variation) (Weis, Abrahamson, and Andersen 1992). Although this would seem to indicate that stabilizing selection is caused by the opposing directional selection forces exerted by parasitoids and birds, the relationship is not so clear-cut. In all of these nineteen cases parasitoids exerted significant upward selection, but birds imposed significantly downward selection only in ten cases, and in several no bird attacks were observed at all. Thus, there was a significant reduction in phenotypic variance not explained by opposing

281

Net Selection Intensity on Gall Diameter

By Population and Generation

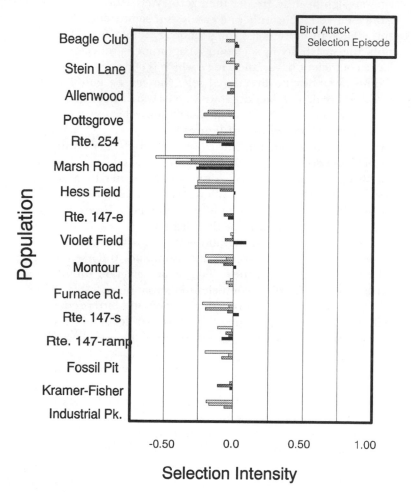

FIGURE 8.11. Directional selection intensities imposed by birds only. As in figure 8.9.

directional-selection episodes. This discrepancy points out the weakness of j': although there is a correction in its calculation to account for variance lost due to directional selection (equation. 8.2), the correction is complete only if the relationship between phenotype and fitness is linear. For *Eurosta*'s gall size this is obviously not the case (fig. 8.7). As an alternative way to establish the existence of stabilizing selection, we examined the fitness functions for sixty-four population-generations to see how often fitness peaks at an intermediate phenotype (Weis, Abrahamson, and Andersen 1992), using a method adapted by Schluter (1988). We found that in half the population-generations *Eurosta* survival was highest in an intermediate-sized gall. Unfortunately, the statistical procedures available to test for significance on an intermediate optimum are not very powerful (Mitchell-Olds and Shaw 1987; Schluter 1988). As a result it is difficult to say with certainty how often selection on gall size is stabilizing. The selection pressure from birds may only occasionally result in true stabilizing selection, but its effect may more often be to diminish the upward directional selection imposed by parasitoids (see below).

Is the variance in selection observed in this data set large or small? Since there are no other studies that have included so many populations and generations, comparisons are difficult. One frame of reference to answer that question is provided in John Endler's 1986 book, *Natural Selection in the Wild*. From the literature, Endler compiled values for 220 directional-selection intensities (due to mortality selection) measured on a variety of traits in seventeen different species. Since Endler was trying to assess the general magnitude of selection independent of direction, the absolute values of selection intensity were reported. Figure 8.12 shows that our values for selection on a single trait in a single species have a frequency distribution much like that for Endler's compilation. The means of the two distributions are very similar (0.32 for *Eurosta* versus 0.34 for Endler's compilation). By contrast, the variance of the *Eurosta* distribution is only about one-fourth that of the compilation. However, this is still an appreciable amount considering the large array of species and characters in the compilation. It will be interesting in future

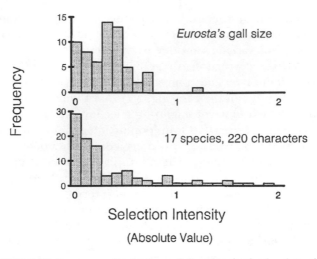

FIGURE 8.12. Frequency distributions of directional selection intensities observed in this study and of the compilation by Endler (1986) for 220 characters in seventeen species (mortality selection).

years, as data accumulate on other systems, to determine how the variability in selection regimes varies among and within species and characters.

Environmental Variation and Variable Selection

Although variation in selection has been observed over several spatial and temporal scales in a variety of other species (e.g., Johnston, Niles, and Rohwer 1972; Boag and Grant 1981; Kalisz 1986; Schluter and Smith 1986; Stewart and Schoen 1987; Gibbs 1988; Scheiner 1989; Kelly 1992) there has been little opportunity to explore the ecological details contributing to variability. In many of the documented cases of selection, the precise ecological causes of selection remain unknown (Endler 1986). In cases where they are known, it has been too difficult to collect data from a sufficient number of populations or across enough generations to make a robust analysis that links variability in ecological factors to variability in selection. In the *Solidago-Eurosta* natural enemy interaction, the agents of selection on gall size, at least during the gallmaker's larval stage, are well understood. It

284

also has been relatively simple to collect a multigeneration, multipopulation database on selection. Thus, we have the requisites for understanding the interplay of ecological factors in determining the magnitude of selection.

Several habitat-specific factors explain some of the population-level variation in selection strength. For instance, the Marsh Road population always showed selection intensities that were either negative or near zero (fig. 8.9). This population occupies a field surrounded by a sizable woodland, which is primary woodpecker habitat, and thus suffered consistently high rates of bird attack (Abrahamson et al. 1989; Weis, Abrahamson, and Andersen 1992). Generally, there is a correlation between bird-attack rate on a gallmaker population and the size of the adjacent woodlands (Weis and Kapelinski 1994), which agrees with the already-cited observations that galls close to a field's woody margin are more likely to be attacked by woodpeckers than those more distant (fig. 8.4) (Confer and Paicos 1986).

A second habitat-specific factor may explain why consistently high and positive selection intensities were found in the Furnace Road population (fig. 8.9). The steep slope of this field and drought-prone, shale-derived soil promote rapid precipitation runoff, leading to dry soil. Ramets growing in this field were smaller than average, and may not have been as responsive to the gall-inducing stimulus, as evidenced by the consistently small gall size in this population (Weis, Abrahamson, and Andersen 1992). The small gall size made most of the gallmakers in that population vulnerable to wasp attack, which can lead to stronger selection.

Weather conditions may explain the most striking incidence of temporal variation we observed in selection intensity, which was observed during the 1988–1989 generation (Sumerford, Abrahamson, and Weis, in prep.). The summer of 1988 was hit by a record drought with the result that mean gall size was under 15 mm, which is considerably less than the 20.5 to 21.5 mm typical of the other years. This would seem to argue for stronger upward directional selection for two reasons: the smaller mean gall size would leave a higher proportion of the gallmaker population vulnerable to *E. gigantea* attack, and the smaller size of the galls would make them less attractive to birds.

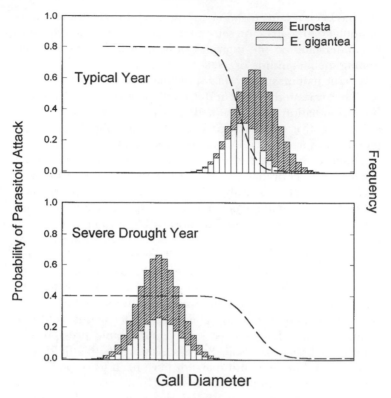

FIGURE 8.13. Schematic diagram showing why the reduction in gall size caused by drought would lead to relaxed selection. When all galls are vulnerable to *E. gigantea,* all have the same expected fitness, that is, there is no opportunity for selection.

However, none of the selection intensities measured in that generation were significantly different from zero, and the selection intensity averaged over all populations that year was slightly negative. The lack of significant selection intensities in the individual populations may in part be due to small sample size, necessitated by a population crash associated with the drought (Sumerford, Abrahamson, and Weis, in prep.). Bird attack was negligible that year, as predicted by its response to gall size, but no detectable upward selection was exerted by parasitoids.

Nonetheless, the lack of selection during the drought year is still consistent with patterns of size-dependent attack by *E. gigantea*.

This seemingly paradoxical result of no selection on gall size is easily explained by the fact that the drought not only reduced mean size, but also reduced the upper limit of the size range from 30 mm to only 23 mm. Nearly all galls fell into the penetrable size range (see fig. 8.4). Since all gallmakers were vulnerable, none were invulnerable (fig. 8.13), and those that escaped parasitism did so not because their galls thwarted parasitoid attack, but because they were never attacked in the first place.

Selection imposed by natural enemies will of course vary with the intensity of attack. Upward selection should tend to be strong when *E. gigantea* is abundant and weak when it is not. Downward selection will be evident only when birds attack in force, and absent when they are scarce. The various enemies in this small community interact, and they may have foraging responses that are sensitive to the density and size distribution of galls. Thus, these elements of enemy community and behavioral ecology may be strong determinants of the observed variation in selection intensity. These influences are the subject of the next chapter.

CHAPTER NINE

The Variable Biotic Environment
and Variable Selection

9.1 THE MANIFOLD EFFECTS OF ENVIRONMENTAL
VARIATION ON SELECTION

Ecologists well understand that the environment acts as an agent of selection and drives the process of adaptation. Trinidadian guppy populations that co-occur with predators are less conspicuously colored and mature at smaller sizes than those in predator-free streams (Endler 1980; Reznick, Bryga, and Endler 1990). Pitch pine populations vary in their degree of cone serotony in proportion to the frequency of fire they have experienced (Givnish 1981). Alleles for resistance to a new insecticide can spread rapidly and widely (Mallet 1989; Raymond et al. 1991). These cases highlight the way in which environment can determine the correlation between phenotype and fitness; small, dull guppies produce more offspring over their lifetime because predators are more successful at capturing large and brightly colored ones.

Selective environments are not static, as illustrated in the last chapter, and can vary in response to any number of factors. A particularly important cause of variable selection occurs when the selective environment changes with the makeup of the selected population itself. This can be the case, for instance, when intraspecific competition selects on resource use. High population density can favor new phenotypes able to capitalize on a resource not used by typical morphs. As natural selection causes the new morph to increase in density, however, the new resource can become limiting, thus eliminating the selective advantage (Mueller 1988). Frequency- and density-dependent selection occur when the strength of selection changes in response to the size or phenotypic composition of the selected population.

When predators, parasites, or pathogens are the agents of selection, their foraging responses or transmission dynamics can change with the size of the prey/host population, and with the proportion of defended and undefended individuals (Roughgarden 1979; Frank 1993; Anderson and May 1992), and this in turn leads to changes in the magnitude, or even the direction, of selection. Frequency-dependent selection, in the form of apostatic selection by predators, has been proposed as a factor that maintains genetic polymorphism (e.g., Greenwood 1984). Thus, an evolutionary response to some initial selective agent can cause a change in that agent, which in turn alters the future trajectory of evolution. Later in this chapter we will examine the data from our long-term-study selection to see if selection intensity on *Eurosta*'s gall size varies in a density- or frequency-dependent fashion.

However, there is a second way in which environment affects natural selection to alter evolutionary trajectories, and this is through phenotypic plasticity. Selection results in evolutionary change only if the favored phenotypes to some degree have different genotypes from the disfavored. Environmental effects on development can obscure an underlying correlation between phenotype and genotype. In the example of intraspecific competition above, imagine the fate of the new mutation that enables use of a new resource when it appears in two different developmental environments. If the mutation occurs in an environment where the new resource is abundant, its beneficial phenotypic effects will be evident, but if it occurs where the new resource is absent, the mutation will be silent at best. Competition is the agent of selection here, and it is equal between the two developmental environments. But in one of those environments the beneficial effects of the mutation are expressed at the phenotypic level while in the other they are not.

In chapter 6 we introduced the idea of phenotypic plasticity. This term from quantitative genetics applies to the situation where a genotype expresses different phenotypes in different environments. Within a population there can be variation among genotypes in their phenotypic sensitivity to the developmental environment. An aphid clone may show robust growth on one host plant but poor growth on another. A different clone from the same population could show the reverse pattern of growth on the two hosts (Moran 1981; Service 1984; Karban 1989; Via 1991; Pilson 1992). When a population occupies a habitat in which both host plants are present, its evolutionary trajectory can become complex, and over the short term can even appear maladaptive (Via and Lande 1985). Here we will consider how *Eurosta*'s phenotypic plasticity in gall size in response to plant variation (chapter 6) might change under selection imposed by natural enemies.

9.2 PLANT VARIABILITY AND SELECTION ON GALL SIZE

Although parasitoids and birds act as selective agents through their gall size-dependent attack rates, from *Eurosta*'s perspective the true target of selection is not gall size itself, but the inducing stimulus it applies to the plant. It is this stimulus which in turn influences final gall size (see fig. 6.1). One challenge is to understand how the stimulus might evolve given the fact that not all plants are equally reactive (see chapters 4 and 6). For instance, the putative necrotic response can lead to rapid gall-maker death on some clones (Anderson et al. 1989; How, Abrahamson, and Zivitz 1994). Even when this response is not triggered, there is considerable variation in how vigorously the plant responds to the gall-inducing stimulus. Greenhouse, garden, and field experiments demonstrating variation in susceptibility and reactivity have been outlined in previous chapters. In short, plant phenotype can be viewed as an element of the gall-maker's developmental environment (Weis 1992). As selection acts on the gall-inducing stimulus, it should favor genes that induce a gall phenotype that increases gallmaker survivorship despite variation in plant reactivity.

Gall Size, Plant Lag Times, and Phenotypic Plasticity

All species face input from the environment during development. Selection should favor developmental programs that preserve fitness given the array of environments that can be expected. As Schmalhausen (1949) pointed out, genotypes do not code for development of a given phenotype; instead, they code for development of a range of phenotypes given a range of environments. Patterns of phenotypic plasticity observed in any one character in any one species may or may not be adaptive, but over the long run selection will favor genotypes with plastic patterns that enhance fitness. The evolution of phenotypic plasticity has become the focus of much theoretical and empirical research (Via and Lande 1985; Parson 1987; Stearns 1988; de Jong 1990a,b; Scheiner and Lyman 1989; Gomulkiewicz and Kirkpatrick 1992; Gavrilets and Scheiner 1993a,b; Schlichting and Piglicci, 1993). It should be of keen interest to evolutionary ecologists to know how real selection regimes can influence the evolution of plasticity patterns; the data we have collected have given us the opportunity to explore the possibilities for *Eurosta*. In chapter 6, we showed that a plant feature we call "lag time" causes plasticity in *Eurosta*'s gall size, and that the degree of the gallmaker's plasticity is genetically variable. In chapter 8 we demonstrated the form and intensity of selection that enemies impose on gall size. By combining these two pieces of information we can evaluate the relative impact of the selection regime on the evolution of gall size per se, and the gallmaker's plasticity in gall size.

Plant lag time is the delay, in days, between oviposition and the first appearance of the gall. Although there is a general tendency for gall size to be smaller on plants with long lag times, greenhouse experiments showed that full-sib families of *Eurosta* differed in their response (Weis and Gorman 1990). In some families, the siblings on fast-reacting plants induced very large galls while the siblings on slow plants induced small ones. Other families showed no evidence of sensitivity to lag time, and in one family, larger galls were induced on slower plants (fig. 9.1). Since siblings are partial replicates of their parents' genotypes, these differences among families indicate genetic variance in sensitivity to lag time (see chapter 6; Via 1984; Falconer 1989). The sensitivity for each family was quantified as a reaction norm.

A reaction norm can be defined as a mathematical function that predicts the expected phenotype of a given genotype as a function of an environmental variable. Formally,

$$p_i = (g_i, e),\qquad (9.1)$$

Gall-Size Reaction Norms

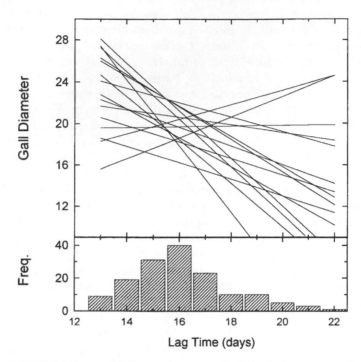

FIGURE 9.1. *Eurosta* gall-size reaction norms. Each line in the upper graph is the regression of gall size on plant lag time for one of the sixteen full-sib families (see chapter 6). The lower graph shows the frequency distribution of plant lag times. (From Weis and Gorman 1990, "Measuring canalizing selection on reaction norms: An exploration of the *Eurosta-Solidago* system," *Evolution* 44:820–831; with permission)

where p_i is the phenotype expected for individuals of genotype g_i at points along an environmental gradient e. In our experiment we found that the reaction norm for gall size over plant lag time could be adequately described by a linear function, Diameter $= B + M$ (Lag Time) where each family had its own unique values for the parameters B and M (fig. 9.1). Values of B and M were estimated by linear regression, where B is the intercept and M is the slope. In estimating the parameters, lag-time values for each gall were coded as deviations from mean lag time, so that the intercept, B, is the basic gall size—the size expected in the average environment. (The parameter B can also be interpreted as the expected gall size averaged across all environments encountered.) Thus, the reaction norm quantifies two things about the expected gall size for a family: first, the size of gall that can be expected on average; and second, the deviation away from that average that will be caused by sensitivity to lag time. The among-family variance in B thus is proportional to the genetic variance in gall size per se, and the among-family variance in M is proportional to the genetic variance in sensitivity of gall size to lag time. These variances are presented in chapter 6. Selection acting on gall size can act on either, or both, of these parameters.

Measuring Selection on Phenotypic Plasticity

Natural selection can act to change reaction norms. Measuring the strength of selection on reaction norms can be accomplished by measuring the deviation of estimated reaction norms from an optimal one. When they are linear functions, reaction norms can deviate from either the optimal elevation or the optimal slope, or both. To understand how we measured selection on *Eurosta*'s gall size reaction norms we present some pictorial equations in figure 9.2. In the figure we calculate the expected fitness of two hypothetical genotypes that differ in reaction norm slope and elevation.

In the first of the pictorial equations we transform the phenotypic reaction norms for genotype a and b into their fitness reaction norms. Genotype a is very sensitive to the environment—its phenotype changes dramatically along the environmental gradient, producing extreme phenotypes in extreme environments. But a produces the optimal phenotype at the center of the gradient. Thus the reaction norm has the optimal intercept, not the slope. By contrast, genotype b produces the same phenotype at all points along the gradient, but this phenotype falls short of the optimal; its reaction norm has the optimal slope, but a suboptimal intercept. The fitness reaction norm can be calculated for these two genotypes by substituting the

Calculating Expected Fitness for Reaction Norms

1) Transforming phenotypic reaction norm into fitness reaction norm.

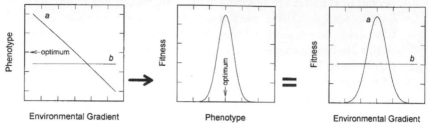

2) Weighting the fitness reaction norm by the frequency with which environments are occupied.

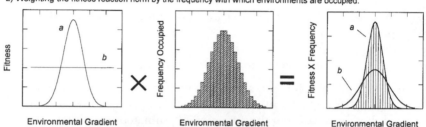

FIGURE 9.2. Pictorial equations to show how the expected fitness is calculated when there is phenotypic plasticity. In the upper equation, the phenotypic reaction norms (phenotype versus environment) are substituted into the phenotype term of the fitness function (fitness versus phenotype). In this substitution, the phenotype terms cancel and one is left with a fitness reaction norm (fitness versus environment). The fitness reaction norm predicts the expected fitness of the two genotypes at any given point on the environmental gradient. However, not all points have an equal probability of being occupied. Thus these fitness reaction norms must be weighted to account for uneven distribution across environments. The second pictorial equation illustrates this weighting. The expected fitness at each point along the environmental gradient is multiplied by the frequency with which that point is occupied (or in other words, fitness is integrated over the environmental frequency distribution). The sum of the weighted fitnesses (or the area under the weighted fitness curve) is the expected fitness, that is, the average fitness of individuals of the given genotype. In this example, genotype *a* has greater expected fitness because it is superior in the most commonly occupied points on the environmental gradient. Although genotype *b* is superior in extreme environments, these are rarely occupied, so its expected fitness is lower. (From Weis and Gorman 1990, "Measuring canalizing selection on reaction norms: An exploration of the *Eurosta-Solidago* system," *Evolution* 44:820–831; with permission)

reaction norm equations (phenotype versus environment) into the phenotype term of the fitness function (fitness versus phenotype). When this is done, the phenotype terms cancel and one is left with the fitness reaction norms (fitness versus environment). The fitness reaction norms reveal that genotype a is highly superior when development occurs somewhere along the middle third of the environmental gradient, but lethal at the environmental extremes. Although genotype b is mediocre at all points along the gradient, it is more fit than a across the lower and upper thirds of the gradient.

Not all points on the environmental gradient have an equal probability of being occupied. The relative fitness of the two genotypes will thus depend on which segments of the gradient the population is normally found. This is illustrated in the second pictorial equation. The expected fitness of the genotypes at each point along the environmental gradient is multiplied by the frequency with which that point is occupied. The sum of the weighted fitnesses (i.e., the area under the curves of the weighted fitness functions) is proportional to the genotype's overall expected fitness. In this example, the area under the curve for genotype a is greater than for b. This makes sense, since a has the optimal phenotype in the middle range of the environmental gradient, which is the region most frequently occupied. Genotype b is superior at the more extreme points on the gradient, but since these points are rarely occupied, the contribution from these points to total fitness is minimal. In this example, reaction-norm intercept is under stronger selection than reaction-norm slope.

We used this method to examine the strength of selection on slope and intercept of *Eurosta* gall-size reaction norms. We did not have different *Eurosta* genotypes, but we did have full-sib families. The expected fitness for each family was calculated by substituting their reaction-norm equations into the gall-size term for the fitness function in figure 8.6. The resulting fitness reaction norms were then weighted by the distribution of lag time in figure 9.1. To quantify the strength of selection on reaction-norm intercept and slope we used a variation on the fitness regression method (Price 1970; Lande and Arnold 1983). The relative strength of selection on the two reaction-norm parameters is proportional to the amount of variance in fitness that each explains. We used the regression model:

$$w_i = \alpha + \delta\, B_i + \gamma\, M_i, \qquad (9.2)$$

where w_i is the relative expected fitness for an individual of family i (that is, the family mean survivorship against enemy attack divided by the population mean survivorship), B_i and M_i are the observed reaction-norm parameters for family i, and α is a constant. The terms δ and γ are the multiple regression coefficients that relate the contributions of intercept and slope, respectively, to fitness. Large, positive values for δ or γ indicate that selection favors an increase in the parameter while a coefficient of zero indicates no selection. Since reaction-norm slope is a measure of phenotypic plasticity, selection on slope is selection on plasticity.

This analysis showed that the selection regime imposed by natural enemies acts strongly on reaction-norm intercept, but weakly on slope (fig. 9.3). Those families that had high reaction-norm intercepts or, in other words, produced large galls averaged over all plants, could expect a greater chance of survival, which makes sense in light of what we know about the persistent

Selection on Reaction-Norm Parameters

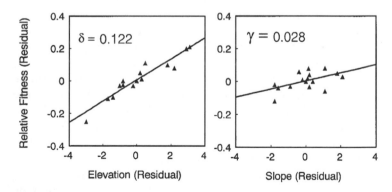

FIGURE 9.3. Partial regression plots of expected fitness over reaction-norm intercept and reaction-norm slope. Variation in reaction-norm intercept explained more of the among-family variance in expected fitness than did variance in slope.

upward directional selection on gall size. However, the largest galls in the experiment were produced by families with the greatest sensitivity to plant lag time (fig. 9.1); did this sensitivity lead to increased fitness? Apparently not. Although weak, selection was against sensitivity, so that the fitness gains on fast plants were inadequate to compensate for losses on slow plants. Note that although the regression coefficient for reaction-norm slope is positive, selection is against sensitivity—the positive coefficient indicates that families with less-negative slopes had greater fitness.

The adverse fitness consequences of sensitivity to plant lag time indicate a canalizing selection regime on gall size. That is, selection favors those genotypes that produce the same phenotype regardless of plant lag time. Although this seems to indicate that selection favors rigidity in the developmental process, in fact the reverse is true. Remember that the true target of selection is the gall-inducing stimulus. Production of a consistent gall size when plants are variable requires that the stimulus level be adjusted to suit the plant actually occupied. Thus the canalizing selection on gall size should favor those insect genotypes with plastic stimulus phenotypes.

Note that although selection is strongest on reaction-norm intercept, that is, the expected gall size averaged over available plants, the genetic variance in intercept is small (chapter 6). This indicates that only a slow evolutionary response of reaction-norm intercept is possible and easily stopped should there be countervailing selection pressures. On the other hand, selection on plasticity is weak, even though the genetic variance in plasticity is large. This pattern is in agreement with predictions that genetic variance in traits closely associated with fitness should show less variance than those with weak fitness effects (Mousseu and Roff 1987; but cf. Price and Schluter 1991).

9.3 COMMUNITY ECOLOGY ON THE NATURAL ENEMIES AND VARIABLE SELECTION, I

The ecological dynamics of a plant-herbivore-natural enemy system can lead to a dynamic of natural selection. Consider for the sake of argument a case in which a herbivore has two host-plant types available, one which is nutritionally superior but attractive to the herbivore's natural enemies. The other is inferior food, but it offers refuge from attack. During generations when enemies are rampant, selection can favor herbivore preference for refuge plants even if they are nutritionally inferior. In other generations, when natural enemies are in decline, the direction of selection can reverse, favoring preference for a more nutritious but less protective host. Add on top of this basic scenario a density-dependent component to enemy attack. During generations when the herbivore population is large, a correspondingly large number will occupy non-refuge plants, and these in turn will suffer a disproportionately greater mortality than they would during low-density generations. Positive density-dependent attack in this case shifts the selective balance toward preference for refuge. But to add another layer of complexity, enemies may compete to prey on the herbivore. Each enemy can have an individualistic response to herbivore density and to the densitiy of the others. Thus, the dynamics of the natural enemy community can induce fluctuations in the strength and direction of selection on the herbivore's host-use characters.

In this section we will deal with the structure of the enemy component-community centered on the *Eurosta* gall and how the attributes of the various enemies cause them to affect selection through the network of interactions they experience. Our approach is admittedly ad hoc and, rather than testing general hypotheses, looks at the specifics of the system to see if more general hypotheses suggest themselves. We used path analysis to see how the intensity of directional selection on gall size varies due to the interactions of the enemies with one another and with gall size and gallmaker density (Weis and Kapelinski 1994).

Path analysis was invented by Sewall Wright (see Wright 1968) and is a generalized form of multiple regression (Li 1975). A path analysis begins with a path diagram (see figs. 4.7 and 9.4), which indicates the supposed causal relationships (arrows) among a series of variables (boxes). In a path model there are independent variables (the causal factors) and dependent variables (the response variables).

When a change in one factor is thought to cause a change in

another, a single-headed arrow is drawn pointing from cause to effect. As is often the case, changes in several different independent variables can cause change in the dependent one, and so arrows from several different causal variables may lead to a response variable. In turn, changes in a causal variable can be caused by change in additional background variables. In such a situation one can follow a chain of causality from the background causal variable through the intermediate causal variable and onto the response variable. When there are several background variables they may be correlated with one another; this correlation among background variables is represented by the curved, double-headed arrows.

In our path model for variable selection (fig. 9.4), one can trace our hypothesized relationships of ecological factors measured in the populations in each generation to the resulting directional selection intensity. The basic causal relationships are from *E. gigantea* and bird attack rates to selection intensity. We supposed that when *E. gigantea* attack rates were high, *Eurosta* mortality in small galls would also be high, giving a selective advantage to large gall size. Similarly, when bird attack was high, the selective advantage would go to larvae-inducing small galls. Thus, there should be a direct (positive) relationship between *E. gigantea* attack and the selection intensity value, and an inverse (negative) relationship between bird attack and the selection intensity value (i.e., when bird attack is high, selection on gall size is more downward). However, the attack rates of these two enemies may themselves be influenced by a number of ecological factors. For instance, *E. gigantea* attack may be higher when mean gall size is low because a greater proportion of the gallmaker population will be vulnerable to attack. When mean gall size is high, more galls may be of an attractive size to birds, and predation may increase. Thus, population mean gall size may affect the intensity of selection indirectly through its effects on the attack rates by *E. gigantea* and by birds. Gall density (number of galls/m²) was hypothesized also to act indirectly on selection, through density-dependent attack by *E. gigantea* and birds. An additional class of interaction was included in the model, based on the observations of Schlichter (1978) and Cane and Kurczewski (1976) that birds find *Mordellistena* and *E. obtusiventris*

299

Ecological Factors Causing Variable Selection:
"Conventional" Model

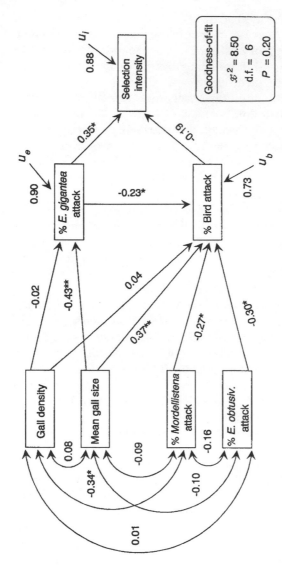

FIGURE 9.4. Path model showing the causal influences of enemy-attack rates, gallmaker population density, and mean gall size on the intensity of natural selection on gall diameter. Paths are signified by arrows. A single-headed arrow indicates that a change in the variable at the arrow's base (independent variable) will cause a change in the variable at its head (dependent variable). The coefficient associated with the causal path is the predicted change in the dependent variable, in standard deviations, by a one standard-deviation change in the independent variable. Variables connected by double-headed, curved arrows may covary through the influence of unmeasured background variables; the coefficient associated with curved arrows is the correlation between the two variables. A nonsignificant goodness-of-fit χ^2 indicates that the correlations among all variables predicted by the path model do not differ from those observed in the data. (*$P < 0.05$; **$P < 0.01$) (From Weis and Kapelinski 1994, "Variable selection on *Eurosta*'s gall size. II: A path analysis of the ecological factors behind selection," *Evolution* 48:734–745;

distasteful. In populations where parasitism rates are high, frequent encounter with parasitized galls could discourage birds from further attack. A similar discouraging effect could arise when birds encounter galls parasitized by *E. gigantea*. Although it has not been suggested that this species is distasteful, its biomass is less than that of *Eurosta* (Stinner and Abrahamson 1979) and thus offers scant reward for the bird's effort. For these reasons, increased attack rates by any of the parasitoid species could increase the selection-intensity value, for even if they do not contribute directly to upward selection, by decreasing bird attack they can curtail downward selection.

With the path diagram in place, path coefficients can be calculated to quantify the strength of each effect on the response variable. Although an explanation of the theory of path coefficients is beyond the scope of this book, a brief description of their calculation and meaning is in order. More complete, and very readable, explanations of path analysis are given by C. C. Li (1975) and R. J. Mitchell (1992, 1993). Basically, a path coefficient is a partial regression coefficient; this means that a path coefficient is a statistical estimate of the change expected in the response variable for a given change in the causal variable, holding all other variables constant. Path coefficients are standardized—they can be read as the change in the response, in standard deviation units, for a one-standard-deviation change in the cause. This standardization scales path coefficients from -1.0 to $+1.0$ (like the correlation coefficient), and has the additional convenience of putting the influences of different variables into the same units, which then facilitates conclusions on their relative strengths.

When the path of causality goes from a background variable through one or more intermediate variables and then on to the response variable, the net effect of that background variable on the response is estimated as a compound path coefficient. To construct a compound coefficient, one identifies the path along the arrows leading from the background to the response and then multiplies together all the coefficients along the way. If more than one path connects background and response, then add together the compound coefficients for each path to get the entire effect coefficient. Data for the years 1984 through 1987

301

from our extended field study of enemy attack rates were used to estimate the path coefficients.

Path coefficients estimated for the model in figure 9.4 showed that most of the hypothesized relationships were supported. For instance, increasing *E. gigantea* attack rate by a standard deviation increased the directional selection intensity by 0.35 standard deviations. A decrease in selection intensity was seen as bird attack rose—meaning that as bird attack went up, the selective advantage to large gall size went down.

The effects of population mean gall size as a background variable were as expected. An increase in gall size had a negative effect on *E. gigantea* and a positive effect on bird attack. When carried through to the selection differential, the net effect of mean gall size on selection was negative (table 9.1), meaning the larger the galls in a population, the weaker the selection for increased size. Unlike mean gall size, gall density had no effect on attack rates; this agrees with findings by Walton (1988) and Cappuccino (1991) that *Eurosta* mortality is density-independent on several spatial scales. Thus there was no direct density-dependent effect on selection.

The incidence of parasitoids had an adverse effect on bird attack. A higher percentage of parasitism by *E. obtusiventris* and *Mordellistena* significantly reduced bird-attack rate. The effect of *E. gigantea* was also negative. However, the total effect of parasitoid attack on selection, mediated through bird attack, was small (table 9.1).

This path model indicates that selection will be most strongly influenced by the direct effects of *E. gigantea* and bird attack, although bird-attack rate can be modified by high parasitism levels. The role of mean gall size on these attack rates was also confirmed; however, the true response by *E. gigantea* and birds to the distribution of gall sizes as they forage through a field may be more complex than suggested by this model.

9.4 COMMUNITY ECOLOGY OF THE NATURAL ENEMIES AND VARIABLE SELECTION, II

The model presented above assumed simple responses of *E. gigantea* and birds to the distribution of gall sizes, and can be

TABLE 9.1.   Effect coefficients of ecological factors on directional selection intensity for the model in figure 9.4.

Ecological Factor	Direct	Indirect E. gigantea	Birds	Total
Eurytoma gigantea attack (%)	0.348	—	0.044	0.392
Bird attack (%)	−0.190	—	—	−0.190
Gall population density (per m²)	—	−0.007	−0.008	−0.016
Population mean gall diameter (mm)	—	−0.15	−0.091	−0.241
Mordellistena attack (%)	—	—	0.051	0.051
Eurytoma obtusiventris attack (%)	—	—	0.057	0.057

Note: Indirect effects by background factors are separated into those effects acting through Eurytoma gigantea parasitism rate and those acting through bird predation rate.

called the "conventional model." However, we further tested the idea that frequency-dependent attack rates by *E. gigantea* and by birds, caused by what we call "false target effects," operate in this system. Such effects would not be detectable in the "conventional model" analysis.

The literature on frequency-dependent predation is large but focuses strongly on apostatic selection. Under this type of selection, the rare morph of a polymorphic prey species is favored because predators tend to specialize on the more abundant food type. In the purest form of apostatic selection, the basic vulnerability of the alternate prey types is equal, and it is only the relative abundance of one morph to the other that determines its expected fitness when exposed to a particular predator species (Greenwood 1984). Similar selective regimes have been posited in models of plant-pathogen interactions (Frank 1993). Apostatic selection is of particular importance because it can maintain genetic polymorphism in the host/prey species. However, it is only one form of frequency-dependent selection that predators, parasitoids, and pathogens can impose. Another form can be imposed when predator/parasitoid foraging effort or efficiency is diminished by "false target" effects.

The term "false target" comes from the operations research literature (Koopman 1980) and refers to objects that resemble the "true targets" of a search. For instance, searching efficiency for submarines will depend on the frequency with which submarine-like signals from false targets are encountered, and how long it takes to determine they are not submarines. In predator-prey interactions, false-target effects can arise when there is a difference in the basic vulnerability of two morphs but some difficulty in distinguishing between them. The undesirable prey morph will always hold a selective advantage. However, the intensity of selection against the desirable morph changes with its frequency relative to the undesirable. An interspecific false-target effect can operate in Batesian mimicry. Suppose a distasteful species is poorly mimicked by another. When mimics are very common, relative to models, predators will encounter the poor mimic often enough to learn how to distinguish it from the distasteful model. However, if models are common and mimics are rare, there will be fewer opportunities for predators to learn the distinction between the two prey species and so even a poor mimic will gain protection. Analogous situations can occur when the two prey classes are alternate genotypes of the same species; the selective disadvantage of vulnerability can decline when the invulnerable genotype becomes common.

False Targets and Functional Responses

For purposes of expediency, the *Eurosta* population can be divided into two parts. Those with galls 21 mm or greater in diameter are for practical purposes invulnerable to attack by *E. gigantea* (fig. 8.4; Weis 1993a). Smaller galls are vulnerable to penetration by *E. gigantea*. Thus 21 mm can be used as a dividing point for "large" and "small" galls. As shown in the last chapter, *E. gigantea* will examine and probe galls much too large for them to penetrate. The only method they seem to use to distinguish large, invulnerable galls from small, suitable ones is exhaustive probing (Weis 1993a). In fact, females will spend on average over 30 minutes probing large galls before giving up in failure, while fewer than 10 minutes are required to successfully parasitize a host in a small gall. This raised the possibility that large, invulnerable galls acted as "false targets" to foraging *E. gigantea* females. The failure rapidly to discriminate between vulnerable

and invulnerable galls would impose a penalty on foraging efficiency when large galls were abundant.

The "false target" effect of large galls was demonstrated in a
laboratory experiment where females had to search for small
galls either in the presence or absence of large galls. In the
"false target" treatment, female wasps were enclosed in cages
that contained eight goldenrod stems bearing small galls and
eight bearing large ones. In the controls, females were offered
eight stems with small galls and eight ungalled stems. The density of small, vulnerable galls was constant between treatments,
but the number of "false targets" was varied. After 24 hours,
galls were dissected to look for newly laid *E. gigantea* eggs. An
average of 2.5 small galls were parasitized when presented with
large galls (no large galls parasitized), but 4.0 were parasitized
when presented with ungalled stems (Weis, Abrahamson, and
McCrea 1985). This showed that large galls distract females.
Thus gallmakers inducing small galls can have a higher expectation of survival if they co-occur with large ones than if they
do not.

These observations suggest that the functional response of
E. gigantea is influenced not only by host density but by the relative frequencies of poorly defended and well-defended hosts.
The extent of the false-target effect will depend on the time required to discriminate between true and false targets. To explore the importance of discrimination time, consider the following adaptation of Holling's disc equation (Holling 1959):

$$N_{te} = \frac{aT\{vN + (1 - v)N\}}{1 + aT_h vN + aT_d(1 - v)N} \tag{9.3}$$

where N_{te} is the number of true targets encountered, N is the
total number of targets available, v is the fraction of targets that
are "true," a is the encounter rate, T is the total time for foraging, T_h is the handling time of true targets, and T_d is the time
required to distinguish false from true targets. Calculations that
evaluated the effect of discrimination time on the encounter
rates with true targets are illustrated in figure 9.5. When discrimination is instantaneous ($T_d = 0$), the effect of increasing
the number of false targets is merely the same as reducing the

number of true targets (fig. 9.5a). However, when discrimination time is larger ($T_d = \frac{1}{2}T_h$), encounter rates decrease further (fig. 9.5b). When it takes twice as long to discriminate false targets as to handle true ones, encounter rates plummet (fig. 9.5c). This model suggests that when the vulnerable and invulnerable prey types are difficult to distinguish, the survivorship of the vulnerable type will depend on its frequency. When rare enough, the vulnerable type's survivorship will converge with the invulnerable's survivorship.

A second kind of false-target effect may influence bird-attack rates. From the bird's perspective, small galls have a lower chance of reward than large ones. They are more likely to contain *E. gigantea* larvae, and are more likely to be empty altogether, due to early larval death (chapter 2). Through repeated encounters, birds may learn to associate small galls with a low probability of reward, and large galls with a higher one. On subsequent foraging trips, birds could be more prone to abandon fields where small galls are abundant. Unrewarding attacks on small galls by young-of-the-year birds could potentially establish behavioral patterns that could last into the next winter and beyond. In this scenario, discrimination of small from large galls could be instantaneous, but the searching intensity could decline with the frequency of encounters with small galls.

Frequency Dependence in the "False Target" Path Model

To see if there are frequency-dependent components to selection, we restructured the data set to analyze a modified path model. This "false target" model (fig. 9.6) treated *Eurosta* as if it comprised two populations, those occupying galls <21 mm and those occupying galls 21 mm or greater. The "mean gall size" and "gall population density" variables in the first path model were replaced with "density of small galls" and "density of large galls." The prediction of the false target model for *E. gigantea* is that attack rate on small, vulnerable galls should decrease as the density of large, distracting galls increases. Similarly, for bird attack, we examined variance in the rate of attack on large galls, when the density of large galls was held constant statistically, in response to variance in small gall density.

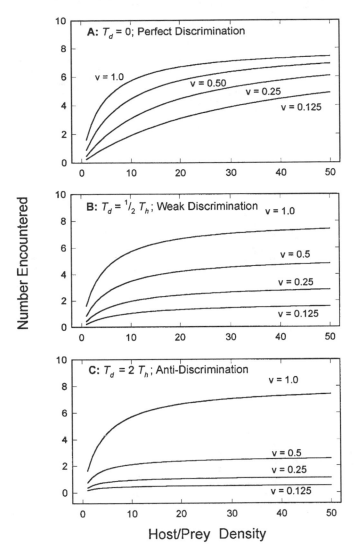

FIGURE 9.5. Functional responses in the presence of "false targets." Variables are as follows: v = proportion of prey that is vulnerable; T_h = handling time of vulnerable prey; T_d = time required to discriminate invulnerable prey. (A) The predator discriminates invulnerable prey instantly, and so encounter rates drop simply by dilution of the prey pool. (B) Discrimination time is half that of handling time. The extra time taken to examine invulnerable prey reduces time to look for vulnerable prey. Decreased encounter rates reflect both dilution and reduced efficiency. (C) Discrimination time exceeds handling time, and so encounter rates plummet.

Ecological Factors Causing Variable Selection: "False Target" Model

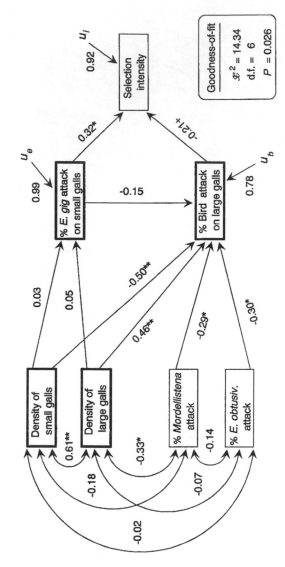

FIGURE 9.6. "False Target" path model showing the causal influence of population densities of small (<21 mm) and large (>21 mm) galls and of enemy-attack rates on directional selection intensity. Variables in the thickened boxes have been modified from the conventional model. (For interpretation of arrows and coefficients, see fig. 9.4.) (From Weis and Kapelinski 1994, "Variable selection on *Eurosta*'s gall size. II: A path analysis of the ecological factors behind selection," *Evolution* 48:734–745; with permission)

Results of the "false target" model analysis suggested a frequency-dependent component of selection for bird attack, but not for *E. gigantea* attack. A significantly negative path coefficient was found from small gall density to bird attack. Since the coefficient for this path quantifies the effect of small gall density on bird-attack rate, holding the density of parasitoid-attacked galls constant, we concluded that this path reflects the response of the birds to small galls *per se*, independent of what they contain. This then is consistent with the possibility that birds learn and remember to avoid fields dense with small, unrewarding galls. At the same time, there is direct density dependence in the attack of birds on large galls that are, on average, more rewarding.

The lack of a false-target effect for *E. gigantea* could be explained if in the field females learned to rapidly discriminate large from small galls. However, other factors could obscure the effects of poor discrimination on parasitization rates. Wasps may not be so limited by time as by egg production, and so the extra 20 minutes spent on invulnerable galls may be of small consequence over their life span, which is over 2 weeks in the laboratory. In addition, numerical responses of wasps to areas of high abundance could compensate for the lost efficiency per individual. Unfortunately, monitoring densities of foraging adult parasitoids is a perennial methodological problem for students of parasitoid biology.

The relative importance of the density- and frequency-dependent components of selection imposed by *E. gigantea* and birds can be evaluated by examining the compound path coefficients from small and large gall density to selection intensity through *E. gigantea* and bird-attack rates. Roughly, the importance of density-dependent selection by each enemy will be the sum of the two compound paths passing from the density variable to the selection intensity (see table 9.2). In the case of both enemies, the sum of the two paths is close to zero: for *E. gigantea* it is $0.001 + 0.016 = 0.017$; for birds it is $(0.115) + (-0.097) = 0.018$. This verifies the overall absence of density-dependent selection seen in the original path model. To evaluate the degree of frequency dependence in selection imposed by the two enemies, consider the absolute value of the difference between the paths coming from small and large gall density: for *E. gigan-*

TABLE 9.2. Effect coefficients of ecological factors on directional selection intensity for the "false target" path model in figure 9.6.

Ecological Factor	Direct	Indirect		Total
		E. gigantea	Birds	
E. gigantea attack on small galls (%)	0.320	—	0.031	0.351
Bird attack on large galls (%)	−0.209	—	—	−0.209
Population density, small galls	—	−0.001	−0.115	−0.016
Population density, large galls	—	−0.016	−0.097	−0.081
Mordellistena attack (%)	—	—	0.061	0.061
Eurytoma obtusiventris attack (%)	—	—	0.062	0.062

Note: Indirect effects by background factors are separated into those effects acting through *Eurytoma gigantea* parasitism rate and those acting through bird predation rate.

tea this is $|0.001 - 0.016| = 0.015$; for birds the difference is $|(0.115) - (-0.097)| = 0.212$. Thus the frequency-dependent component is 15-fold greater in birds than in the parasitoid. Nonetheless, the frequency-dependent component accounts for only about 4% of the variance in selection intensity.

The long-term field study of natural enemy attack has shown that when selection occurs, it is nearly always in the upward direction. Our path analysis of the ecological basis for variable selection has shown that fluctuations in both the phenotypic distribution of gall sizes and in the occurrence of *Eurytoma obtusiventris* and *Mordellistena* can alter the degree to which bird attack can counter the persistent upward selection pressure exerted by *Eurytoma gigantea*. Although density- and frequency-dependent effects were not important in this case, we nonetheless feel that we have developed ideas along these lines that will be helpful elsewhere.

We wish to make a final point on false-target effects. Recall that frequency-dependent selection is an important mechanism for the maintenance of genetic variance. However, frequency dependence through a false-target effect is unlikely to maintain genetic variance on its own. This is because the vulnerable phenotypic classes will always be vulnerable, and thus always at a selective disadvantage. However, this kind of frequency-dependent predation may act to maintain genetic variance when coupled with other processes such as mutation. For instance, the continued action of selection will deplete genetic variance. Mutation will restore some of that variation so that an equilibrium level of genetic variance will be reached when mutation and selection are balanced (Lande 1979). A false-target effect can change this balance because it causes a relaxation of selection as the vulnerable types become rare (fig. 9.7). This relaxation of selection can then shift the equilibrium genetic variance to a higher level. The false-target effect could also contribute to maintenance of a conventional genetic polymorphism. Suppose the false-target morph is strongly defended but pays a small fitness penalty for producing that defense. As the true target morph declines to low frequency by selective predation, the fitness penalty it pays for a lack of defense could fall to the same small value as the fitness penalty paid by the false target for its defense. Allele frequencies would achieve equilibrium. These possibilities bear further attention.

9.5 IS AN EVOLUTIONARY RESPONSE TO SELECTION EVIDENT?

This section examines evidence from our long-term field study for an evolutionary change in gall size. There are several well-documented instances of evolutionary change in natural populations caused by natural selection (see Endler 1986). Unintentional experiments, caused by human disturbance, that demonstrate selection include industrial melanism (Kettlewell 1973; Berry 1990), pesticide resistance in insects (Wood and Bishop 1981; Mallet 1989; Raymond et al. 1991), heavy-metal tolerance in plants (Antonovics, Bradshaw, and Turner 1971) and introduction of novel host plants on insect mouth parts (Carroll and Boyd 1992). Grant and Grant (1993 and references therein) report that a rare fluctuation in climate on the Galapagos Islands changed the distribution of seed types available for *Geospiza fortis* consumption, which in turn selected for narrower beaks. Elegant field experiments have demonstrated changes in coloration (Endler 1980) and life history characters (Reznick and Endler 1982; Reznick, Bryga, and Endler 1990) in

311

False-Target Effect and Frequency Dependence

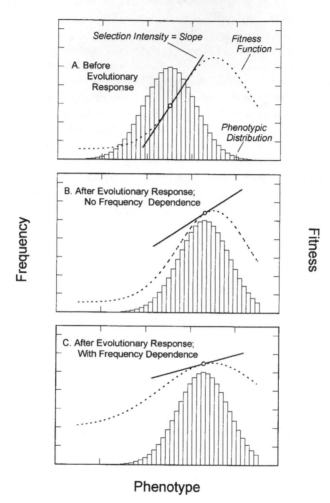

FIGURE 9.7. Changes in fitness function and selection intensity with an evolutionary response to frequency-dependent selection caused by a false-target effect. Phenotype is size. Assume that small individuals are vulnerable to predation and large ones are not. (A) Before evolution has occurred, the population mean phenotype lies away from the optimal phenotype. The intensity of selection is approximated by the slope of the fitness function at

populations of Trinidadian guppies when exposed to several generations of predation. However, these experiments, both intentional and unintentional, are rather extreme examples compared to the "daily and hourly scrutinizing" of selection envisioned by Darwin (1859, p. 133). Experiments have shown more gradual evolutionary responses to ongoing natural selection pressures under laboratory conditions (e.g., Travisano et al. 1995), but what about populations in natural habitats?

Long-term field studies that have looked for selection responses caused by ongoing selection pressures have been less conclusive. One of the most famous examples, and most debated (Provine 1986), is the change in coloration allele frequencies in the Cothill (near Oxford) population of the scarlet tiger moth (*Panaxia dominula*). Fisher and Ford (1947) suggested that after a purported increase in frequency of the dark morph between 1921 and 1936, selection caused a steady decline in frequency of the dark allele over the next 20 years (Jones 1989). Incomplete knowledge of both population size (Wright 1978) and the selective factors involved (Jones 1989) cloud this interpretation, although the steady decline in frequency of the dark allele is consistent with a persistent 10% disadvantage (Wright 1978).

A different kind of result from long-term studies of natural populations has been observed for clutch size in a number of bird species (e.g., Boyce and Perrins 1987; Rockwell, Findlay, and Cooke 1987; Nur 1988; Price and Liou 1989). In each of the

---

the population mean phenotype (see Lande and Arnold 1983). (B) Natural selection has caused the phenotypic distribution to move closer to the optimum. Because the mean phenotype and the optimal phenotype are closer, selection intensity is reduced. However, the absolute fitness for a given phenotype is still the same as it was in (A). (C) Natural selection has caused the phenotypic distribution to move closer to the optimum by the same amount as in (B). However, in this case there is a false-target effect. There are now more large, invulnerable individuals that distract the predator from the small vulnerable ones. As a result, the small ones enjoy an increase in survival. This flattens the fitness function. Thus the intensity of selection is less than in (B) even though the population mean phenotype lies the same distance from the optimum.

313

studies cited, directional selection has consistently favored increase in clutch size over a number of years. Yet no evolutionary increases in clutch size were observed, even though in each case clutch size was shown to be a heritable character. Various kinds of environmental influences on the selection process have been offered to explain stasis in clutch size in the face of upward selection (see below).

Looking for an Evolutionary Response

In this investigation of *Eurosta*'s gall size, we took a similar approach to the clutch-size studies, by measuring both selection intensity and changes in phenotypic distribution over time. If evolution is occurring, one would then expect a general increase in mean gall size over the course of the study. Although the mean size varied significantly among years and populations (fig. 9.8), there is no obvious pattern of size increase from generation to generation when the population means are examined alone.

However, because multiple *Eurosta* populations were studied, we had the opportunity to apply a second, potentially more powerful test for evolutionary change. Using each population as a unit of analysis, we could test for a positive correlation between the intensity of selection imposed on gall size and the magnitude of the ensuing change in gall size.

To address the question of whether gall size evolves at a detectable rate in response to selection, we asked whether the change in mean diameter between year t and year $t + 1$ was proportional to the magnitude of phenotypic selection in year t. This logic follows from the standard quantitative genetic model for response to selection in a single population, $R = h^2 S$, where S is the selection differential, R is the response to selection, measured as the difference between the population mean of the next generation and that of the selected generation, and h^2 is the heritability of the selected trait (see chapter 6). This model predicts that for a given level of heritability, the trait mean should change in direct proportion to the selection differential imposed. One key assumption of this model is that the target trait is not genetically correlated with any other trait under selection (Lande

Mean Gall Diameter across Four Generations

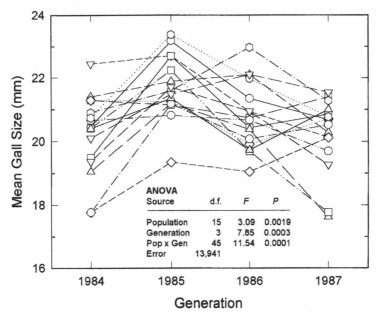

FIGURE 9.8. Population mean gall sizes over four generations from sixteen fields in central Pennsylvania. Although measured selection pressure is being applied, no upward trend in gall size is evident.

1979); when this assumption is violated, selection on the correlated trait can alter, and in some cases negate, an evolutionary response in the trait. Note from this equation that if both the selection differential and response are known, the heritability can be estimated (i.e., the realized heritability [Falconer 1989]).

When replicated populations are considered, the standard selection response model can be modified to

$$\Delta \bar{z}_i = E + h^2 S_i + \epsilon_i, \tag{9.4}$$

where $\Delta \bar{z}_i$ is the change in mean phenotype in the ith population between generations t and $t + n$. This change in mean is attributed to both evolutionary and environmental causes (see

315

Cooke et al. 1990). The evolutionary change in phenotype is the product of h^2 and S_i, the cumulative selection differential over the generations. Other changes in the mean phenotype can be caused by environmental factors that affect all populations equally, E, and environmental factors specific to each population, ϵ_i. Note that this equation is a standard linear regression model in which h^2 is the slope, E is the y-intercept, and ϵ_i is error. Thus, if the selection differentials and the changes in phenotypic mean are known for replicate populations, a statistical test for evolutionary change can be performed.

When applying this model for replicate populations several key regression assumptions are made. First is that h^2 is the same for all populations. In the *Eurosta* study we could not be certain that the populations were true replicates, and so it was quite possible that heritability of gall size would vary among populations, and perhaps between generations. Thus we report the regression slope below as H, which is akin to the mean heritability. Even when h^2 varies, evolutionary responses to selection among the populations should still be proportional to the selection differential applied, so long as h^2 and S_i are uncorrelated. A second key assumption of the regression model is that ϵ_i is uncorrelated with S_i. Since the null hypothesis (no response to selection) would be falsified only by a positive regression slope, the significance of the slope coefficient is assessed by a one-tailed t-test.

Although it is a more sensitive test, the replicate population model failed to detect a statistically significant evolutionary response to selection. When the cumulative change in mean size (mean of the 1987 generation minus the mean of 1983) was regressed over the cumulative selection differential (the sum of the differentials from 1983 to 1986) no relationship is evident (fig. 9.9).

9.6 WHY DON'T WE SEE GALL SIZE EVOLVING?

As reiterated by Williams (1992), it is as important to understand cases where natural selection does not lead to evolution as to understand those where it does. Given the demonstrable se-

Response to Selection?

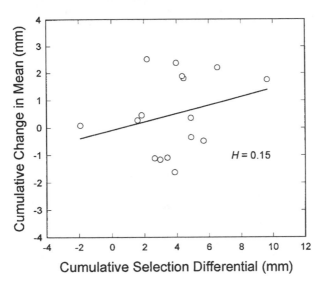

FIGURE 9.9. Regression of the cumulative change in mean gall diameter for the sixteen populations over five generations and the cumulative selection differential applied. A significantly positive regression slope is expected if the populations are responding to selection. (From Weis 1996, "Variable selection on *Eurosta*'s gall size, III: Can an evolutionary response to selection be detected?" *Journal of Evolutionary Biology* 9:623–640; with permission)

lection pressure favoring increase in *Eurosta*'s gall size (Abrahamson et al. 1989; Weis, Abrahamson, and Andersen 1992) and the evidence for genetic variation in *Eurosta* for gall size (Weis and Abrahamson 1986; Weis and Gorman 1990), it is sensible to ask, "Why don't the field data show evidence for gall size evolution?" The possible answers to this question fall into two categories. First, the expectation of evolutionary change is faulty because it is based on overestimates of the relevant parameters, h^2 and S. Second, gall size may be at equilibrium despite the demonstrable effect of size on survival probability. The first of these categories will apply if our simple model of response to selection is an adequate description of the system. The second category includes models that would explain stasis through factors not included in the model.

Errors in Estimation

The lack of a cumulative response to selection may be due to lack of heritable variation in *Eurosta* for gall size. Heritability of gall size estimated in quantitative genetic experiments in the greenhouse (chapter 6) may be too high. When plant genotype was held constant (Weis and Abrahamson 1986), heritability was estimated at 0.40, and when individuals were randomly assigned one of seventy-two plant genotypes, heritability dropped to 0.20 (Weis and Gorman 1990). Both heritability estimates could be inflated because (1) the controlled conditions of the greenhouse reduced environmental variance in gall size, and (2) the use of a full-sib design overestimated insect additive genetic variance in gall size by confounding it with nonadditive genetic variance and with maternal effects (see Weis and Abrahamson 1986).

The heritability estimates in both these experiments exceed the slope of the cumulative regression (0.09; fig. 9.9), which is related to the realized heritability. However, the confidence intervals for all three estimates strongly overlap, and so we cannot say with certainty that the realized heritability estimated in the field is lower than in the greenhouse. Nevertheless, additional environmental factors in the field may dilute the contributions of insect genetic variance to phenotypic variance in gall size. In the "randomized across plant genotype" experiment (Weis and Gorman 1990), 54% of the gall-size variance was explained by an "insect genotype × plant phenotype" interaction effect (insect families differed in their sensitivity to plant lag time) (chapter 6). Other environmental factors could contribute to additional "insect genotype × environment" interaction effects, and thus diminish the proportion of genetic variance available for a selection response.

The expectation of gall-size evolution could be faulty if selection was overestimated. However, the functional relationship between gall size and *Eurosta*'s vulnerability to attack is well understood (chapter 8), and there is no reason to suspect serious bias in estimating the intensity of selection that parasitoids and birds directly impose on gall size (although undetected indirect selection from other sources is possible; see below).

How much change in gall size would be expected if the heritability estimates from the greenhouse experiment are accurate? In a constant year-to-year environment and replicate insect populations we would expect gall size to have increased by an average of 0.71 mm from the 1984 to the 1988 generations (mean cumulative $S = 3.55$ mm, greenhouse $h^2 = 0.20$); the observed mean change was 0.32 mm (s.e. $= 0.34$), which differs neither from the expected change nor from zero. Although the field data failed to show that gall size is evolving in response to selection by natural enemies, the data cannot rule out an evolutionary response that is much slower than selection and genetic-parameter estimates suggest.

Gall Size at Equilibrium

The lack of evidence for evolutionary change in gall size suggests the possibility that gall size is at an evolutionary equilibrium despite continued directional selection and the existence of additive genetic variation. The lack of an evolutionary response despite selection and genetic variance has also been observed for clutch size in birds, and several explanations have been offered (e.g., Price, Kirkpatrick, and Arnold 1988; Cooke et al. 1990). These and other explanations for stasis are considered here.

The first explanation concerns an inadequate understanding of the relationship of fitness to phenotype. Selection in the present study was measured through the statistical association between gall size, a phenotypic trait, and survival. Although the statistical covariance between phenotypic value and fitness (such as the selection differential) has been recognized as a measure of the magnitude of selection on a character (Price 1970; Lande and Arnold 1983), it has been recently pointed out that correlations between phenotypic value and fitness may be due to similar developmental responses by both the measured character and the fitness components to unmeasured environmental factors (Mitchell-Olds and Shaw 1987; Price, Kirkpatrick, and Arnold 1988; Rausher 1992). Price and Liou (1989) offer the following example: female birds that experience superior nutrition may both lay larger clutches and fledge more young than females in inferior condition. A positive statistical relationship between clutch size and fledgling production would be seen (i.e., a positive value of S), but the relationship would not be causal. Instead, this correlation could be due to the underlying causal effects of nutrition on both clutch size and fledgling success. Fitness differences would not be caused by genetic differences in clutch size. Thus, joint dependence of phenotype and fitness on unmeasured environmental factors could make a character appear under strong selection when it is not. For this to occur with *Eurosta*'s gall size, some factor would have to cause both small size and increased mortality due to enemy attack, without any direct causal effect of size on attack. Aside from the few early attacks by *Eurytoma,* no reasonable mechanism could be hypothesized for such a pattern.

A second explanation for stasis in gall size is based on the Red Queen's Hypothesis (van Valen 1973). Selection by parasitoids and birds could be altering frequencies of alleles contributing to gall size, but the genetic improvement in the fly population for gall induction causes the environmental contributions to gall growth to deteriorate by a like amount, such that the mean phenotype for gall size does not change. In the case of clutch size in birds, it has been argued that if population density increases as alleles for larger clutches spread, then the increased intraspecific competition will lower resources available for egg production (Rockwell, Findlay, and Cooke 1987; Cooke et al. 1990; see also Frank and Slatkin 1992).

In *Eurosta*'s case, deterioration in average host-plant quality is an obvious candidate for an environmental effect that could maintain static gall size despite selection. If large galls impose a greater fitness cost on the host plant than small ones, plant genotypes showing lower reactivity to *Eurosta*'s stimulus could increase in frequency at a rate that offsets the gallmaker's genetic improvement. *Eurosta*'s host plant, goldenrod, is a long-lived rhizomatous perennial, and so clonal selection could lead to rapid changes in gene frequencies in plant populations exploited by *Eurosta* populations. This is an attractive hypothesis, but available evidence lends little support. Physiological studies (chapters 3 and 4) have shown conflicting results on the effect of gall infestation on plant clonal growth. A field experiment to measure intensity of clonal selection imposed by the gallmaker has failed to detect any correlation between clonal expansion and either gall size or galling rate (Weis, unpub. data).

A final explanation for stasis is counterbalancing selection on another insect character genetically correlated to gall size. No *Eurosta* character appears to be expressed during the summer and autumn, while it is still growing in the full-sized gall, that would be correlated with gall size. One adult character, emergence date, was examined in a laboratory environment, but no correlation with gall size was detected (Weis, unpub. data). However, the necrotic response of the host plant (Anderson et al. 1989; How, Abrahamson, and Zivitz 1994) could act as a selective force that favors insects inducing small galls.

Recalling the developmental-genetic model of figure 6.1, we can see that the insect genes under selection are those coding for the gall-inducing stimulus. Final gall size is presumably proportional in some way to stimulus strength. At the same time, however, the likelihood that the insect triggers the plant's necrotic response may also depend on stimulus strength. Plant genotypes varied in the frequency with which they showed this response (How, Abrahamson, and Zivitz 1994)—not all larvae died on resistant genotypes—which leaves open the possibility that the response was triggered only when the insect's stimulus exceeded some threshold level. Early stimulus strength may be correlated with final gall size (and so survival) on more susceptible plants (Hess and Abrahamson, unpub. data), and so insect genes for a stong stimulus are favored when expressed on these plants. On resistant plants the threshold for triggering the necrotic response may be lower. A stong early stimulus on these plants could result in gallmaker death, selecting against strong-stimulus genes. When resistant plants are common, the selection pressure from parasitoids to increase early stimulus strength (and thereby increase gall size) could be countered by selection exerted by plant defense to weaken the stimulus. We hope to explore this possibility in the future.

Phenotypic Plasticity and Spurious Evolution

10.1 PHENOTYPIC PLASTICITY AND THE ILLUSION OF SELECTION RESPONSE

The marine gastropod *Thais lamellosa* has a highly variable shell architecture. Populations from localities exposed to the surf have relatively thin and smooth shells. Populations found in more quiet water, by contrast, have thick shells protected by outer ridges and teeth at the shell's aperture. Predation pressures are much stronger in the quieter intertidal waters where the crab *Cancer productus* is also found (Appleton and Palmer 1988). A thick shell can make the snail invulnerable to all but the largest crabs (Palmer 1985a). Consequently, protected snails are found in the habitats where predators are common. To a casual observer, the concordant distribution of shell thickness and predators suggests that *T. lamellosa* is divided into populations that have differentiated in response to local selection pressures. However, an alternative to the selection hypothesis was explored by Appleton and Palmer (1988); they found that population differences could be explained by phenotypic plasticity in shell growth. They reared snails from each type of habitat in the laboratory in three different types of water. Snails reared in regular seawater developed thinner shells with apertural teeth either small or absent. When reared in seawater conditioned with effluent from an aquarium containing *C. productus*, thicker, toothier shells were seen. Most interestingly, the largest teeth were seen on snails reared in seawater conditioned with effluent from crabs that were fed a steady diet of *T. lamellosa*.

As the case of *T. lamellosa* illustrates, plastic developmental responses can result in spatial or temporal patterns in phenotypic variation that resemble a direct response to natural selection. The plastic ability may itself have evolved through selection—a

323

developmental program that constructs a costly defense only when needed could be favored over one that pays the cost regardless of the benefit—but the phenotypic differentiation of the populations occurs without genetic change. Plasticity in shell morphology can raise difficulties in interpreting the gastropod fossil record; apparent instances of "punctuated" morphological changes could in some cases be caused by changes in environmental effects on development rather than changes in gene frequencies (Palmer 1985b).

A case where an induced defense that can resemble an evolutionary response to predation may not be difficult to anticipate, as inducible defenses have fascinated ecologists from at least the time of Wolterek (1909), who coined the term "norm of reaction" to refer to the predator-induced development of spines in *Daphnia* (see Parejko 1992; Luning 1992). Induced plant defenses against insects have received increased attention in the past decade (e.g., Schultz and Baldwin 1982; Brown 1988; Baldwin, Sims, and Kean 1990; Zangerl 1990). With inducible defenses, phenotypic changes occur within a generation. While working on the goldenrod-gallmaker-natural enemy interaction, we have found unexpectedly two instances where phenotypic plasticity can cause phenotypic changes across generations that would be mistaken for evolutionary responses to selection. This chapter serves as a cautionary tale about inferring process from pattern.

10.2 AN ESCALATING ARMS RACE WITHOUT COEVOLUTION

The imagery of an arms race has often been invoked by those investigating the evolution of defense by victim species, and, in enemy species, the evolution of means to overcome defense (Cott 1940; Whitaker and Feeny 1971; Berenbaum and Feeny 1981; Dawkins 1982; Vermeij 1987). This kind of "tit-for-tat" interchange over the generations fits Janzen's (1980) restrictive definition of coevolution, but in a review Thompson (1986) concluded that constraints are likely to prevent continuous and symmetrical escalation in arms races. However, the arms-race analogy retains some heuristic value.

324

While studying selection on gall size, we found patterns of phenotypic plasticity in *Eurytoma gigantea* that could cause average body size to increase should the gallmaker evolve larger galls (Weis, McCrea, and Abrahamson 1989). Increase in parasitoid size would not involve any genetic change in the parasitoid population but instead be caused by the increased nutritional value of large galls. Although the evolutionary change in the host and the corresponding nonevolutionary change in the parasitoid would appear to be an arms race, it would not be coevolutionary since only one species evolves.

Large galls offer more food to *E. gigantea,* and as a result individuals deposited into large galls grow to larger adult size. Larvae of *E. gigantea* feed first on the *Eurosta* larva, but finish their development by feeding on the gall tissues, including both the nutritive and meristematic tissue layers. The amount of tissue consumed can be estimated from the size of the cavity that the parasitoid excavates in its natal gall. In a sample of parasitized galls collected in the fall of 1982, we found the cavity diameter was correlated with outer gall diameter ($r = 0.73$, $n = 19$, $P < 0.005$). This then indicated that large galls had more food for *E. gigantea.*[1] This was further substantiated by the significant correlation between dry body mass of the full-grown parasitoid larvae to natal gall diameter ($r = 0.77$, $n = 19$, $P < 0.001$). The following spring we collected several hundred galls, size sorted them, and collected *E. gigantea* adults as they emerged. The adults were then sexed, and the females dissected and ovipositors measured. As with body mass, the length of a female's ovipositor is correlated with the size of her natal gall (fig. 10.1).

This relationship between natal gall size and resulting daughter size suggested that large galls were thus superior for production of daughters. As shown in chapter 8, attack success by *E.*

---

[1] It might seem paradoxical that gall size has no detectable effect on *Eurosta* body size, but that it does affect the size of this parasitoid. The resolution lies in the fact thta as the gallmaker feeds on the nutritive tissue, cell division replenishes the food supply. This would make outer gall diameter a poor predictor of food availablilty. By contrast, when *E. gigantea* kills the gallmaker, the stimulus for replenishment is removed and so the parasitoid is limited to whatever gall tissue is available at the time of attack. Thus, the available food for this species is correlated with gall size.

CHAPTER 10

Large Females Emerge From Large Galls

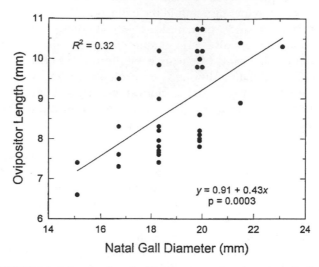

FIGURE 10.1. Ovipositor length of *E. gigantea* as a function of natal gall size. Females developing in large galls emerge with longer ovipositors. (From Weis, McCrea, and Abrahamson 1989, "Can there be an escalating arms race without coevolution? Implications from a host-parasitoid simulation," *Evolutionary Ecology* 3:361–370; with permission)

gigantea is limited by ovipositor length. Successful penetration into the gall's central chamber occurs only when the ovipositor length exceeds wall thickness by over 5%. Wasps with long ovipositors can attack large galls, and because of this, they can attack more galls. Calculations show that when presented with a typical Pennsylvania population of *Eurosta* (mean gall diameter = 21 mm; SD = 3.5) wasps with average ovipositors (8.77 mm) are able to penetrate about half of all galls (Weis 1992). Wasps with the shortest ovipositor length observed (6.6 mm) can penetrate less than 10% of the galls in an average population, while those with the longest observed length (10.8 mm) can penetrate nearly 90% (fig. 10.2).

From the viewpoint of an ovipositing mother, daughters placed in large galls will have greater expected fitness than those placed in small ones. Sex allocation theory (Charnov 1982)

326

Large *E. gigantea* Females
Can Penetrate More Galls

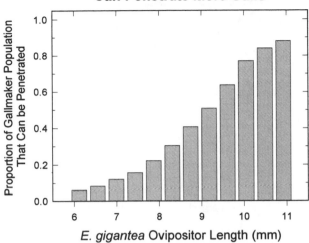

FIGURE 10.2. The proportion of the *Eurosta* population available to *E. gigantea* females with ovipositors of a given length. Galls can be reliably penetrated when the ovipositor length exceeds wall thickness by over 5%.

would predict that if size at maturity has a greater effect on daughter fitness than son fitness, ovipositing females should place daughters in large galls and sons in small ones (see Hurlbutt 1987; Godfray 1993). Results of the 1983 rearings showed this to be the case (fig. 10.3).

Taking things to the population level, if individual ovipositor length depends on individual natal gall size, then mean ovipositor length should depend on mean gall size. We constructed a simulation model to see if the observed distribution of E. *gigantea* ovipositor lengths could be predicted (Weis, McCrea, and Abrahamson 1989) from the distribution of gall sizes. The simulations were based on the estimates we made in the field or laboratory of the following parameters: (1) the observed variance in ovipositor lengths, (2) the mean number of eggs per female at eclosion, (3) the gall diameter mean and variance, (4) the observed regression on mean wall thickness over gall diameter and the within-gall variance in wall

327

E. gigantea **Males Are Allocated to Small Galls**

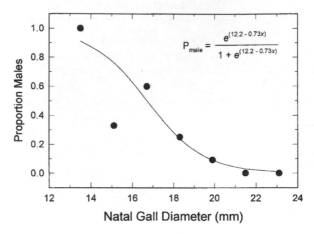

FIGURE 10.3. The sex ratio of wasps emerging from galls of increasing sizes. Males are allocated to small galls and females to large ones. Females benefit from development in large galls because they will have longer ovipositors and can consequently attack a larger proportion of the gallmaker population. (From Weis, McCrea, and Abrahamson 1989, "Can there be an escalating arms race without coevolution? Implications from a host-parasitoid simulation," *Evolutionary Ecology* 3:361–370; with permission)

thickness, (5) the distribution of the number of probes per gall, (6) the logistic regression of sex ratio over natal gall size, and (7) the linear regression of ovipositor length over natal gall size, including the variance about the regression. This model was built on the framework of the simulation presented in chapter 8, which showed that *E. gigantea*'s pattern of gall size-dependent attack on *Eurosta* could be predicted from these parameter estimates.

In each cycle of the model, a simulated "female" with an ovipositor length drawn at random from a normal distribution (with variance equal to observed) was given twelve simulated "galls," one for every egg. The gall sizes were drawn at random, as were the number of probes made. If any "probes" were made at a point where wall thickness was less than 95% of the ovipositor's length, it was scored a success. The sex of the offspring was then determined from the logistic regression in figure 10.3, and if female, its ovipositor length was drawn from a normal distribution determined from the regression and the residual variance about the regression, shown in figure 10.2. A total of one thousand such cycles constituted one "generation" within a run. At the end of each generation, the new mean and variance of ovipositor length were calculated; these statistics were then used to determine the distribution of ovipositor lengths in the next generation. We used the simulation to see how mean gall size would affect mean ovipositor length over successive generations. Note that the simulation assumes no genetic component for ovipositor length. We do not assert that genetic variation is absent, but rather, we used this assumption to see if mean ovipositor length could change across generations due to the environmental effects of natal gall size alone.

The simulation was first used to see if the observed distribution of ovipositor lengths was stable, given the observed distribution of natal gall sizes. Three simulation runs were performed. The first started with mean ovipositor length at its observed value, the second at one standard deviation above the observed, and the third one standard deviation below. Figure 10.4 shows that the simulations predict an equilibrium mean ovipositor length very near the observed mean (8.5 versus 8.77 mm). The two simulated populations started with larger and smaller mean lengths converged upon the equilibrium mean within six generations. The standard deviation in ovipositor lengths increased from the observed value of 1.23 to 1.5 in the first generation and remained stable thereafter. These results indicate that the environmental effect of natal gall size regulates the phenotypic body size distribution of *E. gigantea*.

Next, the simulation was used to see if an increase in mean gall size would result in an increase in mean ovipositor length. Mean gall diameter was increased by 5 mm, and the simulation was run until the mean ovipositor length came to equilibrium. Then, the mean gall size was increased again. Figure 10.5 shows that as mean gall diameter increases, so does mean ovipositor length.

This simulation thus shows the possibility of an escalating arms race between *E. gigantea* and *Eurosta* without coevolution. Suppose gall size were to evolve to larger size in response to the

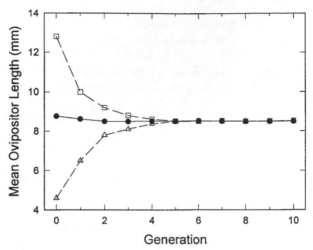

Simulation of Mean Ovipositor Length

FIGURE 10.4. Simulation results showing that the mean ovipositor length of *E. gigantea* returns to its mean value when perturbed. In the simulation, the mean length is ultimately determined by the mean gall size. (From Weis, McCrea, and Abrahamson 1989, "Can there be an escalating arms race without coevolution? Implications from a host-parasitoid simulation," *Evolutionary Ecology* 3:361–370; with permission)

selective pressure the wasp places on the gallmaker. As gall size increased, so would wasp size (fig. 10.5). Because of this, wasps would maintain the selective pressure on gall size, leading to further evolutionary increase. This would not be coevolution; *Eurosta* gall size could increase because of changes in the frequencies of alleles contributing to the gall-inducing stimulus, but the average *E. gigantea* body size would change even if all wasps were genetically identical. Figure 10.5 suggests that the escalating spiral would not turn indefinitely, since a given increase in gall diameter leads to a greater increase in gall-wall thickness than to wasp-ovipositor length. As a result, the proportion of "successful ovipositions" in the simulation fell from over 60% when mean diameter was 15 mm to only 33% when the mean gall was 35 mm, which in turn could diminish the selective pressure for further size increase (chapter 9).

330

Increasing Mean Gall Size
Increases Ovipositor Length

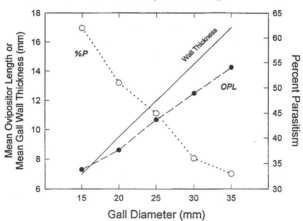

FIGURE 10.5. When population mean gall size was increased, mean ovipositor length also increased. This suggests that if the gallmaker evolved in response to the selection pressure exerted by the parasitoid, phenotypic plasticity in the wasp would cause mean size to increase likewise. (From Weis, McCrea, and Abrahamson 1989, "Can there be an escalating arms race without coevolution? Implications from a host-parasitoid simulation," *Evolutionary Ecology* 3:361–370; with permission)

We do not know of any proven examples where this kind of arms race has occurred. However, selection experiments done by Pimentel and his associates (Pimentel and Al-Hafidh 1965; Pimentel 1968; Pimentel, Levin, and Olson 1978) lend themselves to a similar interpretation. They exposed laboratory colonies of houseflies (*Musca domestica*) to attack by the parasitoid *Nasonia vitripenis* testing the prediction that parasitoid attack will lead to the evolution of host defense. Fly populations responded to selection by evolving defenses, such as shortened development time during the vulnerable pupal stage. The parasitoids that emerged from these evolved hosts were themselves smaller and less fecund. Although the changes in the parasitoid appear as a coevolved response, the possibility is open that phenotypic plasticity in the wasp is solely responsible for its decrease in size (Godfray 1993).

10.3 AN APPARENT RESPONSE TO SELECTION: CORRELATION AND THE DIRECTION OF CAUSATION

In chapter 9, we proposed a way to test for evolutionary change in a trait when replicate populations are available for study. The data required are the intensity of phenotypic selection imposed on a series of populations in generation t, and the change in the trait's mean from generation t to $t + 1$. If regression of the change in mean over the selection intensity is significantly positive, evolutionary change can be inferred. In this section we want to illustrate a potential pitfall of this method; when the trait of interest shows phenotypic plasticity, year-to-year variation in the environment can cause changes in trait mean that appear to be evolutionary but are not.

Nonlinear reaction norms can cause a spurious correlation between selection imposed and subsequent change in mean. Suppose that a species is genetically invariant for a trait, such as body size, and that trait expression is influenced by environment. Such a situation is shown in figure 10.6, where body size increases with environmental quality at a decelerating rate. Two populations are indicated; the "A" population occupies a poor-quality habitat, while "B" occupies a high-quality one. Assume that besides this spatial variation, there is also temporal variation in environmental quality. This can lead to a pattern like that in figure 10.6; the average individual in both populations grows to a larger size in a good year than in a bad year, but the size increment in the poor habitat is much more.

This difference in size increment would appear to be evolutionary if phenotypic selection is stronger in the poor habitat. Suppose the body size in the rich habitat is near an optimum, that is, the size that yields the highest fitness; phenotypic selection will be small. By the same token, if the body size in the poor habitat is far from the optimum, measures of phenotypic selection will be large. When selection intensity and change in body size are measured during the transition from a bad year to a good year, there will be a positive correlation between selection intensity and change in body size, but of course the correlation is not due to evolutionary change. We observed patterns of

Nonlinear Reaction Norms
and Environmental Fluctuation

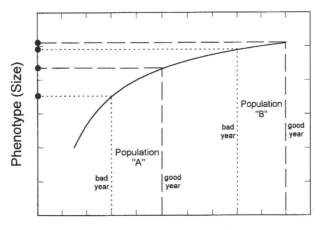

Environmental Quality

FIGURE 10.6. Two genetically uniform populations with a nonlinear reaction norm, and that occupy different quality habitats, would show different responses to annual variation in environmental quality. See text for explanation on how this pattern of phenotypic plasticity could lead to a spurious response to selection.

change in mean gall size during our long-term study that can be explained by a mechanism similar to the one presented here.

The previous chapter presented data indicating that despite consistent selection intensity on *Eurosta* to increase gall size, no evolutionary change was apparent. Figure 9.9 showed the cumulative change in *Eurosta*'s mean gall size over cumulative selection intensity, taken over a five-generation span. No evidence for an evolutionary response was seen. However, a different pattern was seen when the data were examined on a generation-by-generation basis.

Figure 10.7 shows the yearly change in mean gall size regressed over the yearly selection intensities for the four generational transitions examined. Two of the four regressions are significant beyond the 0.05 level (one-tailed test). A third one

Response to Selection?

by Population, per Generation

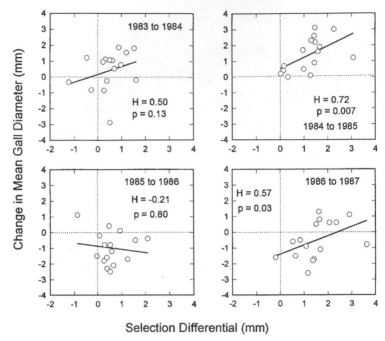

FIGURE 10.7. Regression of the between-generation change in *Eurosta's* gall size and the selection differential imposed. When assumptions of the replicate population model are met, a significant positive regression indicates an evolutionary response to selection. However, violations of this assumption can lead to a spurious conclusion of evolution. (From Weis 1996, "Variable selection on *Eurosta's* gall size, III: Can an evolutionary response to selection be detected?" *Journal of Evolutionary Biology* 9:623–640; with permission)

shows a negative relationship (and this negative relationship would have been marginally significant had the one-tailed test been applied in the opposite direction).

Why do the cumulative and the generation-by-generation data show different results? One possible reason is that the generation-by-generation analysis is indeed showing sporadic evolutionary responses, but that these responses get obscured by statistical noise when cumulative responses across generations

334

are analyzed. However, a careful consideration of the assumptions underlying the statistical model used in the analysis suggest a second, stronger possibility. The regression model was

$$\Delta \bar{z}_i = E + h^2 S_i + \epsilon_i,$$

where $\Delta \bar{z}_i$ is the change in mean gall diameter between years 1 and 2 in population i, and S_i is the selection differential seen in year one. The regression coefficients have specific biological meanings here. The slope, h^2, is related to the mean heritability of the trait under selection. The intercept, E, is the between-generation change in mean size that is caused by those environmental differences between years that affect all populations equally. The error term, ϵ_i, includes measurement error, but it also includes changes in trait means caused by between-year environmental changes that are specific to each population. A basic assumption of regression is that the independent variable (here S_i) is uncorrelated with the error term. The apparent evolutionary response to selection seen in the year-by-year analysis (fig. 10.7) may be due to a violation of this assumption. A strong argument can be made that variation in environmental conditions among fields and between years causes a correlation between selection intensity and change in gall size.

Two features of the long-term study of phenotypic selection on gall size point to a correlation between selection differential and population-specific environmental effects. The first is that the intensity of selection is higher in populations with consistently small mean gall size. Plants in these populations are generally smaller, and so it is likely that their small galls reflect poor growing conditions. Years with better rainfall, for instance, may increase gall size in these fields more than in those with generally better growing conditions. The observed year-to-year variation in mean gall size is consistent with this idea (fig. 9.8). If poor populations benefit more from a good year than a rich one, the variance among the individual population means should be high in bad years and low in good ones. This is the pattern we found; the correlation between the mean of the means and the standard deviation of the means was -0.90 for the five generations studied here. A nonlinear response to environmental quality, as illustrated in figure 10.6, would result in a

Environmental Variation and Spurious Selection Response

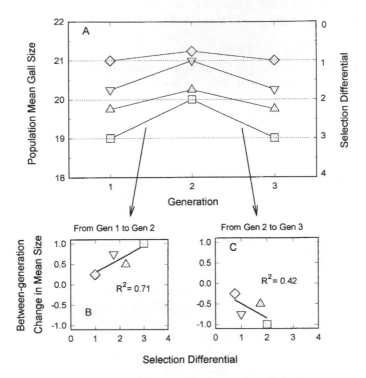

FIGURE 10.8. How population-specific changes in gall size in response to year-to-year change in environment can lead to a spurious selection response. This thought experiment concerns four gallmaker populations, and assumes that there is no genetic variance for gall size, either within or among populations. The four populations occupy habitats that vary in quality, with the poorest habitat producing the smallest galls and the richest the largest galls. Assume that gall size is under a selection pressure, with 22 mm being the optimal size, and that the intensity of selection increases as mean size decreases. (A) The gallmaker populations in poor habitats show greater increases in gall size between the bad year (Generation 1) and the good year (Generation 2) than those in higher-quality habitats. Because gall size is small in poor habitats, selection for increased size is always stronger. (B) During transitions from bad to good years, there will be a positive regression of change in gall size over the intensity of selection measured in the bad year. This is not an evolutionary response. The significant regression occurs because there is a correlation between the habitat-specific portion of the environmentally induced increase in gall size (the extra increase experi-

negative correlation between grand mean and variance among the individual population means.

Figure 10.8 illustrates how these two features of the *Solidago-Eurosta*-natural enemy system can lead to correlations among population-specific environmental effects, selection intensity, and in turn how this correlation could be mistaken for an evolutionary response to selection. In the "bad" year, the population with the smallest mean gall size will also experience the greatest wasp attack and thus the greatest upward selection differential. When a good year follows, the improved environmental conditions will cause this same population to show the greatest increase in gall size. The environmental conditions specific to the poor habitats both contribute to high selection differentials in a "bad" year and marked increases in gall size when a "good" year follows—a pattern that mimics an evolutionary response to selection (fig. 10.8b). In rich habitats, selection will always be lower (because the mean size is closer to the optimum), and the size increment between bad and good years will be smaller. This leads to a positive regression of $\Delta\bar{z}_i$ on S during a transition from a bad to a good year (fig. 10.8b). If this mechanism is in operation, then the opposite pattern should be observed in the transitions from good to bad years—there should be a negative regression of $\Delta\bar{z}_i$ on S (fig. 10.8c). Such an apparent evolutionary change against the direction of selection was observed in the 1985–1986 transition (fig. 10.7).

Over a longer term, the correlation between population-

---

enced by populations in poor habitats) and the stronger phenotypic selection measured there. (C) During the transition from a good year to a bad year, the populations in poor habitats will decrease more than those in rich habitats. This leads to a negative regression of change in size between the good and bad years over selection imposed in the good year. Because the population means are closer to one another in the good year, there is less variance in selection. This has the effect of reducing the variance in the between-year size change that is "explained" by selection differential, and thus reduces the chance of detecting a negative regression in actual data. (From Weis 1996, "Variable selection on *Eurosta*'s gall size, III: Can an evolutionary response to selection be detected?" *Journal of Evolutionary Biology* 9:623–640; with permission)

specific responses to year-to-year environmental changes and cumulative selection differentials should be greatly reduced. This will lead to a smaller bias toward a positive correspondence between selection and size change. Thus, the nonsignificant regression of cumulative change in mean size over cumulative selection differential seen in figure 9.9 is probably a truer indication of *Eurosta*'s evolutionary response to selection.

The overall conclusion to be drawn from this chapter is not particularly original but bears frequent repetition. Dobzhansky (1968) and others since (e.g., Gould and Vrba 1982) have made the point that characters that are adaptive to an organism are not necessarily adaptations (but cf. Williams 1992, appendix A). By extension, cross-generational changes in characters that are adaptive are not necessarily caused by the adaptive process.

Selection on the Hierarchy of Attack and Defense

11.1 ENEMY FORAGING AND VICTIM DEFENSE

Looking in detail at one association, as we have, has given us the occasion to think about many of the evolutionary pressures that enemies place on their victims, and vice versa. In this chapter we address two notions that have impressed us as particularly important in thinking about the structure of natural selection within enemy-victim interactions. First is the fact that for both the enemy and victim species, the many traits they deploy during an encounter are deployed in a sequential hierarchy. This means that characters involved in attack and defense are expressed in a fashion similar to a morphogenic sequence during ontogeny. Changes in early events can alter the impact of later events on the final developmental outcome. In turn, the final products of development determine the selective advantage of changes in early events. In interactions, the hierarchical structure of the several traits can have a crucial influence on the intensity of selection acting on the individual traits. In the *Solidago-Eurosta* interaction, for instance, a goldenrod clone's attractiveness to ovipositing gallmakers may or may not affect its fitness, depending on the clone's reactivity to the hatching larvae—a nonreactive clone pays no penalty for attractiveness. In the parlance of developmental genetics, these sequentially acting traits have epigenetic effects (Atchley, Xu, and Vogl 1994) on fitness.

The second notion, which emerges from this developmental view of attack and defense, is that the expected fitness outcome of a particular enemy-victim encounter is determined from a complex set of environmental influences on the fitness effects of genes. The densities of the interacting populations and the resources available to them will have strong influences on the realized rates of attack, and on post-attack performances of both plant and insect, and hence on fitness. Thus, selection pressures that herbivores place on plants and that plants place on herbivores are bound to vary over space and time.

In order to study how important these manifold effects on selection may be, it would be useful to place them in a conceptual framework. We offer such a framework here, which expands upon a proposal first made by Weis and Campbell (1992).

The hierarchy of attack and defense traits is nearly universal in enemy-victim interactions. To illustrate, consider the parallels between two very dissimilar species interactions—those between bacteria and phage and between ungulates and cats.

At the microscopic scale, *Escherichia coli* is parasitized by bacteriophage viruses, such as T4. Attack begins when the virus attaches by its tail fibers to the host bacteria at an adsorption site on the outer cell surface. After attachment, the virus injects its DNA into the bacterial host and commandeers the host's replication and transcription machinery. The first of the viral genes to be transcribed and translated is an endonuclease that cleaves the host's DNA, resulting in its genetic death. The rest of the host cell machinery remains intact and synthesizes new viral capsule and tail proteins which are then assembled along with new copies of the viral DNA to make progeny viruses. Finally, the host cell lyses and releases the viral offspring, which then restart the cycle on a new host. One can identify five basic steps in a successful viral attack: attachment, viral DNA injection, host DNA destruction, synthesis, and lysis.

Compare this with the attack of a lion on a zebra. The lion first spots and examines a herd, charges, pursues an individual prey, captures, and subdues it, then finally eats it. Thus one can identify up to six basic steps in an attack.

In both of these highly divergent interactions, a successful attack occurs when the enemy species completes a sequence of actions. It is important to note that when one of the steps early in the sequence is inefficient, it may not be possible for the enemy to compensate by increasing the efficiency of later steps. For instance, a virus with a highly efficient mechanism for manipulating the host transcription/translation machinery will not be very successful if its tail fibers are unable to attach to the host surface. Similarly, high performance at early steps will go for naught if there is a failure at later steps, for even the fastest lion will starve if it is also toothless.

Switching now to the victim's perspective, one can see that the sequential nature of the enemy attack routine offers the opportunity for a sequential hierarchy of defenses, or in other words, "defense-in-depth." Many *E. coli* strains are resistant to most T4 strains because of mutations that affect the structure of potential adsorption sites. Should the virus get through, the bacterium has a restriction-modification enzyme system that confers a kind of immunity to attack. When this system is activated the bacteria produces an endonuclease that cleaves the injected viral DNA before it can do any harm. (A second enzyme modifies bacterial DNA by methalation at appropriate sites which then makes it resistant to its own defensive endonuclease.) These two defenses can thwart a viral attack at the two initial steps (those preceding the host's genetic death).

Zebras also show "defense-in-depth." Their disruptive color pattern makes it difficult for a lion to discern the outline of an individual within a herd, which can cause a short, but crucial, delay in the lion's pursuit. Once pursuit begins, zebras are faster than lions over the long run, which usually leads to their escape. Finally, if a lion does catch up, the zebra may be able to fend it off by kicking. Here, the selection intensity on one defense component depends on performance of the other components—if zebras could already outrun any lion, the fitness gained by improving the disruptive coloration would be nil.

The sequential expression of attack and defense determine some features of the selection regime on the individual attack and defense traits, but the ecological context in which the species occur will likewise influence the intensity and direction of selection. Consider the simple case in which a population of the victim species occurs in habitat devoid of the enemy—there will be no benefits to producing defense, and so selection on defense traits will be governed exclusively by their costs. In a similar vein, the selective advantage to predators able to pursue swift prey will be negligible if slow prey are abundant. Thus, the selective advantage of a particular interaction trait can vary across gradients of enemy and victim densities. Selection can also be frequency dependent, with selection changing not only with the frequency of types within a species, but with the frequency of types in the interacting species. Resource availability, and competition for resources, can also enter into the mix of environmental variables that affect the contributions of genes for attack and defense to fitness.

Spatial variation in this mix of environmental factors can contribute to a geographical mosaic of selection regimes as envisioned by John Thompson in his book *The Coevolutionary Process* (Thompson 1994). He coined the term "interaction norm" to denote the variation in outcomes that can occur when ecological factors vary (Thompson 1988). This term is a modification of the "reaction norm" concept from developmental genetics; a reaction norm is the array of phenotypes produced by a given genotype across an array of developmental environments (see chapters 6 and 9). Our conceptual framework for analyzing selection is set in terms of reaction norms for enemy-attack traits and victim defenses. In the following sections we will first explore the structure of selection on "defense-in-depth" in plants using two-dimensional reaction norms. It is a view that grows out of our experience with the *Solidago-Eurosta*-natural enemy system, but is applicable more broadly.

Much effort has been spent on studying the operation of plant resistance to insect attack, less effort toward understanding tolerance of damage, and almost no consideration has been given to the coordinated evolution of the two. As we will show, an important feature of sequentially acting traits like these is their epigenetic effect on fitness. Later we will show how the reaction-norm approach can be applied to the analysis of selection on preference-performance relationships in herbivores. Finally, we will briefly sketch ideas on how multidimensional reaction norms for attack and defense interact. Along the way we hope to show how this framework can be employed to test long-standing hypotheses on natural selection, and how newer issues can be addressed.

342

11.2 "DEFENSE-IN-DEPTH" IN PLANTS: RESISTANCE AND TOLERANCE

The Components of Plant Defense

Mechanisms of plant defense can be divided into two categories. First are those characters that influence the amount of herbivore damage a plant receives. Biochemical, mechanical, and phenological traits all have been shown to influence the probability that an insect will choose an individual plant or influence the amount it eats after the choice is made (reviews include those by Weis and Berenbaum 1989 and Marquis 1992a); we will refer to these as *resistance characters*.[1] Second, work in both natural and agricultural systems has shown that some plant species exhibit physiological and developmental responses that allow regrowth of vegetative or floral structures after damage is sustained; we call these *tolerance characters* (Trumble, Kolodny-Hirsch, and Ting 1993; Rosenthal and Kotanen 1994). Some of the physiological mechanisms that allow plants to tolerate herbivory may also allow it to deal with a broader range of environmental stresses (Chapin 1991), but to aid discussion we will set these aside. Both types of characters, resistance and tolerance, will affect the Darwinian fitness of a plant individual when it encounters an insect herbivore.

Although plant defenses can be put into two categories, there may be multiple levels within each one. For instance, Kogan (1975) has presented a general scheme that showed the sequential action of a foraging herbivore insect. The attack sequence consists of four basic steps (fig. 11.1): (1) host-plant habitat finding, (2) host finding, (3) host recognition and acceptance, and (4) reaction to host suitability. Selection can favor plant characters that interfere with herbivore success at any or all steps. For instance, if an herbivore is limited to searching in one habitat

---

[1] Some authors limit the term "defense" to characters that lessen the likelihood of damage, which we will call *resistance*. In our usage, "defense" includes characters that mitigate the consequences of damage. In an additional point on terminology, other authors include tolerance as a subcategory of resistance, but the two terms denote quite different phenomena.

Host-Plant Attack by Herbivores, and the Opportunities for Plant Resistance

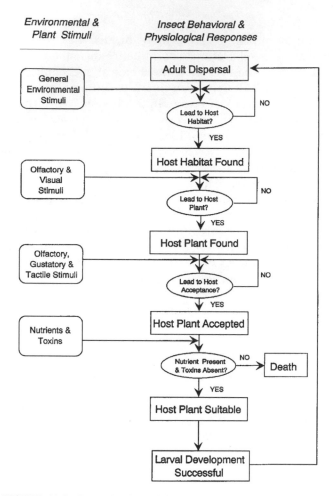

FIGURE 11.1. Generalized model of host-plant selection (from Kogan 1975). Each step in the insect's attack sequence is guided by stimuli associated with the plant or its habitat. Plants have the opportunity to evolve defenses at each step by altering the stimuli. (After Kogan 1975, "Plant resistance in pest management," in *Introduction to Insect Pest Management,* ed. R. L. Metcalf and W. H. Luckmann, 103–145. John Wiley and Sons, New York)

type, selection can favor traits that increase plant dispersal into safe areas. At later steps, very general "green leaf" chemicals as well as highly specific secondary metabolites could play parts in host finding, recognition, and acceptance. Thus, herbivory could select on the type and concentration of chemicals produced. Once an herbivore is settled, its consumption rate could vary with the suitability of the food plant, which will be determined by secondary metabolism, toughness of structural tissue, and concentrations of water and nutrients. Herbivory can act as a selective agent on these as well.

Once the damage is done, plants may be able to survive and reproduce at their reduced size, or compensate for tissues lost through regrowth (Trumble, Kolodny-Hirsch, and Ting 1993). Selection will favor individuals able to achieve high reproductive success despite damage. The modular growth form of plants allows them to accomplish this in several ways. When open flowers are eaten, undamaged floral buds that might otherwise abort can be retained (Hendrix 1988; Lowenberg 1994). When leaves are lost, regrowth can be facilitated by several mechanisms, including increased photosynthetic rates in remaining leaves (Welter 1989), mobilization of stored resources (Kigel 1980), and initiation of new leaves from dormant buds (Belsky 1986). As with resistance, these tolerance mechanisms can act in sequence; the rate at which a dormant bud grows into a new leaf will depend on the supply of photosynthate and nutrients it receives from the rest of the plant.

Understanding the adaptive evolution of plant-defense characters requires that we understand each character as a component within a defense system. However, even the best ecological-genetic studies of selection on plant defenses have concentrated either on resistance (e.g., Maddox and Cappuccino 1986; McCrea and Abrahamson 1987; Maddox and Root 1990; Fritz 1990; Strong, Larsson, and Gullberg 1993) or tolerance (e.g., Holland et al. 1992; Marquis 1992b; Meyer and Root 1993) individually, and some have to a degree confounded the two (e.g., Berenbaum, Zangrel, and Nitao 1986; Simms and Rausher 1989). Simms and Triplett (1995) are among the first to consider the ecological genetics of both resistance and tolerance. It is essential to understand one in order to understand the other. For instance, the frequent failure to find a cost of resistance may be due to the fact that increased resistance comes at the cost of reduced tolerance. This cost cannot be detected unless experimental designs incorporate examination of both resistance and tolerance abilities of plant genotypes. This chapter shows that by casting resistance and tolerance as a system of interacting reactions norms, the contributions that each makes to fitness can be disentangled.

A Developmental-Genetic Model of Defense: The Reaction-Norm Approach

The fitness contributions of sequentially acting traits are not additive. This general principle can be seen by considering selection on enzymes in a biosynthetic pathway. Suppose, for instance, that fitness increases with the flux of material through a pathway mediated by two or more enzymes. Polymorphism at the enzyme loci could lead to genetic variation in fitness, and hence selection. However, genetic variance in the parts may not sum to genetic variance in the whole (Price and Schluter 1991). A mutation that doubles the maximum potential flux through one step will not of necessity double flux through the entire pathway. The flux through the whole system may be controlled by a single limiting step, so that even large increases in the performance at nonlimiting steps will lead to vanishingly small increases in fitness (Hartl, Dykhuizen, and Dean 1985; Dykhuizen, Dean, and Hartl 1987). Similarly, ecological-genetic studies of pollination have shown that seemingly large differences in pollen receipt or dispersal may lead to only small differences in fitness because post-pollination events (e.g., pollen germination, pollen-tube growth) filter their effects on reproductive success (Campbell 1991; Weis and Campbell 1992).

Defense characters also make nonadditive contributions to fitness. Consider this simple thought experiment. Suppose there is a haploid plant population in which there are two defense loci, one for resistance and one for tolerance, and each has two alleles. Also assume there are no costs to defense. The two resistance alleles, R_{10} and R_{20}, will always limit defoliation to 10% and 20%, respectively. Thus, plants with the R_{10} genotype are more

TABLE 11.1.   Epigenetic effects of resistance and tolerance alleles on plant fitness.

A.   Fitness effects of resistance alleles with different tolerance levels.

	Resistance Allele	
Tolerance Fixed at	R_{10}	R_{15}
T_{15}	0	0
T_{25}	−	+

B.   Fitness effects of tolerance alleles with different resistance levels.

	Resistance Fixed at	
Tolerance Allele	R_{10}	R_{15}
T_{15}	0	−
T_{25}	0	+

Note: See text for explanation of alleles and their effects. (A) Table entries indicate the fitness of one resistance allele relative to its alternate when the population is fixed at the tolerance locus. (B) Table entries indicate the fitness of one tolerance allele relative to the alternate when the population is fixed at the resistance locus. (0 denotes no difference in fitness.)

resistant because they suffer *less* damage. The tolerance alleles, T_{15} and T_{25}, are able to sustain up to 15% and 25% defoliation, respectively, without a reduction in fitness. The T_{25} plants are more tolerant because their damage threshold for fitness loss is higher. It might be expected that the R_{10} and T_{25} genes should always be strongly favored, but this is not the case. In fact, the mean fitness advantage of plants with the high-resistance allele will change with the frequency of the high-tolerance allele, and vice versa. Looking at the limiting case (table 11.1), when the population is fixed for T_{25} there is no advantage in limiting defoliation to 10% versus 20% (R_{10} versus R_{20}) because all plants can sustain as much as 25% defoliation without loss of fitness. The difference in resistance is at best selectively neutral. By contrast, when the tolerance locus is fixed for T_{15} there will be an advantage to having the R_{10} allele because it is the only one that keeps defoliation below the threshold for fitness loss. A similar

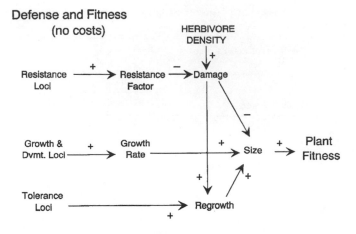

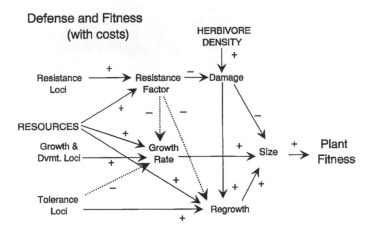

FIGURE 11.2. Developmental genetic model for the effects of defense components on plant fitness. *No defense costs:* In the absence of herbivores, plant fitness is determined by growth and development loci. However, when herbivores are present they will inflict a level of damage determined by their density and by the level of resistance factors (i.e., the products of the resistance loci). This has the potential of reducing plant fitness, but it may also trigger regrowth. The ability to regrow depends on the gene products of tolerance loci. Thus, final plant fitness will depend on genetic factors and

analysis applies to the selective advantage to the high-tolerance genotype. When the population is fixed for R_{10}, plants will not suffer defoliation above their thresholds for fitness loss. However, when R_{20} is fixed, only plants with the high-tolerance genotype will always be above the threshold. When the alleles are at intermediate frequencies, the relative advantage of increased resistance and tolerance will likewise be at intermediate values, but as one defense locus heads toward fixation, the selection coefficient on the other locus becomes vanishingly small. Thus, the hierarchical nature of resistance and tolerance can by itself cause the intensity of selection on them to be negatively correlated. By extension, multiple resistance and tolerance characters can also undergo negatively correlated selection.

The expected fitness of a plant will depend upon many more factors, both genetic and environmental. For this reason defense can be cast in terms of a developmental-genetic model, such as seen in figure 11.2a. This simplified model examines genetic and environmental effects on fitness in a plant species with two defense characters, one that provides resistance and one that provides tolerance (but it could be expanded to an arbitrary number of sequential traits). As with the thought experiment, there are no costs to defense (these will be incorporated below), and for the moment assumes that all individuals experience the same light and nutritional environment and that all phenotypic variance in defensive factors is due to genetic variance. The sole environmental variable in the simple version of the model is herbivore population density.

---

on herbivore density, an environmental factor. *With defense costs:* Costs of defense may arise through resource allocation to defense components or through negative pleiotropic effects of defense alleles on plant growth. For instance, resources allocated to the production of resistance factors like biochemicals or trichomes are unavailable for basic growth or regrowth, and so reduce fitness potential. Furthermore, a developmental program that allows rapid regrowth following damage may compromise basic growth in the undamaged condition. Costs may cause a defended plant to have lower fitness than an undefended plant when herbivores are scarce, since costs can exceed benefits. When herbivores are abundant, benefits may exceed costs.

Resistance loci will code for some resistance factor, such as a repellent secondary metabolite. If so, the phenotype under selection is the tissue concentration of the metabolite because it determines, in part, the expected amount of damage (e.g., percent defoliation, percent ramets galled) the plant receives. The inverse of damage received can be called the *realized resistance*. Damage levels will also be determined by the herbivore population density. All plant phenotypes will have zero damage when herbivores are absent, but when they are present plants with higher levels of a resistance factor will receive relatively less damage. Thus, realized resistance of a plant genotype can be quantified through a function describing the expected level of damage across the herbivore density gradient (fig. 11.3). The overall shape of the resistance function will be generally increasing, but density-dependent herbivore behaviors can induce curvature in the relationship. The resistance function parameters (i.e., terms for slope, curvature) will differ among genotypes based on the level of resistance factor they produce. Since damage received changes with an environmental variable (i.e., herbivore density), one can think of resistance as a reaction norm.

Tolerance can likewise be described as a reaction norm. The expected fitness of a plant can be expected to generally decrease as the amount of damage it receives increases. However, the rate of decrease can vary with genotype. Tolerance loci will influence the level of some physiological or developmental function, such as the ability to increase photosynthetic rates. Although genes affecting tolerance are likely to have other functions in plant growth and development (see below), it will be easier to make several initial points by assuming that they don't. Gene combinations coding for a stronger regrowth response will lose fitness at a slower rate than those with weak responses (fig. 11.4), thus the function parameters describing rate of fitness loss are measures of realized tolerance. The tolerance function parameter for the y-intercept has a special meaning; it is the fitness expected for the plant genotype in the absence of damage. This y-intercept can thus be interpreted as a measure of overall plant vigor. As we will show later, this parameter is important when assessing the costs of defense.

Hypothetical Resistance Reaction Norms

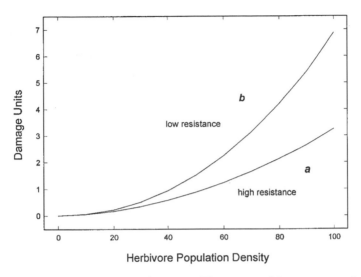

FIGURE 11.3. Resistance reaction norms. The amount of damage expected by genotype will vary with the level of defense produced and by the density of herbivores available for attack. Resistant plants will experience lower damage levels at a given herbivore density, and so its reaction-norm "slope" will be lower.

These separate reaction norms for resistance and tolerance can be combined to yield a single function that characterizes overall defensive abilities. Note that when the resistance reaction norm (damage versus herbivore density) is substituted into the tolerance function (fitness versus damage), the damage terms cancel each other and what remains is fitness as a function of herbivore density. In figure 11.5, we show this substitution for a hypothetical plant population with two resistance (*a* and *b*) and two tolerance (*x* and *y*) genotypes, which can combine in all four combinations (*ax*, *ay*, *bx*, and *by*).

Viewing figure 11.5c, several points on the fitness contributions of resistance and tolerance become apparent. First, the variation in fitness increases with herbivore density. However, it also shows the nonadditive contributions of the two defense components to fitness. Plants with the high-resistance geno-

351

Hypothetical Tolerance Reaction Norms

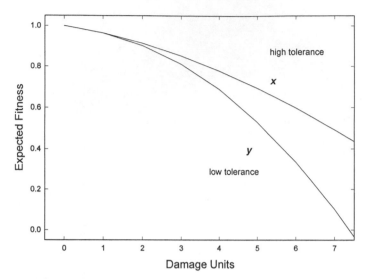

FIGURE 11.4. Tolerance reaction norms. The expected fitness of a genotype will depend on the amount of damage it receives and on its ability to regrow. Tolerant genotypes will regrow more at a given damage level, so that reaction-norm "slopes" will be closer to zero.

type, *a*, in all cases have higher fitness than *b* plants (except when herbivores are absent, when they are equal). Thus, there is strong selection for resistance, especially when herbivores are abundant. However, selection in favor of the more tolerant *x* genotype is predominantly through its fitness effect when combined with the low-resistance *b* genotype. This difference in tolerance selection is an epigenetic effect of resistance on tolerance. When an individual has the low-resistance genotype, *b,* damage levels can be high, and thus differences in tolerance will translate to large differences in fitness. However, genotype *a* restricts how much damage will be done, so there is little damage to be tolerated in the first place. Another way of putting this is that when resistance is low, the entire tolerance reaction norm can be expressed, but high resistance censors

Combining Resistance and Tolerance into Defense Reaction Norms

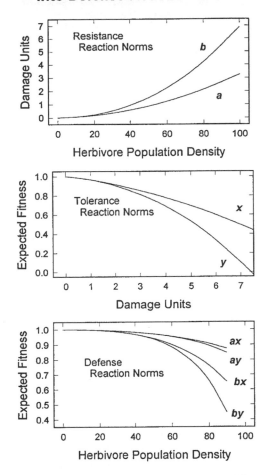

FIGURE 11.5. The defense reaction norm, that is, expected fitness as a function of herbivore density, is a product of resistance and tolerance. By substituting the resistance reaction-norm function (damage versus density) into the damage term of the tolerance reaction function (fitness versus damage) the damage terms cancel and fitness versus density remains. In the plotted example, the more-resistant genotype, *a*, is always strongly favored. By contrast, the more-tolerant genotype, *x*, is strongly favored when in combination with the less-resistant genotype, but only weakly favored when combined with the more-resistant one. This illustrates the epigenetic effects of resistance and tolerance on fitness.

Resistance Censors Expression of
Tolerance Reaction Norms

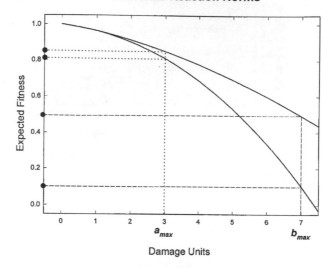

FIGURE 11.6. The censoring effect of resistance on tolerance. Plants with the more-resistant genotype, a, receive little damage, and so only the lower portion of the tolerance reaction norm is expressed. Plants with the less-resistant genotype can experience more damage, and so a greater range of the tolerance reaction norm can be expressed.

expression of the upper end of the tolerance reaction norm (fig. 11.6).

Defense reaction norms are not just conceptual constructs, but measurable properties of real plant genotypes. For instance, resistance can be evaluated by replicating plant genotypes across a series of experimental plots. Insects could then be released in these plots, at graduated densities, and the damage incurred by each genotype could be measured across the density gradient. Tolerance reaction norms are more difficult to measure, especially for insects with highly specific oviposition and establishment requirements (such as most gallmakers, unfortunately). However, tolerance reaction norms to defoliation can be measured by simulated herbivory. Replicates of the genotypes can be treated to graduated damage levels, and growth performance and reproductive output can be measured. Regression methods can then be used to characterize reaction norms, and the expected fitness of the genotypes across the gradient of herbivore density can be calculated by combining the fitted functions, just as in figure 11.5.

The result from the particular hypothetical example laid out in figure 11.5 showed that genes which increase resistance will generally confer a greater fitness advantage over those that increase tolerance. In this example, genetic variance for the two defenses was similar, as were their average levels of effectiveness. We could have chosen more or less extreme differences among genotypes for the two defenses and gotten different outcomes; for instance, if there were minimal differences in the resistance reaction norms, but greater differences in the tolerance reaction norms, selection on tolerance would be stronger. However, the example presented here illustrates the nonadditive effects of resistance and tolerance on fitness.

Since resistance is the first line of defense, it would not be surprising if on average it is under stronger selection. It is interesting to note that experiments with *E. coli* and bacteriophage viruses have suggested that selection on resistance is more intense than on post-attack defenses. Korona and Levin (1993) found that phage strains with novel restriction-modification genotypes could successfully invade a phage-limited bacterial population due to their "immunity." However, the advantage was ephemeral because adsorption site mutations, which conferred complete resistance, soon swept through the population by selection. The interpretation is that resistance made immunity redundant. There is a complication, however, in that the restriction-modification system may have other functions in *E. coli,* so that the relatively greater advantage of resistance could be due to conflicting selection pressure on immunity. This suggests that costs of the defense components can influence the outcome of selection.

Costs in Defense Hierarchies

We have already noted that compared to the work that has been done on resistance or tolerance in their own rights, little consideration has been given to their integration. One approach that has been explored for this integration has been made through plant resource-allocation theory. Coley, Bryant, and Chapin (1985) argued that plant species occupying rich habitats are typically fast growing and replace their short-lived leaves on a regular basis. Because resources are abundant, these plants should be able to regrow if they are damaged. According to their theory, the best strategy in this case is to allocate few resources to resistance (which they call defense) when leaves are ephemeral and easy to replace. By contrast, in poor environments growth is slow and the replacement of damaged tissue is constrained by nutrient limitation. Leaves on these plants are longer lived, and as a result a high initial investment in resistance mechanisms is favored both because the initial cost is amortized over a longer period, and because nutrients are not available for leaf replacement. Thus, the theory rests on the assumption that plants face a trade-off between allocation of resources to costly resistance factors and to growth; the resolution of the trade-off in any one case depends on resource availability. In a similar vein, Herms and Mattson (1992) have argued that investment in resistance mechanisms is constrained by a trade-off between allocation to tissue growth and tissue differentiation. These theories have had success in explaining between-species differences in plant defenses. However, a more explicit consideration of how individual selection acts to generate between-species differences is needed. The developmental-genetic model presented here can be expanded to examine trade-offs imposed by costs of resistance and tolerance.

Defenses can exert a fitness cost by changing allocation patterns (Bazzaz et al. 1987), or by imposing other compromises on the plant's developmental program. For instance, plant growth could be restrained when energy, which could otherwise fuel growth, is instead shunted toward production of resistance compounds (fig. 11.2b). Selection to increase tolerance can also impose a fitness cost by altering allocation patters. A gene that increases allocation to storage organs could benefit the plant by ensuring sufficient reserves to allow regrowth after damage occurs (Vail 1992). However, this benefit could come at the cost of basic growth rate because of reduced allocation to resource-harvesting tissues (leaves and roots). Tolerance can also impose costs through developmental processes indirectly related to allocation. For instance, if regrowth depends on the activation of dormant buds, the activation system (e.g., hormone levels, receptor sight sensitivity, transcription regulation) that yields optimal regrowth may not yield optimal growth in the undamaged condition. Thus, improved tolerance could come at the expense of basic growth.

Costs of defense will be reflected in the defense reaction norms. In fact, one parameter, the y-intercept of the tolerance reaction norm, is key in evaluating costs of both resistance and tolerance. Remember that this y-intercept is the expected fitness of the plant in the absence of damage. Thus, if a new mutation results in increased production of a resistance biochemical or increased allocation to storage, then the decreased allocation to resource harvesting organs (leaves and roots) will be reflected in lowered growth in the undamaged condition.

The implications of costs to resistance for selection can be seen in figure 11.7. Again we have the resistant a and the susceptible b genotypes. Assume also that these two genotypes have the same degree of tolerance—that is, they lose fitness at the same rate with increasing damage. However, because there is a cost to resistance, the tolerance reaction norms are affected. The expected fitness of the resistant genotype, a, is lower at all damage levels (fig. 11.7b), which reflects the cost of allocating resources to resistance factors. As a consequence of this cost, the direction of selection on resistance can change as herbivore population levels rise and fall. When herbivores are rare, resistance can be selected against because the resistant genotypes pay the cost of producing a defense when none is needed. Susceptible genotypes pay no costs, but since herbivores are rare, they receive little damage. When herbivores are abundant, however, the fortunes of the alternate genotypes reverse. The potential for damage is high, and so the cost of defense is overbalanced by the benefit. Thus these reaction norms reflect the pattern conventionally expected for a costly resistance factor (Dirzo and Harper 1982; Berenbaum, Zangerl, and Nitao 1986; Simms and Rausher 1989).

Within the defense reaction-norm framework, cost of resistance is manifested through a positive correlation between the

Resistance Has Fitness Costs

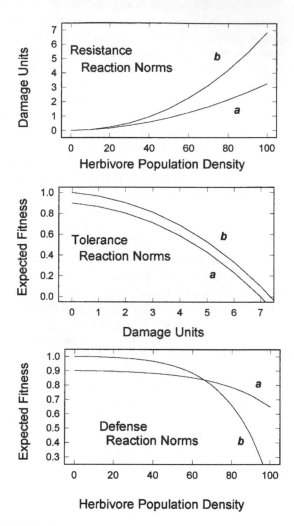

FIGURE 11.7. Allocation to resistance factors can reduce basic growth potential of more-resistant genotypes. This is reflected in reduced fitness in the undamaged condition, as indicated by the lower intercept for the tolerance reaction norm. When herbivores are scarce, the costs of resistance are paid, but no benefits realized. At high herbivore densities the benefits of resistance exceed costs.

steepness of the resistance reaction norm and the y-intercept of the tolerance reaction norm. As seen in figure 11.7, the resistant genotype has a shallow resistance slope and low tolerance intercept while the susceptible one has a high resistance slope and high tolerance intercept. This genetic correlation between elements of the resistance and tolerance reaction norms can be a constraining factor in selection on defense. The costs of resistance can also be manifested through a reduction in slope of the tolerance reaction-norm slope.

Reductions in plant fitness caused by tolerance costs will also be reflected in the y-intercept of the tolerance reaction norm. In figure 11.8, the high-tolerance genotype, x, is less affected by damage than the low-tolerance genotype, y. However, the penalty paid for high tolerance is reduced fitness in the undamaged condition (fig. 11.8b). The fitness penalty is not necessarily due to allocational costs; the balance among the regulatory mechanisms over plant growth and development that enhance regrowth may lead to slower growth under zero-damage conditions. Thus, if damage is low, tolerance is selected against, but high damage selects for high tolerance. Herbivore population density and resistance genotype will interact to determine the amount of damage incurred, and, as a result, the selective advantage of increased tolerance will be context dependent. In the example we have constructed, a benefit to high tolerance is not realized unless herbivore population densities are very high. The positive effects of high tolerance are censored when it co-occurs at high resistance (the ay genotype; fig. 11.8c), but its costs are always expressed, thus tolerance is favored only when herbivores are so abundant that high damage is inevitable. The contributions of the defense components to plant fitness can thus change in direction and magnitude with the wax and wane of herbivore population density.

In this section we have depicted resistance reaction norms without comment on what can give them their shape. Obviously, the expression of resistance factors, such as secondary chemicals, leaf hairs, and leaf toughness, will influence the probability that insects accept and consume plant tissue. Plant nutrient content will also influence plant choice and feeding rates of insects. However, the details of insect natural history will also determine

Tolerance Has Fitness Costs

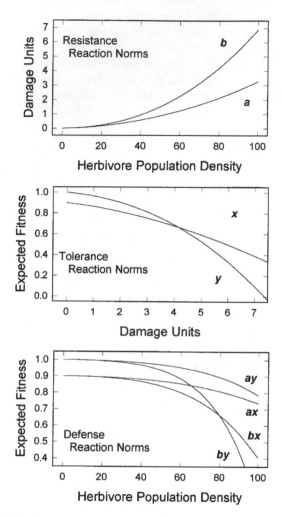

FIGURE 11.8. Tolerant genotypes will not lose much fitness with increasing damage levels. However, if there are costs to tolerance, they will pay a fitness penalty when grown without damage. In the example drawn, the benefits of tolerance exceed costs (average of *ax* and *bx* greater than average of *ay* and *by*) only at the highest herbivore levels. The more-tolerant genotype may have an additional long-term advantage though it lowers variance in fitness.

the resistance reaction norms of its host plant. If insects show aggregative behavior, as with bark beetles, one can expect damage levels to increase exponentially with density, since at low densities few plant individuals will attract enough insect individuals to start an effective infestation. On the other hand, when intraspecific competition leads to density-dependent emigration from a host, resistance reaction norms increase to a plateau. In the next section we present some simple ideas on host choice by insect herbivores. What we will show is that host-plant preference and growth performance can be viewed as insect reaction norms.

11.3 THE SEQUENTIAL HIERARCHY OF ATTACK COMPONENTS: PREFERENCE AND PERFORMANCE

A basic prediction of evolutionary ecology is that habitat-selection behavior should evolve so that animals preferentially settle in habitats where they perform best. On the other hand, selection will shape performance characters to fit the most frequently settled habitats. Selection on these two sets of characters should be reinforcing, such that small differences in habitat suitability could drive selection for stronger preferences, and stronger preferences could in turn drive selection for improved performance in the preferred habitat. Thus, it is expected that the preference of a herbivorous insect for a particular plant type, be it a plant species or a plant genotype within a species, will be correlated with its ability to perform on that type (Futuyma and Moreno 1988). Despite the strong logic of this argument, many empirical studies have not found the predicted correlation (Thompson 1988); our own work on *Eurosta* (chapter 5) has shown only a weak correlation between adult preferences and larval performance. A number of factors can constrain the evolution of preference and performance that will affect their correlation. For instance, correlations may be weak when there are no reliable cues that identify high-quality plants. When adult foraging costs are high, the optimal oviposition strategy can be to place a large number of offspring onto an array of plants rather than to search out the best plants; the aggregate fitness of many mediocre offspring can exceed that of a few superior ones (Weis, Price, and Lynch 1983; Rausher 1988b). Preference and performance characters are likely to be controlled by different genetic loci (see chapters 5 and 7), so correlations between them are likely to arise by linkage disequilibrium rather than pleiotropy, which can limit the stability of the correlation. In the following section we will use the reaction-norm approach to show how ecological conditions can vary the strength of selection leading to a preference-performance correlation. Several genetic models for the evolution of feeding specialization have been constructed (Gould 1984; Futuyma 1987; Rausher 1988b), and so we will not reiterate their findings here. However, what we want to show is that the strength of selection on genes for specialized host preference and performance will vary in response to simple ecological factors. We will concentrate on the effect of host density on selection.

Selection on Preference and Performance: The Reaction-Norm Approach

The structure of selection on the preference and performance shows strong parallels with selection on resistance and tolerance (Weis 1992). Resistance genes are akin to habitat-selection genes; they determine if the plant will develop in a damage-prone or damage-free environment. Likewise, tolerance and performance genes determine the fitness achieved in the occupied habitat. The reaction-norm approach used to examine the structure of selection on defense can be adapted here to think about selection on herbivore-attack characters.

We will consider one scenario that will illustrate the basic points of selection on host preference. The scenario assumes a parthenogenetic insect species in which the adults search for two host-plant types, which occur at equal frequencies (0.50). The female lays eggs on hosts and the offspring feed, mature, and eclose as adults. The two host types are designated as G (good) and NG (not good); the insect may be able to capitalize on some resource offered by G but absent from NG. In the insect population there are alternative genotypes for host-plant preferences, with the *a* genotype showing a preference for G and the *b*

genotype showing no preference. Likewise, there are two genotypes for performance. The y genotype is a performance generalist, that does equally well on both host plants. However, x is a
performance specialist that can take advantage of the extra resource in the G plants; it does as well as the generalist when on
the NG plant, but enjoys a 25% fitness increase on G. The scenario is constructed with no costs to specialization so that the
reaction-norm approach can be seen in a simple situation; costs
of specialization will be added below.

Although one host-plant type may be preferred over the
other, the actual choices made by an insect may be constrained
by absolute abundance of the preferred host (Singer et al. 1989;
Thompson 1994). In this example we assume the two host types
are present at a 50/50 ratio, but that the overall abundance of
plants varies. Under such a condition, one would expect the
preference expressed by genotype a could be absolute (all eggs
deposited on G) if host plants are plentiful. However, if hosts are
sufficiently rare, the less-preferred one could be visited at the
same rate as the preferred. Thus, host choice can be depicted
as a reaction norm, in which the proportion of eggs laid on
one host type (in this case G) is a function of the preference
genotype of the insect and of host-plant population density
(fig. 11.9a).

Performance can also be represented as a reaction norm
(Weis 1992). In this example, the reaction norm shows the mean
fitness of a mother's offspring as a function of the proportion of
those offspring developing on plant type G (fig. 11.9b). Since
the model assumes parthenogenesis, the expected fitness of the
offspring is identical to the fitness associated with a genotype
(i.e., we set aside the complications of recombination to more
clearly illustrate basic points on the structure of the selective regime). The x genotype has higher fitness at all points along the
gradient. That is, because it attains higher fitness on plant G,
and at least half of all eggs will be placed on that plant type, and
so at least half have the opportunity to take advantage of its
beneficial properties. If a larger fraction of eggs are placed on
G, mean fitness increases proportionately. Genotype y, however,
is unable to take advantage of the beneficial properties of G, and

Combining Herbivore Preference and Performance Reaction Norms

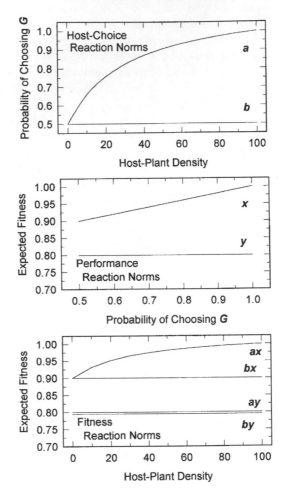

FIGURE 11.9. Insect herbivore preference and performance can also be characterized by reaction norms. Assumptions are described in the text. The actual choices made by an herbivore will depend on its fundamental preferences and by host-plant availability. The choosy insect genotype, *a*, will lay a greater proportion of its eggs on the *G*, the preferred host type, when plants are plentiful, but randomly when hosts are very rare. The alternate genotype, *b*, has no preference. Insects with performance genotype *x* do better

so has the same fitness regardless of how many eggs are placed on that plant type.

The two reaction norms can be combined to yield functions that describe expected fitness as a function of host-plant density of each of the four possible genotypes (fig. 11.9c). Clearly, the performance genotype x (performs better on G) is always strongly favored, and its fitness contribution is synergized when combined with the preference genotype a (G preferred). This graph illustrates the basic prediction that selection on preference and performance are reinforcing, but it also shows that the strength of the reinforcement can vary with environmental conditions, in this case, host abundance.

Although one of the genotypes in this exercise prefers plant type G, it does not express this preference if plants are scarce. This plasticity in choice behavior follows from what Thompson (1994) has called a "preference hierarchy." Singer et al. (1989) have studied this type of behavior in detail in the butterfly *Euphydrius edithi*, which can show preferences between host-plant species, and among individual plants within host-plant species (Ng 1988). In a preference hierarchy, plant abundance, both in terms of densities and frequencies of potential hosts, acts as a constraint on the expression of preference. However, this constraint need not translate into a cost of preference. In the example we constructed there is no cost to preference since the fitness of the preferring genotype is never inferior to the nonpreferring genotype. However, costs of preference, and of differential performance, are possible. The reaction-norm approach can be used to illustrate how these costs can cause the selective advantage of preference and differential performance to shift with population density. In an analogous fashion to defense costs, the costs of attack components will be reflected in the y-intercept of the performance reaction norm.

---

on G than NG and so as the probability of developing on G increase, so does expected fitness. The performance genotype y does equally poor on both plant types. Fitness reaction norms indicate that the G performance-specialist genotype, x, is always favored, but the preference for G is favored only when it occurs in combination with x.

Costs of Preference and Performance

As with defenses, costs can alter the direction of selection on preference and performance characters. These shifts in selection should be tied to host-plant availability, as we show here using the preference-performance reaction-norm approach.

Preference of one host type over another could exert a fitness cost in several ways. For instance, if choice behavior had no plasticity, an absolute preference for G would result in the female bypassing many NG plants when host densities are low. If time is limiting, this could result in fewer total ovipositions and lower fitness. Alternatively, finely tuned sensory mechanisms for recognizing G plants could become unreliable for distinguishing the less preferred, but still serviceable, NG plants from background vegetation. These behavioral constraints impose costs because insects that ignore NG plants give up viable oviposition opportunities when G types are limited—an offspring from an inferior plant is better than no offspring at all. Thus, constraints on foraging may in turn constrain the preference-performance correlation.

We illustrate the consequences of preference costs in figure 11.10. The preferring genotype, a, pays a fitness penalty, which we will assume is manifest as a reduction in the number of eggs successfully deposited because NG plants are ignored. Even though this example is constructed so that all insect genotypes perform better on G, preference for this superior host plant is selected against if plant densities are low. The nonpreferring genotype lays eggs on both plant types, which yields greater oviposition rates when the G types are rare. This result rediscovers the logic of the proverb "beggars can't be choosers."

Another proverb, "a jack-of-all-trades masters none," has entered into discussions on the costs involved in performance specialization (Weis 1992). This notion is embodied in the feeding-specialization hypothesis of Dethier (1954), which suggests that performance generalists can do moderately well on many plant types but excel on none of them, like the proverbial "jack." By contrast, the performance specialist is posited to have been selected to do very well on one, or a few, host plants. The cost of performance specialization is loss of the ability to use all

Preference Has Costs

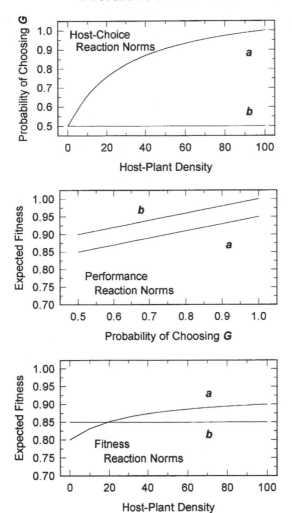

FIGURE 11.10. Preference for the superior host plant can have a cost. The cost occurs when choosy females forgo opportunities for oviposition on lesser-quality plants and as a result lay fewer eggs per lifetime. At low plant densities, choosy insects produce a few high-quality offspring, but less choosy insects produce a mixture of high- and low-quality offspring with a greater aggregate fitness. As a result, selection acts against preference for the superior plant when plants are rare.

other hosts. Supposedly, digestive physiology can optimize to only a narrow range of foods, and so its optimization toward one type leads to suboptimal performance on others. Stated in simple genetic terms, the feeding-specialization hypothesis predicts that novel mutations promoting performance on one host-plant type will also diminish performance on alternative hosts. The evidence in support of the feeding-specialization hypothesis is equivocal (Futuyma and Wasserman 1981; Scriber 1983) and theory indicates that with particular kinds of genetic control on performance, over the long run selection can produce generalists with optimal performance on more than one plant (Via and Lande 1985). However, over the short run, genetically controlled trade-offs in performance on several plants will influence the immediate direction of selection. The selective advantage of an allele that enhances performance on one plant, which at the same time diminishes performance on another, will depend critically on how frequently the superior performance host plant is chosen. Because host-plant choice depends on insect preferences and host-plant abundance, performance evolves in a developmental environment in part defined by these elements of the insect's internal and external environments.

When oviposition reliably occurs on the superior host, genes coding for increased performance on the superior host will be favored. However, realized costs will exceed benefits if the superior host is used less frequently. An example of this is illustrated in figure 11.11, which shows performance reaction norms for two genotypes. One genotype, y, is a performance generalist achieving mediocre fitness levels on both the G and NG plant types. The performance specialist genotype, x, enjoys a 25% fitness advantage over y when deposited onto G, but a 50% fitness penalty when on NG. These two genotypes can then occur with either of the preference genotypes, a or b. Figure 11.11c shows that the performance generalist does equally well, regardless of which preference genotype it occurs with (i.e., both the ay and the by combinations) and of what host-plant density might be— a "jack-of-all-trades." By contrast, the mean fitness of the performance-specialist genotype can vary with either factor. If the x genotype occurs with the no preference genotype, b, it has an equal chance of being deposited on a good or a bad host

Performance Specialization Has Costs

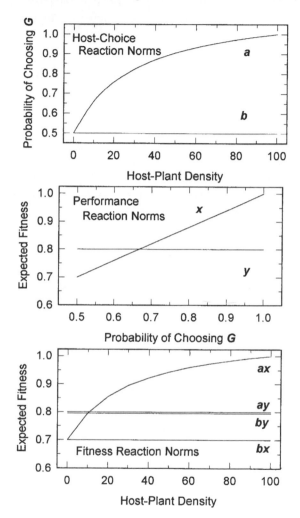

FIGURE 11.11. When performance specialization has a cost, it will be fa-vored by selection only when plants are abundant, and thus the probability of developing on the preferred plant is high.

under all host densities, and thus the cost of differential performance always exceeds the cost. When it occurs with the preference genotype, a, the frequency with which it is deposited on the superior host increases with overall host density.

11.4 MULTIDIMENSIONAL ENVIRONMENTS AND SELECTION ON ATTACK AND DEFENSE

Depicting attack and defense traits as reaction norms offers a way to link the ecological genetics of enemy-victim interactions to phenomena that occur at the population and community levels. This linkage must be established in an explicit framework in order to evaluate the importance of the geographical mosaic of selection within interactions, which Thompson (1994) has argued is the key in understanding coevolution. In this final chapter we have so far considered how the population size of one species in an interaction affects the intensity of selection on the other. Processes that change population size will thus cause selection to be dynamic. However, there are many other factors that will influence selection intensity, and this makes the analysis of selection a multidimensional problem.[2] In this section we will sketch an approach for incorporating multiple ecological factors into reaction norms for attack and defense traits. This approach is akin to the idea of the coevolutionary adaptive landscape, which Thompson (1994) has suggested would be an aid to the development of the evolutionary ecology of interactions.

Examining reaction norms over several environmental dimensions allows the testing of hypotheses on the potency of various ecological factors as constraints or promoters in the evolution of interactions. We consider here only a few issues, namely, density- and frequency-dependence selection on plant resistance, resource availability, and plant defenses; effects of

---

[2] This multidimensional approach differs from the multivariate selection gradient of Lande and Arnold (1983). They present a method designed to evaluate the selection intensity on a set of correlated characters. No explicit information on the nature of the selective agents is implied in their analysis; it is based simply on the covariances among the traits and their covariances with fitness. Here we are proposing a way to evaluate the specific contributions of multiple environmental factors to fitness variance among genotypes.

intraspecific competition among herbivores on preference-performance relationships; and the implication of the "plant-vigor" hypothesis for selection on plant growth. Other very important issues, such as multiple herbivores (Maddox and Root 1990; Linhart 1991) are left for future consideration. However, from this brief look at a few factors, the outlines of the approach should be clear.

Density and Frequency-Dependent Selection on Plant Resistance

The selective pressures on plants generated by their enemies may vary with the absolute and relative abundance of alternate plant genotypes. For instance, models of plant-pathogen coevolution have convincingly shown that novel resistance genotypes are at a selective advantage when at low frequency, since pathogens will not have been selected to overcome the new resistance factor (Antonovics 1992; Frank 1993). As selection drives the resistance genotype to higher frequencies, however, pathogen genotypes that can overcome its effect will increase in number. The selective advantage of that resistance genotype then vanishes. The advantage to rarity per se is due in part to an important feature of pathogen natural history, namely, pathogen attack relies on cell-cell contact with the host. A single mutation, say at a locus coding for a plant receptor site, can confer complete resistance, and, likewise, a single mutation in the pathogen can confer virulence. Because of the specificity of the resistance-virulence association, selection on plant resistance will depend on the genotypic frequencies of the pathogen population. Such specificity is less often the case in plant-herbivore interactions. Taking a bite of leaf does not depend on cell-cell contact in quite the same way as penetration of mesophyll by a fungus. As a result, resistance to animal enemies is prone to be partial. Resistance to insects often works through behavioral stimuli. The plasticity in animal behavior allows an herbivore population to respond to the composition of the plant population (almost) instantly, whereas pathogen populations respond by evolutionary change over several generations. These differences in natural history mean that density- and frequency-dependent selection on plant resistance is unlikely to follow the same course with a plant's animal herbivores as with its pathogens. A more appropriate model is with density and frequency-dependent predation (Greenwood 1984; Endler 1986). Rausher (1978) showed that the swallowtail butterfly, *Battus philenor,* shifts its oviposition preferences to favor host species that were recently encountered. This type of herbivore learning may lead to frequency-dependent host choice. Pollard (1992), discussed aspects of learning and sensory perception of large grazers that could cause their feeding rates on a plant species to vary with density. As mentioned above, the work of Singer and his associates (Singer et al. 1989) and Thompson (1994) on preference hierarchies suggest that oviposition rates on different plant species can change with the frequency that individual insects encounter then. If herbivores show this same type of plasticity in choice among genotypes within species, density and frequency-dependent selection can result.

As suggested above, insects are less discriminating in oviposition as preferred plant types become rare. This behavior not only has impacts on the resistance reaction norm, but causes the intensity of selection on resistance to change with plant density. In figure 11.12 we show resistance reaction norms for two plant genotypes across gradients of herbivore density as above, but also across plant density. Assuming constant genotype frequencies, one sees the potential for density-dependent selection. The level of the resistance factor produced by the two genotypes does not change, but the population size and the behavior of the herbivores cause the realized resistance to change. When plants are abundant, the more resistant (i.e., less preferred) genotype, *a,* receives less damage because herbivores find enough of the preferred genotype, *b.* However, when all plant types are rare, the more resistant genotype can be attacked as often as the susceptible one. This convergence of damage levels, and possibly of fitness, follows from the "beggars can't be choosers" response of the insects. The direction of selection will not necessarily change since the *a* genotype is always more resistant, but the magnitude of the differences between *a* and *b* changes with the ecological conditions. However, when coupled with costs of resistance (fig. 11.7), *a* could shift from the more fit to

Realized Resistance Can Be
Density Dependent

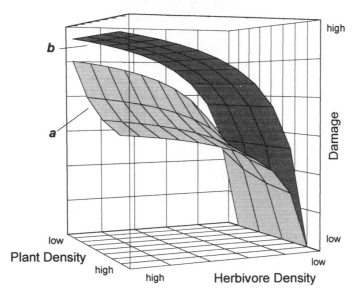

FIGURE 11.12. The damage levels sustained by alternate resistance genotypes can be density-dependent. When herbivore densities are low (right
side of graph), damage is scant for both genotypes. Damage increases for
both as herbivore density increases, but the rate-of-damage increase depends on plant density. When any plants are available, herbivores will be
free to exercise a preference for the less-resistant genotype, *b*. However,
when plants are scarce and herbivores plentiful, insects will also choose the
resistant type ("beggars can't be choosers"). Because herbivores move onto
the resistant genotype, the selective advantage of resistance is diminished.
Thus, the intensity of selection on resistance can be density dependent.

the less fit genotype at low plant densities—it pays the cost of
defense but gets damaged anyway.

Just as behavioral changes in response to absolute density can
cause shifts in selection intensity, so can responses to relative
density of plant types. For instance, low frequency per se could
be advantageous, as in the case of apostatic selection (Greenwood 1984), if herbivores choose whatever plant type is most

common. However, apostatic selection by predators is most likely to occur when the alternative prey types are readily distinguishable and equally palatable. This is not so likely to occur within plant-insect interactions. Resistant and susceptible genotypes within a plant species will be similar for many of the cues that guide herbivore plant choice. For this reason it may take some appreciable time for herbivores to assess the suitability of a host plant. Host-choice mistakes may be common and host suitability variable, and so apostatic selection is probably uncommon. However, other types of frequency-dependent attack may occur widely. The "false target" effect may, for instance, diminish the disadvantage suffered by a susceptible genotype when surrounded by resistant ones (fig. 11.13). The time wasted in examining plants that are ultimately rejected could slow herbivore foraging to the point that they ultimately find fewer suitable hosts. Further, if resistant plants are very common, herbivores could be more likely to emigrate and thereby reduce attack rates on the few susceptible host plants. The opposite pattern is also possible; the relative advantage of resistance diminishes when resistant types are rare but increases when resistance is common. This could happen if numerical responses by herbivores, through immigration, cause local herbivore densities to increase with the frequency of susceptible plants. When they reach high densities, herbivores may "spill over" onto the resistant plants. Alternatively, when resistant types are common, the herbivore population could become concentrated on the few susceptible plants, which would accentuate the advantage of resistance when it is common. Which particular scenario might play out for a particular host plant depends on the behaviors of its herbivores, and if multiple herbivores are selective agents, contrasting frequency-dependent responses could cancel and thus result in frequency independence at some herbivore density combinations. As argued in chapter 9, frequency-dependent effects of these kinds may not of themselves maintain genetic polymorphism, as does apostatic selection. However, these effects may prevent genetic variation from being exhausted by selection when they are coupled with other phenomena, such as costs to defense or with mutation. Clearly, density- and frequency-

Realized Resistance Can Be
Frequency-Dependent

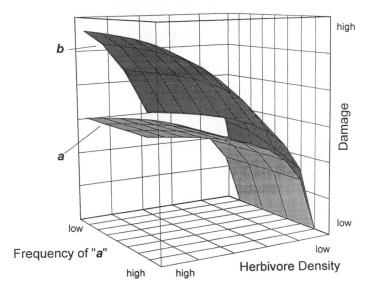

FIGURE 11.13. Damage levels sustained by alternate genotypes can be frequency dependent. If, for instance, herbivores emigrate from areas where they repeatedly encounter resistant plants, the selective disadvantage to susceptibility could decrease when susceptible plants are in the minority. The "false targets" presented by the resistant genotypes thus influence foraging success for "true targets," the susceptible plants.

dependent effects on resistance to insect herbivores are worthy of more study than they have so far received.

Resources and Their Effects on Plant Resistance and Tolerance

Plant ecologists in the past decade have performed a number of studies that examine the effects of resource availability and competition on the production of defensive chemicals and on plant regrowth following damage. Much work was stimulated by the carbon-nutrient balance hypothesis of Bryant, Chapin, and Klein (1983), which states that the supply of light for carbon fixation and the availability of soil nutrients will influence production levels of secondary chemicals. For instance, when there is abundant light but insufficient nitrogen to support growth, photosynthate will be shunted into production of carbon-based secondary chemicals, such as terpenoids and phenolics. Increasing nitrogen will bring carbon and nutrients into balance, and so allocation to growth increases at the expense of secondary chemistry. When nitrogen is abundant but growth is light limited, more assimilated nitrogen will be shunted to production of compounds containing nitrogen, such as alkaloids. A number of studies have supported the carbon-nutrient balance hypothesis (summarized by Fajer, Bowers, and Bazzaz 1992). In an ecological-genetic framework, the carbon-nutrient balance hypothesis is a prediction about reaction norms for secondary chemical production. The biosynthetic response to resources will evolve by natural selection if genotypes vary in their sensitivity to resource supply. At least two cases have shown genotype × nutrient interaction effects on secondary chemistry: Fajer, Bowers, and Bazzaz (1992) showed that some *Plantago lanceolata* genotypes increased levels of an iridoid glycoside under low nutrients more than other genotypes, and Mota-Bravo and Weis (unpub. data) have shown that some *Salix lasiolepis* changed their levels of phenolic glycoside in response to light, while others do not. These reaction norms in secondary chemistry can cause resistance to vary with resources.

Light and nutrient availability thus act as additional dimensions to resistance reaction norms. Although the carbon-nutrient balance hypothesis makes predictions about the population mean reaction norm for plant-defensive chemicals, it makes no predictions of the variation among genotypes in reaction-norm intercept, slope, or curvature. It is important to understand this variation because it can cause selection intensity on resistance to vary across resource gradients. At some light and nutrient levels, a set of plant genotypes may produce nearly identical chemical phenotypes, while at other resource levels the same genotypes could strongly differ in their chemistry (see chapter 5). Predictions on which combinations of resource levels would yield the maximum and the minimum expressed genetic variance (see chapter 6) in resistance would be valuable.

However, details of biosynthesis, evolutionary history, and of course herbivore behavior may interact in complex ways that defy a priori generalizations. Predictions on the intensity of selection on resistance with changes in resources are also made difficult by one additional fact—genetic variation in tolerance can likewise vary with resource levels.

The growth and reproductive responses of plants following herbivore damage have been extensively studied, and several good reviews are available, including those by Crawley (1983), Belsky (1986), Hendrix (1988), and Marquis (1992a). Much has been made of the fact that some plant species, particularly grasses, increase growth rates following leaf removal, a phenomenon known as *compensatory growth* (Belsky 1986). On occasion, plants overcompensate for herbivore attack so that final biomass and/or reproductive output is greater in damaged plants than in undamaged ones. Various interpretations have been offered for this phenomenon, including the suggestion that some plants benefit from being eaten (Paige and Whitham 1987) to the more extreme view that herbivores are mutualists to plants (e.g., Owen and Weigert 1981; McNaughton 1983). A model presented by Vail (1992) showed that when plants predictably suffer moderate herbivore damage, selection can favor developmental and resource allocation programs that maximize reproductive output at their typical (i.e., non-zero) damage levels. Vail's model has been criticized for making too many simplifying assumptions (Belsky et al. 1993), but nonetheless, if plants are adapted to an environment that includes consistent and moderate herbivory, it should not be too surprising if they do better when grown in the normal damaged condition than in the abnormal damage-free condition—species often do better in the environment to which they are adapted than in novel environments. The mean tolerance reaction norm for a plant species so adapted would peak in expected fitness at some intermediate damage level. Be this as it may, it is doubtful that plants generally benefit from herbivory, and improbable that herbivores are plant mutualists (Belsky et al. 1993; Crawley 1993). This discussion of overcompensation deviates from our theme of multidimensional reaction norms, but we bring it up because one of the questions that has recurred in the discourse on compensatory growth is,

"Under what resource conditions is overcompensation likely to occur?" This has stimulated empirical studies on resource effect on regrowth. These have shown that plant response to herbivory can vary from undercompensation to compensation or overcompensation by increasing nutrients (Chapin and McNaughton 1989; Maschinski and Whitham 1989), or decreasing competition (Maschinski and Whitham 1989; Alward and Joern 1993; Edenius, Dannell, and Bergstrom 1993).

These findings can be rephrased to say that resource availability alters the mean tolerance reaction norm. Higher resource levels will tend to increase the reaction-norm intercept, since fertilization increases growth performance in the undamaged state. The reaction-norm slope should become more shallow with increased resources since compensatory growth is enhanced. Following the logic above, resource conditions that favor compensatory growth will tend to diminish selection on resistance; escaping damage will not increase fitness if damage does not affect fitness. Although general patterns on tolerance reaction norms are suggested from studies of resources and compensatory growth, predictions on the genetic variance in reaction-norm parameters do not. In chapter 9 we introduced the idea that stabilizing selection can cause reaction norms to converge in typical environments (see also Weis 1992) but diverge in atypical, stressful ones. Although this pattern frequently has been found in animals (Hoffmann and Parson 1991) it may be less often the case in plants (Hoffmann and Parson 1991; Mazer and Schick 1991).

An extremely interesting issue is how the correlation between resistance and tolerance reaction norms changes with resource availability. Do environmental conditions that increase the expressed genetic variance in resistance also decrease expressed genetic variance in tolerance? If so, selection on the two defense-component variances will be nil in some conditions and intense in others. Expressed genetic variance in resistance and tolerance can be negatively correlated across resource gradients, as can the costs of defenses. For instance, Bergelson (1994) evaluated the cost of resistance in lettuce strains under high- and low-nutrient condition. This is the same as asking if the tolerance reaction-norm intercept changes with resistance level

378

(see sec. 11.2). She found growth performance was lower in resistant strains than in susceptible strains when nutrients were in short supply, but there was no difference when nutrients were added. It would be interesting to see how tolerance changes under these same conditions.

Intraspecific Competition, Herbivore Preference, and Performance

When herbivores accumulate on preferred plant genotypes, intraspecific competition may reduce fitness. For this reason, selection on preference can be density-dependent. Whitham's study of *Pemphigus betae* (Whitham 1980) and its choice of plant parts stands as an analogy. This aphid induces galls on the midribs of cottonwood leaves (*Populus angustifolia*). A gall is initiated in the spring when newly hatched stem mothers emerge from overwintering sites on the tree trunk and ascend to the developing shoots. There, mothers select among the expanding leaves and preferentially settle on those that are destined to become the largest. The aphids then induce gall formation and pass two asexual generations before quitting the gall in favor of their summer host plant. The preference for large leaves is adaptive because the number of asexual granddaughters increases with leaf size. A nonselective aphid would be at a decided disadvantage since the fitness achieved on the largest leaves is several times greater than on an average leaf. However, more than one stem mother can settle on a leaf (although combat between mothers results in the frequent exclusion of interlopers). When two or more galls are formed on a leaf of a given size, competition for the leaf's resources results in fitness losses for both stem mothers. This species seems to have sufficient plasticity in behavior to determine the lesser of the two evils; mothers often choose to settle on smaller leaves rather than share a larger leaf. Resource levels per mother could be less on a crowded, good leaf than on an uncrowded, poor one. One can imagine an extreme situation in which at high density of insect herbivores the fitness advantage shifts toward nonselective genotypes.

An accumulation of feeding insects on a preferred host-plant type could result in strong intraspecific competition on that type. This could result in lower insect growth performance on the preferred plant even if it is fundamentally superior (and the insect population is fixed for alleles that allow superior performance on it). Under these conditions, plant types of lesser fundamental quality could provide a better habitat than the preferred types because they offer refuge from competition. This deterioration of insect performance on the preferred plant type could reverse the direction of selection on host preference when herbivores reach high densities, as shown in figure 11.14. When herbivore densities are low, the choosy insect genotype could enjoy higher fitness (see fig. 11.9) because the resources offered by the preferred host are not saturated. However, at high densities the non-choosy insect genotype could be superior. Of course when all plants are rare, the fitness differences between choosy and non-choosy insect genotypes are smaller. The expected fitness of the non-choosy genotype should be independent of plant density when insect densities are low, but should increase with plant density when insects are very abundant (fig. 11.14). In the former case, competition has no effect on performance; but in the latter case, non-choosers will undergo intense competition with both choosers and non-choosers at low plant densities, but only with non-choosers at very high plant densities.

The scenario for density-dependent selection on host choice assumes that the herbivores choose plants on the basis of their fundamental quality and not on the potential for competition. This is a realistic expectation for many holometabolous insects, where free-living adult insects choose plants for oviposition that are subsequently consumed by the insects' offspring. Unless oviposition-deterring pheromones are laid down with previously deposited eggs, or eggs are recognized visually, numerous females may place eggs on attractive, high-quality plants. In outbreak conditions, preferred plants could be stripped bare before the offspring complete development.

The Plant-Vigor Hypothesis; Herbivore Choice and Plant Fitness

The reaction-norm approach to the ecological genetics of plant defense can be used to investigate the selective consequences of the "plant-vigor hypothesis" for host plants. Price

Selection on Preference Can Be
Density Dependent

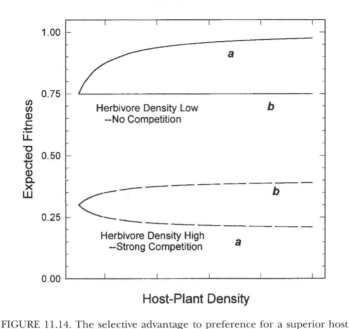

FIGURE 11.14. The selective advantage to preference for a superior host plant could be density dependent. When herbivores are in low abundance, their feeding does not diminish host quality, thus preference for fundamentally high-quality plants will be favored. However, when herbivores are at outbreak levels, the accumulation of damage on the preferred plant can cause its realized quality to fall below the fundamental quality of the poor plant. In this case, non-choosy insects will more often escape the intense competition on the preferred host plants.

(1991) has pointed out that many insect herbivores prefer to feed on vigorously growing plants or plant parts. Such preferences can result in heavier infestation of younger plants than old and of individuals with rich resource supplies than those under nutrient or water stress. Plant genotypes coding for vigorous growth could also be more vulnerable to attack than those coding for slower growth (see chapters 4 and 5). When this occurs, selection on "resistance" becomes selection on plant life-history characters.

In reaction-norm terms, the impact of insect response to plant vigor can be evaluated by comparing the steepness (slope) of the resistance reaction norm and the intercept of the tolerance reaction norm. When insects prefer vigorous plants, genotypes with a high tolerance reaction-norm intercept (high fitness when no damage is incurred) will also have a steep resistance reaction norm (they are more heavily damaged when herbivores are abundant); less-vigorous genotypes will receive lower damage at a given herbivore density. Thus, a positive correlation between resistance slope and tolerance intercept is expected when the plant-vigor hypothesis holds. Furthermore, vigorous plants may better be able to regrow after damage, so positive correlations between resistance and tolerance may also be expected. However, these correlations would also be expected under the cost-of-resistance hypothesis (fig. 11.7), so further experimentation on the mechanisms of host choice is required to distinguish the two. Positive herbivore response to the abundance of new tissue on fast-growing plants—tissue that is low in fiber but high in water and protein—would indicate choice is based on vigor, while negative responses to phytochemicals from slow plants would indicate a cost of resistance. Although one could argue that slow growth itself is an adaptation for resistance, and one that is particularly costly, this semantic construction does not reflect the selective forces in operation. Slow growth of resource harvesting structures could be part of a defensive strategy if it was caused by high allocation to stored reserves that enable regrowth after damage. If this were the case, however, slow growth would be a cost of tolerance (fig. 11.8) rather than an adaptation for resistance.

An interesting consequence of herbivore preference for vigorous plants is that it can break the expected negative correlation between damage and fitness. Intuitively, the plants sustaining the heaviest herbivore attack should achieve the lowest fitness. Arguments on the evolution of compensatory growth (Maschinski and Whitham 1989; Vail 1992) show one way in which this correlation can be broken, but preference for fast-growing plants offers another. Imagine a genotype that has vigorous growth and the capacity to produce one hundred seeds. It is attractive to herbivores that inflict heavy damage so that only fifty seeds are produced. An alternate genotype shows weaker growth and has only a fifty-seed potential. It receives less damage, and in the end produces forty seeds. Although differential attack on the vigorous plant reduces the variance in fitness, the more vogorous one, and hence the more damaged one, still outreproduces the weaker one. Thus, when growth and regrowth vigor respond to a multiplicity of factors, such as competition, fire, and wind damage (Chapin 1991; Belsky et al. 1993), the mean fitness of a plant population can increase at the same time as the mean damage level increases.

CHAPTER TWELVE

Goldenrod, Gallmakers, and the Evolutionary Ecology of Plant-Insect Interactions

12.1 THE *SOLIDAGO-EUROSTA*-NATURAL ENEMY SYSTEM IN CONTEXT

Mark Twain, in a brief essay entitled "The Whole Human Race," claimed his expansive knowledge of human nature was based on the thorough examination of a single example, namely himself (Anderson 1972). This book has reviewed our own examination of one example, the system of interactions among *Solidago, Eurosta,* and the associated parasitoids and predators. However, this is only one in a startling variety of enemy-victim associations, and so what generalizations can be made? Over the course of these studies our focus has shifted toward the task of mapping out the details and mechanisms of this tritrophic-level interaction from the viewpoint of basic evolutionary theory. Do these antagonistic interactions generate selection pressures? Is there genetic variation in the traits involved in the interaction? Does variation in attack and defense occur at more than one level in the interaction sequence? Is there evidence for short-term response to selection? Do host-plant preferences lead to reproductive isolation? Can speciation occur in sympatry prior to complete reproductive isolation? Looking in detail at one association, as we have, has given us the occasion to think in detail about many of the evolutionary pressures that all enemies place on their victims, and vice versa. Although testing broad ecological hypotheses concerning plant-herbivore interactions per se has been a secondary goal, we would nevertheless like to close with a chapter that examines the *Solidago-Eurosta*-natural enemy system in the context of several ideas that have stimulated plant-herbivore research over the past decade. Our studies have addressed some of these ideas better than others, and suggestions for further work on goldenrod and its gallmakers will be made. In many ways we have only now reached the point where we sufficiently understand the detailed natural history of these species and their interaction to ask the really important questions.

12.2 COEVOLUTION OF PLANTS AND INSECT HERBIVORES

The concept of coevolution is really a family of hypotheses about the evolution of interacting species (Thompson 1989), with the two most commonly understood of these being reciprocal adaptation and cospeciation. The notion that reciprocal adaptation leads to a coevolutionary arms race between interacting lineages has motivated much work, especially on the chemical ecology of host selection and plant defense (e.g., Berenbaum, Zangerl, and Nitao 1986). The cospeciation hypothesis has recently been put to test through phylogenetic analysis of plant and insect taxa (e.g., Futuyma and McCafferty 1990).

The notion that the *Eurosta solidaginis* clade has been split into lineages through speciation of the host plant can be ruled out. Although the exact phylogenetic positions of *S. altissima* and *S. gigantea* are uncertain, it is clear that since the former is a hexaploid it cannot be ancestral to the latter, which is typically tetraploid (chapter 2). Semple and Ringius (1983) suggest these species derive from the diploid *S. canadensis*, which only rarely hosts *Eurosta*. Genetic evidence points to a recent split of the *S. gigantea*-associated *Eurosta* lineage from the *S. altissima*-associated clade (chapter 7) and so although a shift to a closely related host has occurred, cospeciation is not indicated.

At a larger scale, the genus *Eurosta* is spread across the genus *Solidago* in no particularly clear pattern. *Eurosta* species and subspecies, as determined from sequences of the mitochondrial cytochrome oxidase I and II genes (fig. 2.3), are spread across a variety of goldenrod species with little apparent regard to plant phylogeny.

A far more difficult question to answer is whether *E. solidaginis* and *S. altissima* have undergone episodes of reciprocal adaptation. Spatial variation in the interaction can give evidence for local adaptation and counteradaption. Our associates Timothy Craig, Joanna Itami, and John Horner have begun to look at geographic differences, and we anticipate interesting new information will come to light as a result of their efforts. But based on what we know so far, many of the conditions for reciprocal adaptation are present in the Pennsylvania populations that have been studied most intensively. Both the plant and insect populations show genetic variance for many of the important traits involved in the interaction. For instance, plant genotypes differ in their attractiveness to *Eurosta,* their propensity to form galls, the frequency with which they mount a hypersensitive response, and the size of the gall they ultimately produce (chapters 4, 5, 7). The insect population exhibits genetic variation in gall size, and in their sensitivity to plant responsiveness (chapter 6). The differentiation of host-plant preferences of the *S. altissima* and *S. gigantea*-associated races testifies to the existence of genetic variation in those behaviors at least at some time in the not too distant past (chapter 7). Thus both species seem capable of evolutionary responses should the two species place selection on one another.

Evidence for reciprocal selection between goldenrod and the gallmaker is less certain. Although there is much evidence that the presence of galls on an individual *Solidago* ramet can reduce growth and reproductive output (chapter 3), in the field these effects are washed out when integrated over entire genets. The limited evidence we have indicates that neither gall load nor gall size exerts selection on *S. altissima* (chapter 3). Since *Eurosta* attacks more vigorous plant genotypes, its negative effects may simply serve to reduce the fitness differences between inherently stronger and weaker plants—in effect, it may reduce the opportunity for selection on the plant by reducing the variance in fitness. Furthermore, selection on growth vigor could have many sources, such as intra- and interspecific competition. Thus, susceptibility to the gallmaker may evolve as a correlated response to selection by these other agents.

Ample evidence exists that *Eurosta* is under strong selection from natural enemies to increase gall size (chapters 8, 9). Whether goldenrod exerts selection on *Eurosta* remains a matter of speculation. The failure of gall size to evolve in response to the parasitoids' size-dependent attack is evidence for unmeasured, counterbalancing selective forces; the plant's hypersensitive response is a prime suspect (chapter 9).

12.3 HOST SPECIFICITY AND HERBIVORE SPECIATION

Many biologists have argued that the speciation of populations in sympatry is too rare to be important for the multiplication of species (Mayr and Provine 1980; Futuyma 1986; Mayr 1988; Carson 1989). It is argued that gene flow and recombination counteract disruptive selection, which favors individuals possessing distinct sets of genes and gene combinations adaptive to different sections of the habitat. Recombination under sympatry would exclude the potential for divergence within an interbreeding population. Geographic barriers, it is argued, are required to restrict gene flow and eliminate the homogenizing effect of recombination.

Other biologists have countered that the requirement of allopatry to reduce or eliminate recombination, and consequently to enable genetic differentiation, is relaxed in some groups of organisms (Bush 1975a,b, 1994; Rice 1984, 1987; Tauber and Tauber 1989; Johnson et al. 1996). In these organisms, the homogenizing effect of recombination is minimized because gene flow is restricted by factors such as strict habitat-based mating and/or strong disruptive selection.

The debate surrounding speciation modes and processes continues in spite of the growing number of theoretical and empirical studies that suggest sympatric speciation has an appreciable role in nature (Johnson et al. 1996 and references therein). Theoretical models, for example, have suggested that genetically controlled, habitat-based assortative mating can partition individuals into separate mating pools and hence can facilitate genetic divergence. A fundamental conclusion of these models is "that speciation, when defined as the differentiation of taxa into lineages irrevocably committed to distinct evolutionary fates, can occur in the presence of gene flow" (Bush 1995).

Understanding the roles of gene flow and reproductive isolation is central to resolving the sympatric speciation vs. allopatric speciation debate. At issue is whether speciation can occur prior to full reproductive isolation, or alternatively whether reproductive isolation is an absolute requirement of genetic divergence. In the former case, reproductive isolation is an end product of speciation while in the latter case, reproductive isolation is the cause of speciation (Bush 1994, 1995; Claridge 1995). Perhaps most important to understanding speciation processes is determining how initial gene flow reductions occur. Are initial reductions of gene flow the consequence of extrinsic, geographical (or spatial) barriers or does reduced gene flow result from intrinsic, genetic barriers within one or more populations? This is the fundamental difference between theories of allopatric and sympatric speciation.

In the absence of a time machine, we cannot be absolutely certain of the incipient steps in the speciation of divergent populations. However, studies such as ours (chapter 7) do offer important insights into how speciation proceeds. We have shown for *Eurosta* host races that partial reproductive isolation results from its strong host-plant association and allochronic emergence. Furthermore, the poor performance of hybrids of the two *Eurosta* host races creates strong disruptive selection for host use. Thus, our results add to those already available (e.g., Feder and Bush 1989a,b, 1991; Feder, Chilcote, and Bush 1989a,b; Bush 1994; Berlocher 1995; Feder, Reynolds, and Wang 1995; Payne and Berlocher 1995) in suggesting that "speciation (the splitting of lineages) may occur long before complete reproductive isolation evolves" (Bush 1995). The consequence can be that "reproductive isolation is only the end product of the speciation process, not its cause" (Bush 1995).

This realization questions the usefulness of the biological species concept. Its strict application excludes sympatric host-race formation and subsequent sympatric speciation—processes that are documented to occur in nature. Similarly, the biological species concept eliminates many "good" species that form hybrids. For example, because they extensively hybridize few, if any, plant systematists could support the suggestion that all species in the red oak group should be collapsed into only one variable species. The continuum of reproductive isolation in natural populations calls into question the fundamental nature of species.

The debate surrounding speciation modes and processes will continue until we resolve not only the details of its incipient steps but also the fundamental nature of species.

12.4 THE THIRD TROPHIC LEVEL AND PLANT DEFENSE

A review paper by Price et al. (1980) on plants, herbivores, and herbivore natural enemies pointed out the rich potential for interactions spanning three trophic levels. Much work since that time has demonstrated that the vulnerability of insect herbivores to their natural enemies is frequently influenced by properties of the host plant (see reviews by Whitman 1988; Hare 1992). It has been shown that plant chemicals not only influence insect herbivores, but in some cases can attract parasitoids (Williams, Elzen, and Vinson 1988). One of the more intriguing possibilities to be investigated is that plants may incorporate natural enemies into their defense repertoire.

Selection may favor plant traits that facilitate enemy attack on the herbivore. For this to work, the enemies must curtail herbivore damage. For instance, Janzen's work on *Acacia* (Janzen 1966) demonstrated that the ants attracted and sustained by extrafloral nectaries and Beltian bodies on the host tree provided protection from diverse natural enemies. In the Netherlands, biological control of the greenhouse whitefly was successful on glabrous cucumber varieties but not on those with trichome-covered leaves. Parasitoids foraged much more efficiently on the smooth leaves and so were able to drive down pest populations (Hulspas-Jordan and van Lenteren 1978) and thus increase yield. However, the timing of enemy attack relative to herbivore feeding is crucial in determining if the plant actually benefits— what is bad for the herbivore is not necessarily good for the plant. How often do plant features that enhance enemy attack also enhance plant fitness?

Early data on the *Solidago-Eurosta* system suggested that parasitoids, particularly *Eurytoma gigantea,* could reduce the gall-maker's impact on goldenrod by arresting gall growth. This was based on the observation that parasitized galls are smaller (Stinner and Abrahamson 1979; chapter 3). However, phenological studies (Weis and Abrahamson 1985; chapter 8) showed that attack occurs after the gall reaches full size, and thus after the damage is done. Galls were parasitized because they were small, rather than small because they were parasitized. For goldenrod, facilitating parasitoid attack on the gallmaker would not improve fitness. Still, plant reactivity does influence gallmaker vulnerability; some goldenrod genotypes are more likely to produce small, penetrable galls than others (chapters 4, 6). Thus, the goldenrod-gallmaker-parasitoid interaction joins the list of instances in which plant traits can facilitate enemy attack, yet not improve plant fitness.

The individualistic details of a tritrophic-level interaction will determine if natural enemies benefit the host plant. The details for any given plant-enemy pair can conceivably change over space or time, resulting in what John Thompson (1994) has termed a "geographic mosaic of outcomes." Local conditions can lead to unexpected effects of enemies on plant fitness, as has been demonstrated for the leguminous vine *Vicia sativa*. Extrafloral nectaries attract ants, and when ants are abundant they effectively clear the plant of external feeders. However, larvae of the *Cydia* moth feed in the developing pods out of the ant's reach. The pod feeders are especially destructive in areas of high ant abundance because ants also exclude *Cydia*'s parasitoids (Koptur and Lawton 1988).

The empirical work on tritrophic-level systems, including the one centered around *Eurosta*, points to the restrictive conditions required for plants to incorporate facilitation of natural enemies into the defense repertoire. Enemy attack must quickly rid the plant of the herbivore threat. For this reason, Koptur (1991) suggests that plants are less likely to benefit by facilitating parasitoids than ants since parasitized herbivores often continue to feed on the host plant. In some instances parasitized larvae feed longer and consume more than their healthy counterparts (Slansky 1986). When isolated demes of sedentary insects pass many generations on a host plant, plant facilitation of mobile enemies could enhance fitness, as with the hairless cucumbers. In addition to swift action against the herbivore, facilitation should be more effective if both the herbivore and its natural enemy are consistently abundant—the potential for damage must be high enough to warrant facilitation, and enemy attack must be frequent enough to pay the cost of facilitation. Finally, selection for enemy-attracting traits will be compromised when they also attract herbivores (Williams, Elzen, and Vinson 1988). The occurrence of extrafloral nectaries in diverse plant groups suggests that defense systems incorporating ants may often meet the restrictions (Koptur 1992). It remains to be seen if parasitoid attraction imposes a significant selective force on plant odors or other characters that can flag the presence of herbivores.

12.5 COSTS OF RESISTANCE: ECOLOGICAL, ALLOCATIONAL, AND PLANT-VIGOR HYPOTHESES

There is much evidence from a variety of plant species for genetic variation in resistance and susceptibility both within and among plant populations (see reviews by Karban 1992; Weis and Campbell 1992; Thompson 1994). Are there ecological reasons why selection does not act on this genetic variation to further improve defense capabilities? Several suggestions have been made as to why selection does not take defense to higher levels. First, there may be trade-offs between defense against one enemy and defense against another (Simms and Rausher 1989; Linhart 1991). This "ecological cost" hypothesis can be extended to include trade-offs between defense against herbivores and competitive ability (Windle and Franz 1979). A second suggestion is that defense levels are set by an allocational trade-off between the costs and benefits of defense (Simms 1992; Zangerl and Bazzaz 1992); those genotypes that allocate more to defense suffer less herbivory, but at the cost of reduced allocation to growth and hence to fitness (Herms and Mattson 1992). As we suggested in chapter 11, the intensity of selection for or against defense can change with herbivore population density, with defended genotypes at a disadvantage when herbivores are rare, but with the advantage when herbivores are abundant. A third hypothesis extends from the plant-vigor hypothesis (Price 1991). Peter Price has noted that many herbivores preferentially attack the largest and fastest-growing plants and plant parts available. These may be superior food, since they are frequently higher in nutrients and lower in secondary chemicals (Coley, Bryant, and Chapin 1985). Although the plant-vigor hypothesis

391

concerns herbivore population dynamics as a function of the availability of vigorous plants and plant parts, Craig et al. (1988) have suggested that because of these preferences, insect herbivory can act as a force that counteracts upward selective pressure on vigor. As with the allocational cost-of-defense argument, the plant-vigor argument predicts that reduced allocation to growth will be correlated with a reduction in herbivory. These hypotheses, all of which are based on ecological constraints, are not mutually exclusive. None of them explicitly incorporates genetic variation in tolerance, even though increased resistance could conceivably come at the cost of reduced tolerance, and vice versa. Do patterns of genetic variation in the *Solidago-Eurosta* system support any of them?

The results of Maddox and Root (1990) argue against ecological costs paid through resistance to other herbivores. Susceptibility to *Eurosta* was positively genetically correlated with susceptibility to several goldenrod insects (fig. 4.4), including xylem feeders, leaf miners, and other gallmakers, but uncorrelated to others. Thus, an evolved increase in resistance to *Eurosta* would not decrease resistance to other herbivores. However, the fitness consequences of attack were not measured in this experiment, and so it is possible that negative genetic correlations in tolerance to different types of damage could constrain selection on goldenrod defense. Ecological costs expressed through lost competitive ability are also possible, as we will argue below.

We were unable to successfully perform a definitive test of the allocational cost-of-defense argument. A large quantitative genetic field experiment was initiated in the spring of 1988 for this purpose, but it failed due to transplant mortality from the record drought of that year. However, we have several bits of evidence that argue against an allocational cost of resistance, at least as it is conventionally understood. Cost arguments have been framed in terms of resource allocation to secondary chemicals, trichomes, or similar defensive structures. Increased allocation to these structures is presumed to decrease the potential level of damage, hence it can confer a benefit. When herbivores are absent from the habitat, the cost of allocation will be paid, but no benefit will accrue (since no damage can occur in the absence of herbivores). The signature of costly resistance is a negative genetic correlation between plant performance in the herbivore-free environment and in environments where herbivores are present at typical levels (Dirzo and Harper 1992; Berenbaum, Zangerl, and Nitao 1986; Simms and Rausher 1989). If any such correlation should exist for *Solidago* in the absence and presence of *Eurosta*, it is unlikely to be caused by allocation to secondary chemicals or trichomes. Experiments described in chapters 3 and 5 indicate that increased levels of secondary chemicals may actually increase the potential for damage. The density of trichomes had no effect on *Eurosta* host choice (chapter 5). *Solidago altissima* genotypes with a history of resistance had actually lower constitutive levels of phenolics than those that were historically susceptible (chapter 3). Similarly, *Eurosta* attraction to methanol-soluble secondary chemicals (presumably terpenoids) from *S. altissima* was positively related to their concentration (table 5.4; Abrahamson et al. 1994), at least over the range of concentrations tested. Granted, these experiments tested concentrations of entire chemical classes, and herbivores may respond more to the absolute and relative concentrations of individual components within the class (Berenbaum, Zangerl, and Nitao 1986). Nevertheless, these simple experiments failed to show that conventional plant defenses are effective against this gallmaking herbivore. Thus, the conventional cost-of-defense framework does not seem to apply.

Costs of resistance are still possible here, but through avenues other than allocation to defensive structures. The peculiar nature of gall induction makes it quite likely that genes which influence susceptibility to gallmakers have other essential plant functions. Weis, Walton, and Crego (1988) argued that reactivity to the gall-inducing stimulus may be a pleiotropic effect of loci that govern normal growth and development. Plant genotypes that respond strongly to the gallmaker by producing fast-growing, protective galls may simply be genotypes with rapid growth. Conversely, those plant genotypes with weak response to the gallmakers' stimulus may simply be slow growers. Table 4.3 shows that when offered to *Eurosta* in a no-choice situation, susceptible *Solidago* genotypes (those with high galling rates in the field and experimental garden) produce more and larger galls than resistant (low galling rates) genotypes. Susceptible genotypes tend to grow more vigorously in the early season, up to the time that *Eurosta* oviposits, as we showed in a common garden experiment (fig. 5.9), and these taller plants are more likely to be galled. *Eurosta* females clearly preferred genotypes with fast growth rates (fig. 5.2). These findings point to a pleiotropic relationship between susceptibility and plant vigor. The signature for allocational costs of resistance, that is, a negative genetic correlation between fitness in an herbivore-free environment and one with herbivores present, would also be seen in cases where susceptibility is a pleiotropic effect of loci promoting vigorous growth. However, it can be argued that causality is reversed from the allocational-cost scenario—instead of resistant plants growing poorly, poorly growing plants are resistant. At some gallmaker density it would be expected that the costs and benefits of resistance are balanced (chapter 11) and so selection on resistance is nil. No selection for resistance (decreased gall load) was found in natural clones, when clonal expansion rate was used as the surrogate for fitness (chapter 3). It would be helpful if in the future the cost-of-resistance experiment could be tried again (preferably in a non-drought year).

The role of the hypersensitive response relative to plant vigor remains unclear. Since it occurs in less-vigorous plant genotypes, it could be argued that it exerts a large cost on goldenrod growth. As an inducible defense, however, it would not be expected to exert an allocational cost in the herbivore-free environment. Yet the evidence indicates that resistant plants start later and grow slower in the period before *Eurosta* females are active (chapters 4, 5), and hence before a hypersensitive response can be triggered. This could indicate that hypersensitivity entails some reorganization of the regulatory mechanisms for plant growth and development, and that these changes then exert a fitness cost. However, there is another possible mechanism for the slow growth of hypersensitive genotypes. Suppose that the response is not directly triggered by the gallmaker's stimulus, but instead it is triggered when the growth of stimulated plant cells is too far ahead of the surrounding plant tissue. If such is the case, there will be a bigger disparity between stimulated cells in low-vigor plants than in high-vigor plants. Thus, plants would be hypersensitive because they are slow growers instead of being slow growers because they are hypersensitive. This is a testable hypothesis; the rate of hypersensitive responses in susceptible plants should be increased by growing them in nutrient-poor conditions, and decreased in resistant genotypes by growing them in nutrient-rich conditions.

Explaining goldenrod's genetic variation in resistance and susceptibility as a pleiotropic effect of variation in plant vigor does not answer completely the question as to why resistance does not evolve further, but rather displaces it by one step. Why is there genetic variation in vigor? Of course segregational load is one possible explanation. An intriguing possibility that merits more work is that variation in susceptibility and resistance are maintained through their correlations to competitive ability. To understand the possibilities in the *Solidago-Eurosta* system, recall that goldenrod is a successional species. Virtually all successful recruits germinate during the third and fourth years after disturbance (Hartnett and Bazzaz 1983b). In the following years, the population expands by clonal growth only. Unless another disturbance occurs, local populations go extinct when succession advances to later stages. Thus, goldenrods experience very different competitive environments over their lifetime—against weedy annuals in the first years, against perennials and one another soon thereafter, and finally shrubs and trees.

Perhaps the variation seen in vigor revealed by the experiments described in this book (which were performed in conditions that resemble early succession) reflects a cost of competitive ability. Those genotypes that perform best as competition intensifies later in life may do relatively poorly under the less stringent conditions early on. Conversely, the vigorous genotypes that start off with a bang may fade when the going gets tough. The unpredictable distribution of disturbance in both space and time could further contribute to the maintenance of variation. That disturbance regime has undoubtedly changed much in the past 35,000 years, with retreat of the glaciers, introduction of agriculture by Native Americans, and finally intensification of agriculture and other types of disruptions that came with European settlement. Increases in the frequency of disturbance may have selected for plants that thrive under early successional conditions, which could by extension have changed infestation by *Eurosta* from an unusual to a common event for goldenrod.

12.6 THE RELATIONSHIP OF OVIPOSITION PREFERENCE AND OFFSPRING PERFORMANCE

One would expect that ovipositing females would position their offspring on host plants such that their progeny would attain the greatest fitness. It seems intuitive that natural selection would favor those females making the most appropriate oviposition-site choices. Females should evolve behavioral mechanisms that enable them to select the best sites for their offspring and to avoid poor-quality sites.

Many positive relationships between adult oviposition preference and offspring performance are known (e.g., Whitham 1992; Hanks, Paine, and Millar 1993; Brody and Waser 1995; Preszler and Price 1995). A positive correlation is typically interpreted as showing that some host-plant characteristic (e.g., vigor, stem length, bud size) is related to oviposition choice, which in turn is correlated with larval survival and/or growth. Because the fitness of herbivores varies according to host-plant quality (e.g., Heard 1995b), females selecting higher-quality host plants should gain fitness over nonselective females. Such oviposition behavior could promote specialization on a subset of the available host plants. These types of relationships are central to the theories of coevolution and plant-animal interactions.

However, positive preference-performance correlations will not result unless there is some genetic relationship between oviposition-preference behavior and larval survival. For example, if preference and performance are controlled by two different genes, we might expect that selection would favor linkage disequilibrium of those loci. Those herbivore genotypes possessing specific combinations of alleles at linked loci related to oviposition choice and larval performance would have a selective advantage over those that do not. However, there may be barriers to the development of linkage disequilibrium. Larval sur-

396

vival, for example, appears to be controlled by different alleles at many autosomal loci (Craig, Itami, and Horner, unpub.) and it is likely that the genes involved in host-plant choice are different from those that influence larval survival on a particular host plant (Thompson 1989; Horner and Abrahamson 1992). The sheer numbers of genes involved in larval survival and adult oviposition preference may prevent linkage disequilibrium, and hence may eliminate any potential for correlation.

It is also possible that the adaptations related to oviposition preferences of herbivores can be broken up as their host plant's distribution and abundance changes with alteration of the landscape. Preference-performance relationships developed over thousands of years of evolution could be weakened appreciably if the host-plant abundance changes. For example, both *S. altissima* and *S. gigantea* likely have undergone huge expansions in abundance during the past three hundred years. Populations of these old-field species were limited by the extensive forest cover that existed prior to European colonization. As the populations of these host species expanded, the relative frequencies of susceptible and resistant genotypes likely shifted, given the differential growth rates of susceptible and resistant clones. This, in turn, could alter the equilibrium of herbivore preference-performance adaptations to host-plant genotypes. The time available following the disruption of host-plant and herbivore populations may be insufficient to enable the reestablishment of an equilibrium between adult oviposition preferences and offspring survival on various host-plant genotypes.

The array of examples (including ours with *E. solidaginis*) in which preference and performance are weakly correlated suggests that there are probably many factors that limit the coevolution of plants and animals. For preference-performance relationships, these factors may include, in addition to the above, the short adult life of many insect herbivores which may curtail their developing choosy oviposition behavior, the limits to the discriminatory ability of adult females, the lack of host-plant characters that provide reliable cues for suitability, the influence of natural-enemy attack on preferred and less-preferred host-plant genotypes, as well as others. The frequency with which we

encounter weak correlations between adult oviposition preference and offspring performance begs for an explanation to this
paradox (chapter 5).

12.7 EPILOGUE: MODEL SYSTEMS AND THE STUDY OF
EVOLVING INTERACTIONS

We have been able to study many facets in the evolutionary
ecology of plant-insect interactions by using the *Solidago-Eurosta*-
natural enemy system. We have gained insights on the intricacy
of herbivore effects on ramet- and genet-level plant growth, the
ecology of host-choice behavior, the mechanisms for host-race
formation and sympatric speciation, and on the ecological basis
of natural selection. The breadth of the questions addressed has
contributed to the depth of understanding of the factors driving
the evolution of specialized interactions. However, focus on a
single system of necessity limits the generality of the conclusions
drawn, simply because it will be only one example of many such
systems. Furthermore, any model system will present practical
limitations on experimentation (for instance, the inability to im-
plant *Eurosta* eggs into plants of our choosing hampers further
investigation of *Solidago*'s tolerance of gallmaker attack and of its
hypersensitive response; goldenrod's longevity hampers formal
estimates of fitness using survivorship and lifetime fecundity).
What then should be the role of model systems in the study of
plant-insect interactions?

A broad comparative approach can draw general conclusions,
but if the cases being compared are superficially understood,
the conclusions will be suspect. Detailed knowledge on the multiple facets of a few well-chosen systems can provide the bench-
marks against which the conclusions of comparative studies can
be calibrated. Perhaps it is time for ecologists and evolutionists,
entomologists and botanists, to focus their efforts on a relatively
small number of plant-insect associations. Berenbaum and Zangrel (1992) have listed six criteria to be considered when selecting a system to investigate chemical mediation of plant-insect
coevolution: (1) plant and insect natural histories should be well
known, (2) the systematic status (at the species level) of plant
and insect should be well supported, (3) plant chemistry should

be easily quantified and well known, (4) plant breeding system should make it suitable for genetic analysis (e.g., easily out-crossed, diploid, high fecundity), (5) life spans should be short enough for artificial selection experiments, and (6) the insect host range and the plant's herbivore community should be small. These attributes are desirable not just for the study of plant chemistry, but for all aspects of plant-insect interactions. To this list we would also add the following: (1) the vertebrate herbivores, parasitic nematodes, and the microbial pathogens in the system must be known, and their contributions to the inter-action investigated, (2) plants and insects should be easily grown in field plots at controlled densities and genotype frequencies, (3) background information on the physiology and develop-mental biology of plant and insect should be available, (4) tech-niques to study the insect's sensory physiology should be estab-lished, (5) the phylogeny of plant and insect at the generic level and higher should be well understood to facilitate comparisons with close relatives, and (6) both insects and plants should be related to the model systems used by molecular geneticists so that gene libraries, PCR primers and genetic maps can be readily adapted. Surely any model system will fall short of one or more of these criteria. However, progress toward understanding the evolutionary ecology of plant-insect interactions may be quicker if investigators emulate their colleagues in molecular biology and high-energy physics by entering into large collaborations to study the questions that single investigators cannot.

References

Abrahamson, W. G. 1980. Demography and vegetative reproduction. In *Demography and the Evolution of Plant Populations*, ed. O. T. Solbrig, 89–106. Blackwell Scientific, Oxford, England. (Chapter 3)

———. 1989. Plant-animal interactions: An overview. In *Plant-Animal Interactions*, ed. W. G. Abrahamson, 1–22. McGraw-Hill, New York. (Chapter 3)

Abrahamson, W. G., and H. Caswell. 1982. On the comparative allocation of biomass, energy, and nutrients in plants. *Ecology* 63:982–991. (Chapter 3)

Abrahamson, W. G., and M. Gadgil. 1973. Growth form and reproductive effort in goldenrods. *American Naturalist* 107:651–661. (Chapter 3)

Abrahamson, W. G., and K. D. McCrea. 1985. Seasonal nutrient dynamics of *Solidago altissima* (Compositae). *Bulletin of the Torrey Botanical Club* 112: 414–420. (Chapters 2, 3, 4)

———. 1986a. The impacts of galls and gallmakers on plants. *Proceedings of the Entomological Society of Washington* 88:364–367. (Chapter 3)

———. 1986b. Nutrient and biomass allocation in *Solidago altissima:* Effects of two stem gallmakers, fertilization, and ramet isolation. *Oecologia* 68: 174–180. (Chapters 3, 8)

Abrahamson, W. G., and A. E. Weis. 1987. Nutritional ecology of arthropod gallmakers. In *Nutritional Ecology of Insects, Mites, and Spiders*, ed. F. Slansky, Jr., and J. G. Rodriguez, 235–258. John Wiley and Sons, New York. (Chapters 2, 3, 4, 5, 7)

Abrahamson, W. G., S. S. Anderson, and K. D. McCrea. 1988. Effects of manipulation of plant carbon nutrient balance on tall goldenrod resistance to a gallmaking herbivore. *Oecologia* 77:302–306. (Chapters 4, 5)

———. 1991. Clonal integration: Nutrient sharing between sister ramets of *Solidago altissima* (Compositae). *American Journal of Botany* 78:1508–1514. (Chapters 3, 4)

Abrahamson, W. G., P. O. Armbruster, and G. D. Maddox. 1983. Numerical relationships of the *Solidago altissima* stem gall insect-parasitoid guild food chain. *Oecologia* 58:351–357. (Chapters 3, 4, 7)

Abrahamson, W. G., K. D. McCrea, and S. S. Anderson. 1989. Host preference and recognition by the goldenrod ball gallmaker *Eurosta solidaginis* (Diptera: Tephritidae). *American Midland Naturalist* 121:322–330. (Chapters 2, 5, 7)

Abrahamson, W. G., J. F. Sattler, K. D. McCrea, and A. E. Weis. 1989. Variation in selection pressures on the goldenrod gall fly and the competitive interactions of its natural enemies. *Oecologia* 79:15–22. (Chapters 2, 3, 4, 5, 7, 8, 9)

401

REFERENCES

Abrahamson, W. G., K. D. McCrea, A. J. Whitwell, and L. A. Vernieri. 1991. The role of phenolics in goldenrod ball gall resistance and formation. *Biochemical Systematics and Ecology* 19:615–622. (Chapters 3, 7)

Abrahamson, W. G., J. M. Brown, S. K. Roth, D. V. Sumerford, J. D. Horner, M. D. Hess, S. T. How, T. P. Craig, R. A. Packer, and J. K. Itami. 1994. Gallmaker speciation: An assessment of the roles of host-plant characters and phenology, gallmaker competition, and natural enemies. In *Gall-Forming Insects,* ed. P. Price, W. Mattson, and Y. Baranchikov, 208–222. USDA Forest Service, North Central Experiment Station. General Technical Report NC-174. (Chapters 2, 5, 7, 12)

Addicott, F. T. 1970. Plant hormones in the control of abscission. *Biological Reviews* 45:485–524. (Chapter 3)

Alpert, P. 1991. Nitrogen sharing among ramets increases clonal growth in *Fragaria chiloensis. Ecology* 72:69–80. (Chapter 3)

Alpert, P., and H. A. Mooney. 1986. Resource sharing among ramets in the clonal herb, *Fragaria chiloensis. Oecologia* 70:227–233. (Chapter 3)

Alward, R. D., and A. Joern. 1993. Plasticity and overcompensation in grass responses to herbivory. *Oecologia* (Berlin) 95:358–364. (Chapter 11)

Ananthakrishnan, T. N. 1984. *Biology of Gall Insects.* Oxford and IBH, New Delhi, India. (Chapter 3)

Anderson, F., ed. 1972. *A Pen Warmed Up in Hell: Mark Twain in Protest.* HarperCollins, New York. (Chapter 12)

Anderson, R. M., and R. M. May. 1992. *Infectious Diseases of Humans: Dynamics and Control.* Oxford University Press, Oxford, England. (Chapter 9)

Anderson, S. S., K. D. McCrea, W. G. Abrahamson, and L. M. Hartzel. 1989. Host genotype choice by the ball gallmaker *Eurosta solidaginis* (Diptera: Tephritidae). *Ecology* 70:1048–1054. (Chapters 2, 3, 4, 5, 7, 9)

Antonovics, J. 1992. Toward community genetics. In *Plant Resistance to Herbivores and Pathogens: Ecology, Evolution and Genetics,* ed. R. S. Fritz and E. L Simms, 426–449. University of Chicago Press, Chicago. (Chapter 11)

Antonovics, J., A. D. Bradshaw, and R. G. Turner. 1971. Heavy metal tolerance in plants. *Advances in Ecological Research* 7:1–85. (Chapters 1, 9)

Appleton, R. D., and A. R. Palmer. 1988. Water-borne stimuli released by predatory crabs and damaged prey induce more predator-resistant shells in a marine gastropod. *Proceedings of the National Academy of Science* 85:4387–4391. (Chapter 10)

Arnold, S. J., and M. J. Wade. 1984. On the measurement of natural and sexual selection: Theory. *Evolution* 38:709–719. (Chapter 8)

Askew, R. R. 1961. On the biology of the inhabitants of oak galls of Cynipidae (Hymenoptera) in Britain. *Transactions of the Society for British Entomology* 14:237–268. (Chapters 1, 2)

———. 1975. The organization of chalcid-dominated parasitoid communities centered upon endophytic hosts. In *Evolutionary Strategies of Parasitic Insects and Mites,* ed. P. W. Price, 130–153. Plenum Press, New York. (Chapter 7)

Atchley, W. R., S. Z. Xu, and C. Vogl. 1994. Developmental quantitative genetic models of evolutionary change. *Developmental Genetics* 15:92–103. (Chapters 1, 11)

Averill, A. L., and R. J. Prokopy. 1987. Interspecific competition in the tephritid fruit fly *Rhagoletis pomonella*. *Ecology* 68:878–886. (Chapter 5)

Baehrecke, E. H., S. B. Vinson, and H. J. Williams. 1990. Foraging behavior of *Campoletis sonorensis* in response to *Heliothis virescens* and cotton plants. *Entomologia experimentalis et applicata* 55:47–57. (Chapter 8)

Baldwin, I. T., C. L. Sims, and S. E. Kean. 1990. The reproductive consequences associated with inducible alkaloidal responses in tobacco. *Ecology* 71:252–262. (Chapter 10)

Barbosa, P., P. Gross, and J. Kemper. 1991. Influence of allelochemicals on tobacco hornworm and its parasitoid, *Cotesia congregata*. *Ecology* 72:1567–1575. (Chapter 8)

Baust, J. G., and R. E. Lee. 1982. Environmental triggers to cryoprotectant modulation in separate populations of the gall fly, *Eurosta solidaginis* (Fitch). *Journal of Insect Physiology* 28:431–436. (Chapter 2)

Bazzalo, M. D., E. M. Heber, M. A. Del Pero Martinez, and O. H. Caso. 1985. Phenolic compounds in stems of sunflower plants inoculated with *Sclerotinia sclerotiorum* and their inhibitory effects on the fungus. *Phytopathologische Zeitschrift* 112:322. (Chapters 3, 4)

Bazzaz, F. A., N. R. Chiariello, P. D. Coley, and L. F. Pitelka. 1987. Allocating resources to reproduction and defense. *BioScience* 37:58–67. (Chapter 11)

Beaudry, J. R. 1963. Studies on *Solidago* L. VI. Additional chromosome numbers of taxa of the genus *Solidago*. *Canadian Journal of Genetics and Cytology* 5:150–174. (Chapter 2)

Beaudry, J. R., and D. L. Chabot. 1957. Studies on *Solidago* L. I. *Solidago altissima* and *S. canadensis*. In *Montréal Université Institut Botanique, Contribution* 70:65–72. (Chapter 2)

Beck, E. G. 1947. Some studies on the *Solidago* gall caused by *Eurosta solidaginis* Fitch. Ph.D. diss., University of Michigan, Ann Arbor. (Chapters 2, 3)

Belsky, A. J. 1986. Does herbivory benefit plants? A review of the evidence. *American Naturalist* 127:870–892. (Chapter 11)

Belsky, A. J., W. P. Carson, C. L. Jensen, and G. A. Fox. 1993. Overcompensation by plants: Herbivore optimization or red herring? *Evolutionary Ecology* 7:109–121. (Chapter 11)

Bentur, J. S., and M. B. Kaslode. 1996. Hypersensitive reaction and induced resistance in rice against the Asian rice gall midge *Orseolia oryzae*. *Entomologia Experimentalis et Applicata* 78:77–81. (Chapters 3, 4)

Berenbaum, M. R., and P. P. Feeny. 1981. Toxicity of angular furanocoumarins to swallowtails: Escalation in the coevolutionary arms race. *Science* 212:927–929. (Chapter 10)

Berenbaum, M. R., and A. R. Zangerl. 1992. Quantification of chemical coevolution. In *Plant Resistance to Herbivores and Pathogens: Ecology, Evolution and Genetics*, ed. R. S. Fritz and E. L. Simms, 69–87. University of Chicago Press, Chicago. (Chapter 12)

Berenbaum, M. R., A. R. Zangerl, and J. K. Nitao. 1986. Constraints on chemical coevolution: Wild parsnip and the parsnip webworm. *Evolution* 40:1215–1228. (Chapters 1, 11, 12)

Bergelson, J. 1994. The effects of genotype and the environment on costs of resistance in lettuce. *American Naturalist* 143:349–359. (Chapter 11)

Berlocher, S. H. 1995. Population structure of *Rhagoletis mendax*, the blueberry maggot. *Heredity* 74:542–555. (Chapters 7, 12)

Berry, R. J. 1990. Industrial melanism and the peppered moths (*Biston betularia* (L.)). *Biological Journal of the Linnean Society of London* 39:301–322. (Chapter 9)

Bertalanffy, L. V. 1957. Quantitative laws in metabolism and growth. *Quarterly Review of Biology* 32:217–231. (Chapter 3)

Billett, E. E., and J. H. Burnett. 1978. The host parasite physiology of the maize smut fungus *Ustilago maydis*. Part II: Translocation of C-14 labeled assimilates in smutted maize plants. *Physiology Plant Pathology* 12:103–112. (Chapter 3)

Birch, M. L., J. W. Brewer, and O. Rohfritsch. 1992. In *Biology of Insect-Induced Gall*, ed. J. D. Shorthouse and O. Rohfritsch, 171–184. Oxford University Press, New York. (Chapter 3)

Bloom, A. J., F. S. Chapin III, and H. A. Mooney. 1985. Resource limitation in plants—an economic analogy. *Annual Review of Ecology and Systematics* 16:363–392. (Chapter 3)

Boag, P. T., and P. R. Grant. 1981. Intense natural selection in a population of Darwin's finches (Geospinzinae). *Science* 214:82–85. (Chapter 8)

Bosio, C. F., K. D. McCrea, J. K. Nitao, and W. G. Abrahamson. 1990. Defense chemistry of *Solidago altissima:* Effects on the generalist herbivore *Trichoplusia ni* (Lepidoptera: Noctuidae). *Environmental Entomology* 19:465–468. (Chapter 4)

Boyce, M. S., and C. M. Perrins. 1987. Optimizing clutch size in a fluctuating environment. *Ecology* 68:142–153. (Chapter 9)

Bradbury, I. K. 1973. The strategy and tactics of *Solidago canadensis* L. in abandoned pastures. Ph.D. diss., University of Guelph, Ontario, Canada. (Chapter 2)

Bradbury, I. K., and G. Hofstra. 1975. The partitioning of net energy resources in two populations of *Solidago canadensis* during a single developmental cycle in southern Ontario. *Canadian Journal of Botany* 54:2449–2456. (Chapter 3)

Bresticker, D. H., and W. G. Abrahamson. 1984. Phenotypic variation among clones of *Solidago altissima* and correlations to total biomass. *Bulletin of the Ecological Society of America* 65:202. (Chapter 3)

Brody, A. K. 1992. Oviposition choices by a pre-dispersal seed predator (*Hylemya* sp.). II. A positive association between female choice and fruit set. *Oecologia* 91:63–67. (Chapter 5)

Brody, A. K., and N. M. Waser. 1995. Oviposition patterns and larval success of a pre-dispersal seed predator attacking two confamilial host plants. *Oikos* 74:447–452. (Chapter 12)

Bronner, R. 1977. Contribution a l'étude histochimique des tissus nourriciers des zoocecidies. *Marcellia* 40:1–134. (Chapter 3)

Bross, L. S., A. E. Weis, and L. Hanzley. 1992. Ultrastructure of cells of the goldenrod *Solidago altissima* ball gall induced by *Eurosta-solidaginis*. *Cytobios* 71:51–55. (Chapters 2, 8)

Brower, L. P., and J.V.Z. Brower. 1964. Birds, butterflies and plant poisons: A study in ecological chemistry. *Zoologica* 49:137–159. (Chapter 8)

Brown, D. G. 1988. The cost of plant defense: An experimental analysis with induced proteinase inhibitors in tomato. *Oecologia* 76:467–470. (Chapter 10)

Brown, J. M., W. G. Abrahamson, and P. A. Way. 1996. Mitochondrial DNA phylogeography of host races of the goldenrod ball gallmaker, *Eurosta solidaginis* (Diptera: Tephritidae). *Evolution* 50:777–786. (Chapters 2, 7)

Brown, J. M., W. G. Abrahamson, R. A. Packer, and P. A. Way. 1995. The role of natural-enemy escape in a gallmaker host-plant shift. *Oecologia* 104: 52–60. (Chapters 2, 5, 7)

Brues, C. T. 1946. *Insect Dietary.* Harvard University Press, Cambridge, MA. (Chapter 3)

Bryant, J. P. 1987. Feltleaf willow snowshoe hare interactions: Plant carbon/nutrient balance and floodplain succession. *Ecology* 68:1319–1327. (Chapter 4)

Bryant J. P., F. S. Chapin III, and D. R. Klein. 1983. Carbon/nutrient balance of boreal plants in relation to vertebrate herbivory. *Oikos* 40:357–368. (Chapters 3, 4, 11)

Bryant, J. P., G. D. Wieland, P. B. Reichardt, V. I. Lewis, and M. C. McCarthy. 1983. Pinosylvin methyl ether deters snowshoe hare feeding on green alder. *Science* 222:1023–1025. (Chapter 4)

Burstein, M., and D. Wool. 1992. Great tits exploit aphid galls as a source of food. *Ornis Scandinavica* 23:107–109. (Chapter 1)

———. 1993. Gall aphids do not select optimal galling sites (*Smynthurodes betae;* Pemphigidae). *Ecological Entomology* 18:155–164. (Chapter 5)

Bush, G. L. 1969a. Mating behavior host specificity and the ecological significance of sibling species in frugivorous flies of the genus *Rhagoletis* (Diptera: Tephritidae). *American Naturalist* 103:669–672. (Chapter 7)

———. 1969b. Sympatric host race formation and speciation in frugivorous flies of the genus *Rhagoletis* (Diptera, Tephritidae). *Evolution* 23:237–251. (Chapter 7)

———. 1974. The mechanism of sympatric host race formation in the true fruit flies (Tephritidae). In *Genetic Mechanisms of Speciation in Insects,* ed. M.J.D. White, 3–213. Australia and New Zealand Book Company, Sydney. (Chapter 7)

———. 1975a. Sympatric speciation in phytophagous insects. In *Evolutionary Strategies of Parasitic Insects and Mites,* ed. P. W. Price, 187–206. Plenum Press, New York. (Chapters 7, 12)

———. 1975b. Modes of animal speciation. *Annual Review of Ecology and Systematics* 6:339–364. (Chapters 7, 12)

———. 1982. Host shifts, genetic models of sympatric speciation and the origin of parasitic insect species. *Proceedings of the Fifth International Symposium on Insect-Plant Relations,* Pudo, Wageningen, Netherlands, 297–306. (Chapter 7)

———. 1994. Sympatric speciation in animals: New wine in old bottles. *Trends in Ecology and Evolution* 9:285–288. (Chapters 7, 12)

————. 1995. Reply from G. L. Bush. *Trends in Ecology and Evolution* 10:30. (Chapters 7, 12)

Bush, G. L., and D. J. Howard. 1986. Allopatric and non-allopatric speciation; assumptions and evidence. In *Evolutionary Processes and Theory*, ed. S. Karlin and E. Nevo, 411–438. Academic Press, Orlando, FL. (Chapter 7)

Butlin, R. 1987. A new approach to sympatric speciation. *Trends in Ecology and Evolution* 2:310–311. (Chapter 7)

————. 1990. Divergence in emergence time of host races due to differential gene flow. *Heredity* 65:47–50. (Chapter 7)

Cain, A. J., and W. B. Provine. 1992. Genes and ecology in history. In *Genes in Ecology*, ed. R. J. Berry, T. J. Crawford, and G. M. Hewitt. Blackwell Scientific, Oxford, England. (Chapter 1)

Cain, M. L. 1990a. Models of clonal growth in *Solidago altissima*. *Journal of Ecology* 78:27–46. (Chapters 2, 3)

————. 1990b. Patterns of *Solidago altissima* shoot growth and mortality: The role of below-ground ramet connections. *Oecologia* 82:201–209. (Chapters 2, 3, 5)

Campbell, D. R. 1991. Effects of floral traits on sequential components of fitness in *Ipomopsis aggregata*. *American Naturalist* 137:713–737. (Chapter 11)

Cane, J. T., and F. E. Kurczewski. 1976. Mortality factors affecting *Eurosta solidaginis* (Diptera; Tephritidae). *Journal of the New York Entomological Society* 84:275–292. (Chapters 2, 8, 9)

Cappuccino, N. 1991. Density dependence in the mortality of phytophagous insects on goldenrod (*Solidago altissima*). *Environmental Entomology* 20:1121–1128. (Chapters 2, 9)

————. 1992. The nature of population stability in *Eurosta solidaginis* a non-outbreaking herbivore of goldenrod. *Ecology* 73:1792–1802. (Chapter 2)

Caraco, T., and C. K. Kelly. 1991. On the adaptive value of physiological integration in clonal plants. *Ecology* 72:81–93. (Chapter 3)

Carango, P., K. D. McCrea, W. G. Abrahamson, and M. I. Chernin. 1988. Induction of a 58,000 dalton protein during goldenrod gall formation. *Biochemical Biophysical Research Communications* 152:1348–1352. (Chapters 2, 3)

Carroll, S. P., and C. Boyd. 1992. Host race radiation in the soapberry bug: Natural history with the history. *Evolution* 46:1052–1069. (Chapter 9)

Carson, H. L. 1975. The genetics of speciation at the diploid level. *American Naturalist* 109:83–92. (Chapters 7, 12)

————. 1989. Genetic imbalance, realigned selection, and the origin of species. In *Genetics, Speciation, and the Founder Principle*, ed. L. V. Giddings, K. Y. Kaneshire, and W. W. Anderson, 345–362. Oxford University Press, New York. (Chapter 7)

Carter, D. J., and J. C. Deeming. 1980. *Azygophleps albovittata* (Lepidoptera: Cossidae) attacking groundnuts in northern Nigeria, with descriptions of the immature and imaginal stages. *Bulletin of Entomological Research* 70:399–406. (Chapter 4)

Chapin, F. S. 1991. Integrated responses of plants to stress. *BioScience* 41:29–36. (Chapter 11)

Chapin, F. S., and S. J. McNaughton. 1989. Lack of compensatory growth under phosphorous deficiency in grazing-adapted grasses from the Serengeti Plains. *Oecologia* (Berlin) 79:551–557. (Chapter 11)

Chapin, F. S., D. A. Johnson, and J. D. McKendrick. 1980. Seasonal movement of nutrients in plants of differing growth form in an Alaskan tundra ecosystem: Implications for herbivory. *Journal of Ecology* 68:189–209. (Chapter 3)

Charnov, E. L. 1982. *The Theory of Sex Allocation.* Princeton University Press, Princeton, NJ. (Chapter 10)

Claridge, M. 1995. Species and speciation. *Trends in Ecology and Evolution* 10:38. (Chapters 7, 12)

Cockerell, T.D.K. 1890. Galls. *Nature* (London) 42:344. (Chapters 1, 6)

Coley, P. D., J. P. Bryant, and F. S. Chapin III. 1985. Resource availability and plant antiherbivore defense. *Science* 230:895–899. (Chapters 1, 4, 11, 12)

Collinge, S. K., and S. M. Louda. 1988. Herbivory by leaf miners in response to experimental shading of a native crucifer. *Oecologia* 75:559–566. (Chapter 4)

Collins, M., M. J. Crawley, and G. C. McGavin. 1983. Survivorship of the sexual and agamic generations of *Andricus quercuscalicis* on *Quercus cerris* and *Q. robur. Ecological Entomology* 8:133–138. (Chapter 3)

Colvill, K. E., and C. Marshall. 1981. The patterns of growth assimilation of $^{14}CO_2$ and distribution of ^{14}C-assimilate within vegetative plants of *Lolium perenne* at low and high density. *Annals of Applied Biology* 99:179–180. (Chapter 3)

Confer, J. L., and P. Paicos. 1985. Downy woodpecker predation on goldenrod galls. *Field Ornithology* 56:56–64. (Chapters 2, 8)

Confer, J. L., C. J. Hibbard, and D. Ebbets. 1986. Downy woodpecker reward rates from goldenrod gall insects. *The Kingbird* 36:188–192. (Chapter 2)

Cook, R. E. 1985. Growth and development in clonal plant populations. In *Population Biology and Evolution of Clonal Organisms*, ed. J.B.C. Jackson, L. W. Buss, and R. E. Cook, 259–296. Yale University Press, New Haven, CT. (Chapter 3)

Cooke, F., P. D. Taylor, C. M. Francis, and R. F. Rockwell. 1990. Directional selection and clutch size in birds. *American Naturalist* 136:261–267. (Chapter 9)

Cooper-Driver, G. A., and P. W. LeQuesne. 1987. Diterpenes as insect antifeedants and growth inhibitors—role in *Solidago* species. In *Allelochemicals: Role in Agriculture and Forestry*, ed. G. R. Waller, 534–550. American Chemical Society Symposium 330, ACS, Washington, DC. (Chapter 4)

Coquillett, D. W. 1910. The type species of the North American genera of Diptera. *United States National Museum Proceedings* 37:534. (Chapter 2)

Cornell, H. V. 1983. The secondary chemistry and complex morphology of galls formed by the Cynipinae (Hymenoptera): Why and how? *American Midland Naturalist* 110:225–232. (Chapters 3, 6)

———. 1990. Survivorship, life history, and concealment: A comparison of leaf miners and gall formers. *American Naturalist* 136:581–597. (Chapters 1, 2)

Cott, H. B. 1940. *Adaptive Coloration in Animals.* Oxford University Press, Oxford, England. (Chapter 10)

Courtney, S. P. 1981. Coevolution of pierid butterflies and their cruciferous food plants. III. *Anthocharis cardamines* (L.). Survival, development and oviposition on different host plants. *Oecologia* 51:91–96. (Chapter 5)

———. 1982. Coevolution of pierid butterflies and their cruciferous food plants. IV. Crucifer apparency and *Anthocharis cardamines* (L.). *Oecologia* 52:258–265. (Chapter 5)

Courtney, S. P., and T. T. Kibota. 1990. Mother doesn't know best: Selection of hosts by ovipositing insects. In *Insect-Plant Interactions,* vol. 2, ed. E. A. Bernays, 161–188. CRC Press, Boca Raton, FL. (Chapter 5)

Craig, T. P., J. K. Itami, and P. W. Price. 1988. A strong relationship between oviposition preference and larval performance on a shoot-galling sawfly. *Ecology* 70:1691–1699. (Chapter 5)

———. 1989. A strong relationship between oviposition preference and larval performance in a shoot-galling sawfly. *Ecology* 70:1691–1699. (Chapters 4, 5)

———. 1990. The window of vulnerability of a shoot-galling sawfly to attack by a parasitoid. *Ecology* 71:1471–1482. (Chapters 1, 5, 8)

Craig T. P., P. W. Price, and J. K. Itami. 1986. Resource regulation by a stem-galling sawfly on the arroyo willow. *Ecology* 67:419–425. (Chapters 3, 4, 5)

Craig T. P., P. W. Price, K. M. Clancy, G. L. Waring, and C. F. Sacchi. 1988. Forces preventing coevolution in the three-trophic-level system: Willow, a gall forming herbivore, and parasitoid. In *Chemical Mediation of Coevolution,* ed. K. C. Spencer, 57–80. Academic Press, New York. (Chapters 4, 12)

Craig, T. P., J. K. Itami, W. G. Abrahamson, and J. D. Horner. 1993. Behavioral evidence for host-race formation in *Eurosta solidaginis. Evolution* 47:1696–1710. (Chapters 2, 5, 7)

Craig, T. P., J. K. Itami, J. D. Horner, and W. G. Abrahamson. 1994. Host shifts and speciation in gall-forming insects. In *Gall-Forming Insects,* ed. P. Price, W. Mattson, and Y. Baranchikov, 194–207. USDA Forest Service, North Central Experiment Station, General Technical Report NC-174. (Chapter 7)

Craig, T. P., C. Shantz, J. K. Itami, and W. G. Abrahamson. In prep. Multiple determinants of oviposition preference in *Eurosta solidaginis.* (Chapters 5, 7)

Crawley, M. J. 1983. *Herbivory. The Dynamics of Animal-Plant Interactions.* Blackwell Scientific, Oxford, England. (Chapter 11)

———. 1993. On the consequences of herbivory. *Evolutionary Ecology* 7:124–125. (Chapter 11)

REFERENCES

Croat, T. 1972. *Solidago canadensis* complex of the Great Plains. *Brittonia* 24: 317–326. (Chapter 2)

Curran, C. H. 1923. Two varieties of *Eurosta solidaginis* Fitch (Trypetidae, Dipt.). *Entomology News* 34:302. (Chapter 2)

——. 1925. Note on the rearing of *Eurosta solidaginis* and variety *fascipennis* (Trypaneidae, Dipt.). *Canadian Entomologist* 57:128–129. (Chapter 2)

Darwin, C. D. 1859. *On the Origin of Species. A Facsimile.* Printed 1966. Harvard University Press, Cambridge, MA. (Chapters 6, 9)

Dawkins, R. 1982. *The Extended Phenotype: The Gene as a Unit of Selection.* Freeman, San Francisco. (Chapters 2, 6, 10)

——. 1986. *The Blind Watchmaker.* Norton, New York. (Chapter 1)

de Jong, G. 1990a. Genotype-by-environment interaction and the genetic covariance between environments: Multilocus genetics. *Genetica* 81: 171–177. (Chapters 6, 9)

——. 1990b. Quantitative genetics of reaction norms. *Journal of Evolutionary Biology* 3:447–468. (Chapters 6, 9)

Dethier, V. G. 1954. Evolution of feeding preferences in phytophagous insects. *Evolution* 8:33–54. (Chapter 11)

deWit, C. T. 1960. On competition. *Versl. Landbouwk. Onderz. Rijkslandb. Proefstn.* 66:1–82. (Chapter 2)

Diehl, S. R., and G. L. Bush. 1984. An evolutionary and applied perspective of insect biotypes. *Annual Review of Entomology* 29:471–504. (Chapter 7)

——. 1989. The role of habitat preference in adaptation and speciation. In *Speciation and Its Consequences,* ed. D. Otte and J. A. Endler, 345–365. Sinauer Associates, Sunderland, MA. (Chapter 7)

Dieleman, F. L. 1969. Effects of gall midge infestation on plant growth and growth regulating substances. *Entomologia experimentales et applicata* 12: 745–749. (Chapter 3)

Dirzo, R., and J. L. Harper. 1992. Experimental studies on slug-plant interactions. IV. The performance of cyanogenic and acyanogenic morphs of *Trifolium repens* in the field. *Journal of Ecology* 70:119–138. (Chapters 11, 12)

Dobzhansky, T. 1968. On some fundamental concepts of Darwinian biology. In *Evolutionary Biology,* vol. 2, ed. T. Dobzhansky, M. K. Hecht, and W. C. Steere, 1–34. Appleton-Century-Crofts, New York. (Chapter 10)

——. 1973. Nothing in biology makes sense except in the light of evolution. *American Biology Teacher* 35:125–129. (Chapter 1)

Dodge, K. L., P. W. Price, J. Kettunen, and J. Tahvanainen. 1990. Preference and performance of the leaf beetle *Disonycha pluriligata* (Coleoptera: Chrysomelidae) in Arizona and comparisons with beetles in Finland. *Environmental Entomology* 19:905–910. (Chapter 5)

Dreger-Jauffret, F., and J. D. Shorthouse. 1992. Diversity of gall-inducing insects and their galls. In *Biology of Insect-induced Galls,* ed. J. D. Shorthouse and P. Rohfritsch, 8–33. Oxford University Press, New York. (Chapter 3)

Dykhuizen, D. E., A. M. Dean, and D. L. Hartl. 1987. Metabolic flux and fitness. *Genetics* 115:25–31. (Chapter 11)

409

REFERENCES

Ecale, C. L., and E. A. Backus. 1995. Mechanical and salivary aspects of potato leafhopper probing in alfalfa stems. *Entomologia Experimentalis et Applicata* 77:121–132. (Chapter 3)

Edenius, L., K. Danell, and R. Bergstrom. 1993. Impact of herbivory and competition on compensatory growth in woody plants: Winter browsing by moose on Scots pine. *Oikos* 66:286–292. (Chapter 11)

Edmunds G. F., and D. N. Alstad. 1978. Coevolution of insect herbivores and conifers. *Science* 199:941–945. (Chapter 1)

Edson, K. M., S. B. Vinson, D. B. Stoltz, and M. D. Summers. 1981. Virus in a parasitoid wasp: Suppression of the cellular immune response in the parasitoid's host. *Science* 211:582–583. (Chapter 3)

Ehrlich, P. R., and P. H. Raven. 1964. Butterflies and plants: A study in co-evolution. *Evolution* 18:586–608. (Chapter 7)

Elzen, G. W. 1983. Minireview: Cytokinins and insect galls. *Comparative Biochemistry Physiology* 76:17–19. (Chapter 3)

Endler, J. A. 1980. Natural selection on color patterns in *Poecilia reticulata*. *Evolution* 34:76–91. (Chapter 9)

———. 1986. *Natural Selection in the Wild*. Princeton University Press, Princeton, NJ. (Chapters 1, 8, 9, 11)

Fajer, E. D., M. D. Bowers, and F. A. Bazzaz. 1992. The effect of nutrients and enriched CO_2 environments on production of carbon-based allelochemicals in *Plantago:* A test of the carbon-nutrient balance hypothesis. *American Naturalist* 140:707–723. (Chapter 11)

Falconer, D. S. 1989. *Introduction to Quantitative Genetics*. 3d ed. Longman, Harlow, Sussex, England. (Chapters 6, 8, 9)

Farrell, B. D., C. Mitter, and D. J. Futuyma. 1992. Diversification at the insect-plant interface. *BioScience* 42:34–42. (Chapter 1)

Feder, J. L. 1995. The effects of parasitoids on sympatric host races of *Rhagoletis pomonella* (Diptera: Tephritidae). *Ecology* 76:801–803. (Chapter 7)

Feder, J. L., and G. L. Bush. 1989a. A field test of differential host-plant usage between 2 sibling species of *Rhagoletis pomonella* fruit flies (Diptera: Tephritidae) and its consequences for sympatric models of speciation. *Evolution* 43:1813–1819. (Chapters 7, 12)

———. 1989b. Gene frequency clines for host races of *Rhagoletis pomonella* in the midwestern United States. *Heredity* 63:245–266. (Chapters 7, 12)

———. 1991. Genetic variation among apple and hawthorn host races of *Rhagoletis pomonella* across an ecological transition zone in the midwestern United States. *Entomologia experimentalis et applicata* 59:249–265. (Chapters 7, 12)

Feder, J. L., C. A. Chilcote, and G. L. Bush. 1988. Genetic differentiation between sympatric host races of the apple maggot fly *Rhagoletis pomonella*. *Nature* 336:61–64. (Chapter 7)

———. 1989a. Inheritance and linkage relationships of allozymes in the apple maggot fly. *Journal of Heredity* 80:277–283. (Chapters 7, 12)

———. 1989b. Are the apple maggot, *Rhagoletis pomonella*, and blueberry maggot, *R. mendax*, distinct species? Implications for sympatric speciation. *Entomologia experimentalis et applicata* 51:113–123. (Chapters 7, 12)

—. 1990a. Regional, local and microgeographic allele frequency variation between apple and hawthorn populations of *Rhagoletis pomonella* in western Michigan. *Evolution* 44:595–608. (Chapter 7)

—. 1990b. The geographic pattern of genetic differentiation between host association populations of *Rhagoletis pomonella* (Diptera: Tephritidae) in the eastern United States and Canada. *Evolution* 44:570–594. (Chapter 7)

Feder, J. L., K. Reynolds, W. Go, and E. C. Wang. 1995. Intra- and interspecific competition and host race formation in the apple maggot fly, *Rhagoletis pomonella* (Diptera: Tephritidae). *Oecologia* 101:416–425. (Chapters 7, 12)

Feeny, P. 1976. Plant apparency and chemical defense. *Recent Advances in Phytochemistry* 10:1–40. (Chapter 1)

Feeny, P., S. Stadler, I. Ahman, and M. Carter. 1989. Effects of plant odor on oviposition by the black swallowtail butterfly, *Papilio polyxenes* (Lepidoptera: Papilionidae). *Journal of Insect Behavior* 2:803–827. (Chapter 5).

Felsenstein, J. 1981. Skepticism towards Santa Rosalia, or why are there so few kinds of animals? *Evolution* 35:124–138. (Chapter 7)

Felt, E. P. 1917. *Key to American Insect Galls*. New York State Museum Bulletin 200, Albany, NY. (Chapter 7)

—. 1940. *Plant Galls and Gall Makers*. Hafner, New York. (Chapters 2, 3, 7)

Fernald, M. L. 1950. *Gray's Manual of Botany*. 8th ed. Van Nostrand, Reinhold, New York. (Chapter 2)

Fernandes, G. W. 1990. Hypersensitivity: A neglected plant resistance mechanism against insect herbivores. *Environmental Entomology* 19:1173–1182. (Chapters 3, 4)

Fisher, R. A. 1918. The correlation between relative on the supposition of Mendelian inheritance. *Transaction of the Royal Society of Edinburgh* 52:399–433. (Chapter 6)

Fisher, R. A., and E. B. Ford. 1947. The spread of a gene in natural conditions in a colony of the moth *Panaxia dominula* L. *Heredity* 1:143–174. (Chapter 9)

Foote, R. H., F. L. Blanc, and A. L. Norrbom. 1993. *Handbook of the Fruit Flies (Diptera: Tephritidae) of America North of Mexico*. Comstock, Ithaca, NY. (Chapter 2)

Fowler, S. V., and J. H. Lawton. 1985. Rapidly induced defenses and talking trees: The Devil's advocate position. *American Naturalist* 126:181–195. (Chapter 4)

Fox, C. W. 1993. A quantitative genetic analysis of oviposition preference and larval performance on two hosts in the bruchid beetle, *Callosobruchus maculatus*. *Evolution* 47:166–175. (Chapter 5)

Fox, C. W., and R. G. Lalonde. 1993. Host confusion and the evolution of insect diet breadths. *Oikos* 67:577–581. (Chapter 5)

Frank, S. A. 1993. Coevolutionary genetics of plants and pathogens. *Evolutionary Ecology* 7:45–75. (Chapters 9, 11)

Frank, S. A., and M. Slatkin. 1992. Fisher's fundamental theorem of natural selection. *Trends in Ecology and Evolution* 7:92–95. (Chapters 1, 9)

411

REFERENCES

Friedman, D., and P. Alpert. 1991. Reciprocal transport between ramets increases growth of *Fragaria chiloensis* when light and nitrogen occur in separate patches but only if patches are rich. *Oecologia* 86:76–80. (Chapter 3)

Fritz, R. S. 1990. Effects of genetic and environmental variation on resistance of willow to sawflies. *Oecologia* 82:325–332. (Chapters 4, 11)

Fritz, R. S., and E. L. Simms, eds. 1992. *Plant Resistance to Herbivores and Pathogens*. University of Chicago Press, Chicago. (Chapters 1, 3, 4)

Fritz, R. S., W. S. Gaud, C. F. Sacchi, and P. W. Price. 1987. Patterns of intra- and interspecific association of gall-forming sawflies in relation to shoot size on their willow host plant. *Oecologia (Berl.)* 73:159–169. (Chapter 5)

Futuyma, D. J. 1983. Evolutionary interactions among herbivorous insects and plants. In *Coevolution,* ed. D. J. Futuyma and M. Slatkin, 207–231. Sinauer Associates, Sunderland, MA. (Chapter 7)

———. 1986. *Evolutionary Ecology.* 2d ed. Sinauer Associates, Sunderland, MA. (Chapters 7, 12)

———. 1987. The role of behavior in host-associated divergency in herbivorous insects. In *Evolutionary Genetics of Invertebrate Behavior,* ed. M. D. Huettel, 295–302. Plenum Press, New York. (Chapters 7, 11)

Futuyma, D. J., and S. S. McCafferty. 1990. Phylogeny and the evolution of host plant associations in the leaf beetle genus *Ophraella* (Coleoptera, Chrysomelidae). *Evolution* 44:1885–1913. (Chapters 1, 12)

Futuyma, D. J., and G. Moreno. 1988. The evolution of ecological specialization. *Annual Review of Ecology and Systematics* 19:207–234. (Chapter 11)

Futuyma, D. J., and S. C. Peterson. 1985. Genetic variation in the use of resources by insects. *Annual Review of Entomology* 30:217–238. (Chapter 7)

Futuyma, D. J., and T. E. Phillipi. 1987. Genetic variation and covariation in responses to host plants by *Alsophila pometaria*. *Evolution* 41:269–279. (Chapter 7)

Futuyma, D. J., and S. S. Wasserman. 1981. Food plant specialization and feeding efficiency in the tent caterpillars, *Malacosoma disstria* Hubner and *M. americanum* (F.). *Entomologia experimentalis et applicata* 30:106–110. (Chapter 11)

Gavrilets, S., and S. M. Scheiner. 1993a. The genetics of phenotypic plasticity. 5. Evolution of reaction norm shape. *Journal of Evolutionary Biology* 6:31–48. (Chapters 6, 9)

———. 1993b. The genetics of phenotypic plasticity. 6. Theoretical predictions for directional selection. *Journal of Evolutionary Biology* 6:49–68. (Chapters 6, 9)

Gay, P. E., P. J. Grubb, and H. J. Hudson. 1982. Seasonal changes in the concentrations of nitrogen, phosphorus and potassium, and in the density of mycorrhiza, in biennial and matrix-forming perennial species of closed chalkland turf. *Journal of Ecology* 70:571–593. (Chapter 3)

Geervliet, J.B.F., L.E.M. Vet, and M. Dicke. 1994. Volatiles from damaged plants as major cues in long-range host-searching by the specialist parasitoid *Cotesia rubecula*. *Entomologia experimentalis et applicata* 73:289–297. (Chapter 8)

Gershenzon, J. 1984. Changes in the level of plant secondary metabolites under water and nutrient stress. In *Phytochemical Adaptation to Stress*, ed. B. N. Timmerman, C. Steelink, and F. A. Loewus, 273–320. Plenum Press, New York. (Chapter 4)

Gibbs, H. L. 1988. Heritability and selection on clutch size in Darwin's medium ground finch (*Geospiza fortis*). *Evolution* 42:750–762. (Chapter 8)

Giddens, J. H., F. Perkins, and L. C. Walker. 1962. Movement of nutrients in coastal Bermuda grass. *Agronomy Journal* 54:379–391. (Chapter 3)

Givens, K. T. 1982. Polyploidy, reproductive ecology, and stem gall insect interactions in the *Solidago canadensis* goldenrod complex. M.S. thesis, Bucknell University, Lewisburg, PA. (Chapter 2)

Givnish, T. J. 1981. Serotiny, geography, and fire in the pine barrens of New Jersey. *Evolution* 35:101–123. (Chapter 9)

Gleason, H. A., and A. Cronquist. 1992. *Manual of Vascular Plants of Northeastern United States and Adjacent Canada*. 2d ed. New York Botanical Garden, Bronx. (Chapter 2)

Godfray, H.C.J. 1993. *Parasitoids: Behavioral and Evolutionary Ecology*. Princeton University Press, Princeton, NJ. (Chapter 10)

Goldberg, D. E., and P. A. Werner. 1983. The effects of size of opening in vegetation and litter cover on seedling establishment of goldenrods (*Solidago* spp.). *Oecologia* 60:149–155. (Chapter 2)

Gomulkiewicz, R., and M. Kirkpatrick. 1992. Quantitative genetics and the evolution of reaction norms. *Evolution* 46:390–411. (Chapters 6, 9)

Gould, F. 1984. Role of behavior in the evolution of insect adaptation to insecticides and resistant host plants. *Bulletin of the Entomological Society of America* 30:33–41. (Chapter 11)

Gould, S. J., and R. C. Lewontin. 1979. The spandrels of San Marco and the panglossian paradigm: A critique of the adaptationist program. *Proceedings of the Royal Society of London*, ser. B 205:581–598. (Chapter 6)

Gould, S. J., and E. S. Vrba. 1982. Exaptation—a missing term in the science of form. *Paleontology* 8:4–15. (Chapter 10)

Grant, B. R., and P. R. Grant. 1993. Evolution of Darwin's finches caused by a rare climatic event. *Proceedings of the Royal Society of London*, ser. B 251:111–117. (Chapter 9)

Greenwald, L. L., K. D. McCrea, and W. G. Abrahamson. 1985. The effects of soil moisture and density on the competitive interactions between *Solidago canadensis* and *S. altissima*. *Bulletin of the Ecological Society of America* 66:182. (Chapter 2)

Greenwood, J.J.D. 1984. The functional basis of frequency-dependent food selection. *Biological Journal of the Linnean Society of London* 23:177–199. (Chapters 9, 11)

Gross, R. S., and P. W. Price. 1988. Plant influences on parasitism of two leafminers: A test of enemy-free space. *Ecology* 69:1506–1516. (Chapter 8)

Gross, R. S., and P. A. Werner. 1983. Relationships among flowering phe-

413

nology, insect visitors and seed set of individuals: Experimental studies on four cooccurring species of goldenrod (*Solidago:* Compositae). *Ecological Monographs* 53:95–117. (Chapter 2)

Hanks, L. M., T. D. Paine, and J. G. Millar. 1993. Host species preference and larval performance in the wood-boring beetle *Phorocantha semipuncta* F. *Oecologia* 95:22–29. (Chapters 5, 12)

Hare, J. D. 1992. Effects of plant variation on herbivore-natural enemy interactions. In *Plant Resistance to Herbivores and Pathogens*, ed. R. S. Fritz and E. L. Simms, 278–298. University of Chicago Press, Chicago. (Chapters 8, 12)

Harper, J. L. 1977. *Population Biology of Plants*. Academic Press, London. (Chapter 3)

Harrington, W. H. 1895. Occupants of the galls of *Eurosta solidaginis*, Fitch. *Canadian Entomologist* 27:197–198. (Chapter 2)

Harris, P. 1980. Effects of *Urophora affinis* Frfld. and *U. quadrifasciata* (Meig.) (Diptera: Tephritidae) on *Centaurea diffusa* Lam. and *C. maculosa* Lam. (Compositae). *Zeitschrift für Angewandte Entomologie* 90:190–201. (Chapter 3)

Hartl, D. L., D. E. Dykhuizen, and A. M. Dean. 1985. Limits of adaptation: The evolution of selective neutrality. *Genetics* 111:655–674. (Chapter 11)

Hartley, S. E. 1992. The insect galls on willow. *Proceedings of the Royal Society of Edinburgh* 98B:91–104. (Chapters 3, 4, 7)

Hartley, S. E., and J. H. Lawton. 1992. Host-plant manipulation by gall-insects: A test of the nutrition hypothesis. *Journal of Animal Ecology* 61: 113–119. (Chapter 7)

Hartnett, D. C., and W. G. Abrahamson. 1979. The effects of stem gall insects on life history patterns in *Solidago canadensis*. *Ecology* 60:910–917. (Chapters 2, 3, 8)

Hartnett, D. C., and F. A. Bazzaz. 1983a. Physiological integration among intraclonal ramets in *Solidago canadensis*. *Ecology* 64:779–788. (Chapter 3)

———. 1983b. The genet and ramet population dynamics of *Solidago canadensis* in an abandoned field. *Journal of Ecology* 73:407–413. (Chapter 12)

———. 1985. The integration of neighbourhood effects by clonal genets in *Solidago canadensis*. *Journal of Ecology* 73:415–427. (Chapter 3)

Hartvigsen, G., D. A. Wait, and J. S. Coleman. 1995. Tri-trophic interactions influenced by resource availability: Predator effects on plant performance depend on plant resources. *Oikos* 74:763–768. (Chapter 4)

Harvey, P.H., and M. D. Pagel. 1991. *The Comparative Method in Evolutionary Biology*. Oxford University Press, New York. (Chapter 1)

Haukioja, E., and S. Neuvonen. 1985. Induced long-term resistance of birch foliage against defoliators: Defensive or incidental? *Ecology* 66:1303–1308. (Chapter 4)

Hawkins B. A., and R. J. Gagne. 1989. Determinants of assemblage size for the parasitoids of Cecidomyiidae (Diptera). *Oecologia* 81:75–88. (Chapters 1, 2)

Hawkins, B. A., and R. D. Goeden. 1982. Biology of a gall-forming *Tetrastichus* (Hymenoptera: Eulophidae) associated with gall midges on saltbush in southern California. *Annals of the Entomological Society of America* 75:444–447.

Heady, S. E., R. G. Lambert, and C. V. Covell, Jr. 1982. Determination of free amino acids in larval insect and gall tissues of the goldenrod, *Solidago canadensis* L. *Comparative Biochemistry and Physiology* 73:641–644. (Chapter 3)

Heard, T. A. 1995a. Oviposition and feeding preferences of a flower-feeding weevil, *Coelocephalapion aculeatum,* in relation to conspecific damage to its host-plant. *Entomologia Experimentalis et Applicata* 76:203–209. (Chapter 5)

———. 1995b. Oviposition preferences and larval performance of a flower-feeding weevil, *Coelocephalapion aculeatum,* in relation to host development. *Entomologia Experimentalis et Applicata* 76:195–201. (Chapters 5, 12)

Heinrich, B. 1976. The foraging specializations of individual bumblebees. *Ecological Monographs* 46:105–128. (Chapter 2)

Heinrich, B., and S. L. Collins. 1985. Caterpillar leaf damage and the game of hide-and-seek with birds. *Ecology* 64:592–602. (Chapter 8)

Henderson, C. R., Jr. 1982. Analysis of covariance in the mixed model: Higher-level, nonhomogeneous and random regressions. *Biometrics* 38:623–640. (Chapter 6)

Hendrix, S. D. 1988. Herbivory and its impact on plant reproduction. In *Plant Reproductive Ecology: Patterns and Strategies,* ed. J. Lovett Doust and L. Lovett Doust, 246–263. Oxford University Press, Oxford, England. (Chapter 11)

Herms, D. A., and W. J. Mattson. 1992. The dilemma of plants: To grow or defend? *Quarterly Review of Biology* 67:283–335. (Chapters 1, 11, 12)

Hess, M. D. 1993. The potential role of intraspecific larval competition in host shifts of the herbivorous ball-gallmaker *Eurosta solidaginis* (Diptera: Tephritidae). M.S. thesis, Bucknell University, Lewisburg, PA. (Chapter 2)

Hess, M. D., W. G. Abrahamson, and J. M. Brown. 1996. Intraspecific competition in the goldenrod ball-gallmaker: Larval mortality, adult fitness, ovipositional and host plant response. *American Midland Naturalist* 136:121–133. (Chapters 2, 3, 4, 5, 7)

Hickman, J. C., and L. F. Pitelka. 1975. Dry weight indicated energy allocation in ecological strategy analysis of plants. *Oecologia* 21:117–121. (Chapter 3)

Hirose, T. 1975. Relations between turnover rate, resource utility and structure of some plant populations: A study in the matter budgets. *Journal of the Faculty of Science, University of Tokyo, III.* 11:355–407. (Chapter 2)

Hirose, T., and M. Monsi. 1975. On a meaning of life form of plants in relation to their nitrogen utilization. In *Nitrogen Fixation and Nitrogen Cycle,* ed. H. Takahashi, 87–94. Japanese Committee for the International Biological Program, vol. 12, University of Tokyo Press, Japan. (Chapter 3)

Hoagland, D. R., and D. I. Arnon. 1950. The water-culture method for growing plants without soil. *California Agricultural Experimental Station Circular* 347 (revised). (Chapter 3)

Hoffman, C. A. 1985. Gall-forming response of *Curcubita foctadissima* to infestation by the squash stem borer, *Melitta curcubitae. Bulletin of the Ecological Society of America* 66:195. (Chapter 1)

Hoffmann, A. A., and P. A. Parson. 1991. *Evolutionary Genetics and Environmental Stress.* Oxford University Press, Oxford, England. (Chapter 11)

Holland, E. A., W. J. Parton, J. K. Detling, and D. L. Coppock. 1992. Physiological responses of plant populations to herbivory and their consequences for ecosystem nutrient flow. *American Naturalist* 140:685–706. (Chapter 11)

Hollenbach, H. G. 1984. Inheritance of seedling traits in *Solidago altissima.* Senior Honors thesis, Bucknell University, Lewisburg, PA. (Chapter 6)

Holling, C. S. 1959. The components of predation as revealed by a study of small-mammal predation on the European pine sawfly. *Canadian Entomologist* 91:293–320. (Chapter 9)

Hori, K. 1992. Insect secretions and their effect on plant growth, with special reference to Hemipterans. In *Biology of Insect-Induced Galls,* ed. J. D. Shorthouse and O. Rohfritsch, 157–170. Oxford University Press, New York. (Chapters 2, 3)

Horner J. D., and W. G. Abrahamson. 1992. Influence of plant genotype and environment on oviposition preference and offspring survival in a gallmaking herbivore. *Oecologia* 90:323–332. (Chapters 3, 4, 5, 7, 12)

———. Submitted. Influence of plant genotype and early-season water deficits on oviposition preference and offspring performance in a gallmaking herbivore. *Oikos.* (Chapters 4, 5, 7)

Houle, D. 1992. Comparing evolvability and variability of quantitative traits. *Genetics* 130:195–204. (Chapter 6)

Hovanitz, W. 1959. Insects and plant galls. *Scientific American* 201:151–162. (Chapter 3)

How, S. T., W. G. Abrahamson, and T. P. Craig. 1993. Role of host plant phenology in host use by *Eurosta solidaginis* (Diptera: Tephritidae) on *Solidago* (Compositae). *Environmental Entomology* 22:388–396. (Chapters 3, 4, 5, 7)

How, S. T., W. G. Abrahamson, and M. J. Zivitz. 1994. Disintegration of clonal connections in *Solidago altissima* (Compositae). *Bulletin of the Torrey Botanical Club* 121:338–344. (Chapters 3, 9)

Hull, R. 1974. Integrated control of pests and diseases in sugar beet. In *Biology in Pest and Disease Control,* ed. D. P. Jones and M. E. Solomon, 269–276. Blackwell Scientific, Oxford, England. (Chapter 4)

Hulspas-Jordan, P. M., and J. C. van Lenteren. 1978. The relationship between host-plant leaf structure and parasitization efficiency of the parasitic wasp *Encarsia formosa* Gahan (Hymenoptera: Aphelinidae). *Med. Fac. Landbouww. Rijkuniv. Ghent* 43:431–440. (Chapter 12)

Hunter, M. D., and J. C. Schultz. 1993. Induced plant defenses breached: Phytochemical induction protects an herbivore from disease. *Oecologia* 94:1195–1203. (Chapter 8)

Hurlbutt, B. 1987. Sexual size dimorphism in parasitoid wasps. *Biological Journal of the Linnean Society* 30:63–89. (Chapter 10)

Hutchings, M. J. 1988. Differential foraging for resources and structural plasticity in plants. *Trends in Ecology and Evolution* 3:200–204. (Chapter 3)

Iason, G. R., and A. J. Hester. 1993. The response of heather (*Calluna vulgaris*) to shade and nutrients—predictions of the carbon-nutrient balance hypothesis. *Journal of Ecology* 81:75–80. (Chapter 4)

Jaenike, J. 1981. Criteria for ascertaining the existence of host races. *American Naturalist* 117:830–834. (Chapter 7)

———. 1990. Host specialization in phytophagous insects. *Annual Review of Ecology and Systematics* 21:243–273. (Chapter 7)

Janzen, D. H. 1966. Coevolution of mutualism between ants and acacias in Central America. *Evolution* 20:249–275. (Chapter 12)

———. 1980. When is it coevolution? *Evolution* 34:611–612. (Chapter 10)

Johnson, C. D., and D. H. Siemens. 1991a. Expanded oviposition range by a seed beetle (Coleoptera: Bruchid) in proximity to a normal host. *Environmental Entomology* 20:1577–1582. (Chapter 7)

———. 1991b. Interactions between a new species of *Acanthoscelides* and a species of Verbenaceae. A new host family for Bruchidae (Coleoptera). *Annals of the Entomological Society of America* 84:165–169. (Chapter 7)

Johnson, P. A., F. C. Hoppensteadt, J. J. Smith, and G. L. Bush. 1996. Conditions for sympatric speciation: A diploid model incorporating habitat fidelity and non-habitat assortative mating. *Evolutionary Ecology* 10:187–205. (Chapters 7, 12)

Johnston, R. F., D. M. Niles, and S. A. Rohwer. 1972. Hermon Bumpus and natural selection in the house sparrow *Passer domesticus*. *Evolution* 26:20–31. (Chapter 8)

Jones, D. A. 1989. 50 years of studying the scarlet tiger moth. *Trends in Ecology and Evolution* 4:298–301. (Chapter 9)

Jones, C. G., and J. H. Lawton. 1991. Plant chemistry and insect species richness of British umbellifers. *Journal of Animal Ecology* 60:767–778. (Chapter 7)

Joshi, S. C., P. Tandon, and A.L.S. Rajee. 1985. Changes in certain oxidative enzymes and phenolics in *Camellia sinenses–Elaeocarpus lancifolius* leaf-roll galls. *Cecidologia Internationale* 6:51–58. (Chapter 4)

Judd, W. W. 1953. Insects reared from goldenrod galls caused by *Eurosta solidaginis* Fitch (Diptera: Trypetidae) in southern Ontario. *Canadian Entomologist* 85:294–296. (Chapter 2)

Kalisz, S. 1986. Variable selection on the timing of germination on *Collinsia verna* (Scrophulariaceae). *Evolution* 40:479–491. (Chapter 8)

417

REFERENCES

Karban, R. 1989. Fine scale adaptation of herbivorous thrips to individual host plants. *Nature* 340:60–61. (Chapters 1, 9)

———. 1990. Herbivore outbreaks on only young trees: Testing hypothesis about aging and induced resistance. *Oikos* 59:27–32. (Chapter 4)

———. 1992. Plant variation: Its effects on populations of herbivorous insects. In *Plant Resistance to Herbivores and Pathogens*, ed. R. S. Fritz and E. L. Simms, 195–215. University of Chicago Press, Chicago. (Chapter 12)

Karban, R., and J. R. Carey 1984. Induced resistance of cotton seedlings to mites. *Science* 225:53–54. (Chapter 4)

Karban, R., and S. Courtney. 1987. Intraspecific host plant choice: Lack of consequences for *Streptanthus tortuosus* (Cruciferae) and *Euchloe hyantis* (Lepidoptera: Pieridae). *Oikos* 48:243–248. (Chapter 5)

Keating, S. T., M. D. Hunter, and J. C. Schultz. 1990. Leaf phenolic inhibition of gypsy moth nuclear polyhedrosis virus: Role of polyhedral inclusion body aggregation. *Journal of Chemical Ecology* 16:1445–1457. (Chapter 8)

Keep, E., and J. B. Briggs. 1971. A survey of *Ribes* species for aphid resistance. *Annals of Applied Biology* 68:23–30. (Chapter 4)

Keese, M. C., and T. K. Wood. 1991. Host-plant mediated geographic-variation in the life-history of *Platycotis vittata* (Homoptera, Membracidae). *Ecological Entomology* 16:63–72. (Chapter 7)

Kelly, C. A. 1992. Spatial and temporal variation in selection on correlated life-history traits and plant size in *Chamaecrista fasciculata*. *Evolution* 46:1658–1673. (Chapter 8)

Kester, K. M., and P. Barbosa. 1994. Behavioral responses to host foodplants of two populations of the insect parasitoid *Cotesia congregata* (Say). *Oecologia* 99:151–157. (Chapter 8)

Kettlewell, B. 1973. *The Evolution of Melanism: The Study of a Recurring Necessity with Special Reference to Industrial Melanism in the Lepidoptera*. Clarendon Press, Oxford, England. (Chapters 1, 9)

Kigel, J. 1980. Analysis of regrowth patterns and carbohydrate levels in *Lolium multiflorum* Lam. *Annals of Botany* 45:91–101. (Chapter 11)

Kirst, G. O., and H. Rapp. 1974. Zur Physiologie der Galle von *Mikiola fagi* Htg. auf Blättern von *Fagus silvatica* L.: 2. Transport ^{14}C-markierter Assimilate aus dem befallenen Blatt und aus Nachbarblättern in die Galle. *Biochemie und Physiologie der Pflanzen (BPP)* 165:445–456. (Chapter 3)

Knerer, G., and C. E. Atwood. 1973. Diprionid sawflies: Polymorphism and speciation. *Science* 179:1090–1099. (Chapter 7)

Knight, R. L. 1954. The genetics of jassid resistance in cotton. *Journal of Genetics* 52:199–207. (Chapter 4)

Kogan, M. 1975. Plant resistance in pest management. In *Introduction to Insect Pest Management*, ed. R. L. Metcalf and W. H. Luckmann, 103–145. John Wiley and Sons, New York. (Chapter 11)

Koopman, B. O. 1980. *Search and Screening*. Pergamon Press, New York. (Chapter 9)

REFERENCES

Koptur, S. 1991. Extrafloral nectaries of herbs and trees: Modeling the interaction with ants and parasitoids. In *Ant-Plant Interactions,* ed. C. R. Huxley and D. F. Cutler, 213–230. Oxford University Press, Oxford. (Chapter 12)

———. 1992. Extrafloral mediated interactions between insects and plants. In *Insect-Plant Interactions,* ed. E. A. Bernays, 81–129. Vol. 4. CRC Press, Boca Raton, FL. (Chapter 12)

Koptur, S., and J. H. Lawton. 1988. Interactions among vetches bearing extrafloral nectaries, their biotic protective agents and herbivores. *Ecology* 69:278–283. (Chapter 12)

Korona, R., and B. R. Levin. 1993. Phage-mediated selection and the evolution and maintenance of restriction-modification. *Evolution* 47:556–575. (Chapter 11)

Kosuge, T. 1969. The role of phenolics in host response to infection. *Annual Review of Phytopathology* 7:195–222. (Chapters 3, 4)

Kouki, J. 1993. Female's preference for oviposition site and larval performance in the water-lily beetle, *Galerucella nymphaeae* (Coleoptera: Chrysomelidae). *Oecologia* 93:42–47. (Chapter 5)

Krebs, R. A., J.S.F. Barker, and T. P. Armstrong. 1992. Coexistence of ecologically similar colonising species. III. *Drosophila aldrichi* and *D. buzzatii:* Larval performance on, and adult preference for, three *Opuntia* cactus species. *Oecologia* 92:362–372. (Chapter 5)

Küster, E. 1930. Anatomie der Gallen. *Schroders Handbook of Entomology* 1:1–197. (Chapter 3)

Lack, D. L. 1947. *Darwin's Finches.* University Press, Cambridge, England. (Chapter 1)

Lande, R. 1979. Quantitative genetic analysis of multivariate evolution, applied to brain:body size allometry. *Evolution* 33:402–416. (Chapters 8, 9)

Lande, R., and S. J. Arnold. 1983. The measurement of selection on correlated characters. *Evolution* 37:1210–1226. (Chapters 3, 9, 11)

Larsson, S., and B. Ekbom. 1995. Oviposition mistakes in herbivorous insects: Confusion or a step towards a new host plant? *Oikos* 72:155–160. (Chapters 5, 7)

Larsson, S., and D. R. Strong. 1992. Oviposition choice and larval survival of *Dasineura marginemtorquens* (Diptera: Cecidomyiidae) on resistant and susceptible *Salix vimimalis. Ecological Entomology* 17:227–232. (Chapter 5)

Layne, J. R., Jr. 1993. Winter microclimate of goldenrod spherical galls and its effects on the gall inhabitant *Eurosta solidaginis* (Diptera: Tephritidae). *Journal of Thermal Biology* 18:125–130. (Chapter 2)

Layne, J. R., Jr., R. E. Lee, Jr., and J. L. Huang. 1990. Inoculation triggers freezing at high subzero temperatures in a freeze-tolerant frog (*Rana sylvatica*) and insect (*Eurosta solidaginis*). *Canadian Journal of Zoology* 68:506–510. (Chapter 2)

Leddy, P. M., T. D. Paine, and T. S. Bellows, Jr. 1993. Ovipositional preference of *Siphoninus phillyrea* and its fitness on seven host plant species. *Entomologia experimental et applications* 68:822–827. (Chapter 5)

Lee, R. E., Jr., J. J. McGrath, R. T. Morason, and R. M. Taddeo. 1993. Survival of intracellular freezing, lipidcoalescence and osmotic fragility in body fat cells of the freeze-tolerant gall fly *Eurosta solidaginis*. *Journal of Insect Physiology* 39:445–450. (Chapter 2)

LeQuesne, P., G. A. Cooper-Driver, M. Villani, M. N. Do, P. A. Morrow, and D. A. Tonkyn. 1986. Biologically active diterpenoids from *Solidago* species-plant-insect-interactions. In *New Trends in Natural Products Chemistry 1986: Studies in Organic Chemistry*, vol. 26, ed. A. Rahman and P. W. LeQuesne, 271–281. Elsevier Science, Amsterdam. (Chapter 4)

Leroi, A. M., M. R. Rose, and G. V. Lauder. 1994. What does the comparative method reveal about adaptation. *American Naturalist* 143:381–402. (Chapter 1)

Levins, R. 1968. *Evolution in Changing Environments: Some Theoretical Explorations*. Princeton University Press, Princeton, NJ. (Chapter 2)

Lewis, W. H., and M.P.F. Elvin-Lewis. 1977. *Medical Botany: Plants Affecting Man's Health*. John Wiley and Sons, New York. (Chapter 2)

Lewontin, R. C. 1986. Gene, organism and environment. In *Evolution from Molecule to Man*, ed. D. S. Bendall, 273–286. Columbia University Press, New York. (Chapter 6)

Li, C. C. 1975. *Path Analysis, a Primer*. Boxwood Press, Pacific Grove, CA. (Chapter 9)

Lichter, J. P., A. E. Weis, and C. R. Dimmick. 1990. Growth and survivorship differences in *Eurosta* (Diptera; Tephritidae) gall sympatric host plants. *Environmental Entomology* 19:972–977. (Chapter 7)

Lincoln, D. E. 1985. Host-plant protein and phenolic resin effects on larval growth and survival of a butterfly. *Journal of Chemical Ecology* 11:1459–1467. (Chapter 4)

Lincoln, D. E., and D. Couvet. 1989. The effect of carbon supply on allocation to allelochemicals and caterpillar consumption of peppermint. *Oecologia* 78:112–114. (Chapter 4)

Linhart, Y. B. 1991. Disease, parasitism and herbivory: Multidimensional challenges in plant evolution. *Trends in Ecology and Evolution* 6:392–396. (Chapter 11, 12)

Lovett Doust, J. 1980. A comparative study of life history and resource allocation in selected Umbelliferae. *Biological Journal of the Linnean Society* 13:139–154. (Chapter 3)

———. 1981. Population dynamics and local specialization in a clonal perennial (*Ranunculus repens*). I. The dynamics of ramets in contrasting habitats. *Journal of Ecology* 69:743–755. (Chapter 3)

Lovett Doust, L., and J. L. Harper. 1980. The resource costs of gender and maternal support in an andromonoecious umbellifer, *Smyrnium olusatrum* L. *New Phytology* 85:251–264. (Chapter 3)

Lowenberg, G. J. 1994. Effects of floral herbivory on maternal reproduction in *Sanicula arctopoides* (Apiaceae). *Ecology* 75:359–369. (Chapter 11)

Lu, T., M. A. Menelaou, D. Vargas, F. R. Fronczek, and N. H. Fischer. 1993.

Polyacetylenes and diterpenes from *S. canadensis. Phytochemistry* 32: 1483–1488. (Chapter 4)

Luning, J. 1992. Phenotypic plasticity of *Daphnia pulex* in the presence of invertebrate predators: Morphological and life history responses. *Oecologia* 92:383–390. (Chapter 10)

Maddox, G. D., and N. Cappuccino. 1986. Genetic determination of plant susceptibility to an herbivorous insect depends on environmental context. *Evolution* 40:863–866. (Chapters 4, 11)

Maddox, G. D., and R. B. Root. 1987. Resistance to 16 diverse species of 6 herbivorous insects within a population of goldenrod, *Solidago altissima:* Genetic variation and heritability. *Oecologia* 72:8–14. (Chapters 2, 3, 4, 7)

———. 1990. Structure of the encounter between goldenrod (*Solidago altissima*) and its diverse insect fauna. *Ecology* 71:2115–2124. (Chapters 2, 4, 11, 12)

Maddox, G. D., R. E. Cook, P. H. Winberger, and S. Gardescu. 1989. Clone structure in four *Solidago altissima* (Asteraceae) populations: Rhizome connections within genotypes. *American Journal of Botany* 76:318–326. (Chapter 3)

Mallet, J. 1989. The evolution of insecticide resistance: Have the insects won? *Trends in Ecology and Evolution* 4:336–340. (Chapter 9)

Mani, M. S. 1964. *Ecology of Plant Galls.* Dr. W. Junk, The Hague. (Chapters 2, 7)

Mapes, C. C., and P. J. Davies. 1992. Hormonal requirements of ball gall and stem tissues of *Solidago altissima* in tissue culture. *Plant Physiology Supplement* 99:38. (Chapter 3)

Marquis, R. J. 1990. Genotypic variation in leaf damage in *Piper arieanum* (Piperaceae) by a multi-species assemblage of herbivores. *Evolution* 44: 104–120. (Chapter 1)

———. 1992a. The selective impact of herbivores. In *Plant Resistance to Herbivores and Pathogens: Ecology, Evolution and Genetics,* ed. R. S. Fritz and E. L. Simms, 301–325. University of Chicago Press, Chicago. (Chapter 11)

———. 1992b. A bite is a bite is a bite: Constraints on response to folivory in *Piper-arieianum* (Piperaceae). *Ecology* 73:143–152. (Chapter 11)

Maschinski, J., and T. G. Whitham. 1989. The continuum of plant responses to herbivory: The influence of plant association, nutrient availability and timing. *American Naturalist* 134:1–19. (Chapter 11)

May, R. M., and R. M. Anderson. 1983. Epidemiology and genetics in the coevolution of parasites and hosts. *Proceedings of the Royal Society of London* B219:281–313. (Chapter 3)

Maynard Smith, J. 1966. Sympatric speciation. *American Naturalist* 100:637–650. (Chapter 7)

Mayr, E. 1942. *Systematics and the Origin of Species from the Viewpoint of a Zoologist.* Columbia University Press, New York. (Chapter 7)

———. 1963. *Animal Species and Evolution.* Belknap Press of Harvard University Press, Cambridge, MA. (Chapter 7)

REFERENCES

————. 1970. *Populations, Species and Evolution.* Belknap Press of Harvard University Press, Cambridge, MA. (Chapter 6)

————. 1988. *Towards a New Philosophy of Biology: Observations of an Evolutionist.* Harvard University Press, Cambridge, MA. (Chapters 7, 12)

Mayr, E., and W. B. Provine, eds. 1980. *The Evolutionary Synthesis: Perspectives on the Unification of Biology.* Harvard University Press, Cambridge, MA. (Chapters 7, 12)

Mazer, S. J., and C. T. Schick. 1991. Constancy of population parameters for life history and floral traits in *Raphanus-sativus* L. 1. Norms of reaction and the nature of genotype by environment interactions. *Heredity* 67: 143–156. (Chapter 11)

McCalla, D. R., M. K. Genthe, and W. Hovanitz. 1962. Chemical nature of an insect gall growth factor. *Plant Physiology* 37:98–103. (Chapter 3)

McColloch, J. W. 1923. The Hessian fly in Kansas. Technical Bulletin 11, Agricultural Experiment Station, Kansas State Agricultural College, Manhattan, KS. (Chapter 3)

McCune, B. 1993. Multivariate analysis on the PC-ORD system. Oregon State University, Corvalis, OR. (Chapter 2)

McCrea, K. D., and W. G. Abrahamson. 1985. Evolutionary impacts of the goldenrod ball gallmaker on *Solidago altissima* clones. *Oecologia* 68:20–22. (Chapter 3)

————. 1987. Variation in herbivore infestation: Historical vs. genetic factors. *Ecology* 68:822–827. (Chapters 3, 4, 5, 7, 11)

McCrea, K. D., W. G. Abrahamson, and A. E. Weis. 1985. Goldenrod ball gall effects on *Solidago altissima:* ^{14}C translocation and growth. *Ecology* 66: 1902–1907. (Chapter 3)

McNaughton, S. J. 1983. Compensatory plant growth as a response to herbivory. *Oikos* 40:329–336. (Chapter 11)

McPheron, B. A., D. C. Smith, and S. H. Berlocher. 1988. Genetic differences between host races of *Rhagoletis pomonella.* Nature 336:64–66. (Chapter 7)

Mecum, L. K. 1994. Downy woodpecker (*Picoides pubescens*) predation on the goldenrod gallmaker *E. solidaginis.* M.S. thesis, Bucknell University, Lewisburg, PA. (Chapters 2, 8)

Melville, M. R., and J. K. Morton. 1982. A biosystematic study of the *Solidago canadensis* (Compositae) complex. I. The Ontario populations. *Canadian Journal of Botany* 60:976–997. (Chapter 2)

Menken, S. B. 1981. Host races and sympatric speciation in small ermine moths, *Yponomeutidae. Entomologia experimentalis et applications* 30:280–292. (Chapter 7)

Messina, F. J. 1978. Mirid fauna associated with old-field goldenrods (*Solidago:* Compositae) in Ithaca, N.Y. *Journal of the New York Entomological Society* 46:137–143. (Chapter 2)

Messina, F. J., and R. B. Root. 1980. Association between leaf beetles and meadow goldenrods (*Solidago* spp.) in central New York. *Annals of the Entomological Society of America* 73:641–646. (Chapter 2)

Meyer, G. A. 1993. A comparison of the impacts of leaf- and sap-feeding insects on growth and allocation of goldenrod. *Ecology* 74:1101–1116. (Chapter 2)

Meyer, G. A., and R. B. Root. 1993. Effects of herbivorous insects and soil fertility on reproduction of goldenrod. *Ecology* 74:1117–1128. (Chapters 2, 11)

Meyer, J. 1987. *Plant Galls and Gall Inducers.* Borntraeger, Berlin. (Chapter 3)

Meyer, J., and H. J. Maresquelle. 1983. *Anatomie des Galles.* Gebruder Borntraeger, Berlin. (Chapter 3)

Miles, P. W. 1968. Insect secretions in plants. *Annual Review of Phytopathology* 6:137–164. (Chapter 3)

Miller, J. R., and K. L. Strickler. 1984. Finding and accepting host plants. In *Chemical Ecology of Insects,* ed. W. J. Bell and R. T. Carde, 127–157. Chapman and Hall, London. (Chapters 4, 5)

Miller, W. E. 1959. Natural history notes on the goldenrod ball gall fly, *Eurosta solidaginis* (Fitch), and on its parasites, *Eurytoma obtusiventris* Gahan and *E. gigantea* Walsh. *Journal of the Tennessee Academy of Science* 34:246–251. (Chapter 2)

Mills, R. R. 1969. Effect of plant and insect hormones on the formation of the goldenrod gall. *National Cancer Institute Monograph* 31:487–491. (Chapter 3)

Milne, L. J. 1940. Autoecology of the golden-rod ball gall fly. *Ecology* 21:101–105. (Chapter 2)

Ming, Y. 1989. A revision of the genus *Eurosta* Loew with a scanning microscopic study of taxonomic characters (Diptera: Tephritidae). M.S. thesis, Washington State University, Pullman, WA. (Chapters 2, 7)

Mitchell, R. J. 1992. Testing evolutionary and ecological hypotheses using path analysis and structural equation modelling. *Functional Ecology* 6:123–129. (Chapter 9)

———. 1993. Using path analysis to study plant-pollinator interactions. In *Design and Analysis of Ecological Experiments,* ed. S. M. Scheiner and J. Gurevitch, 211–231. Chapman and Hall, New York. (Chapter 9)

Mitchell-Olds, T. J., and R. G. Shaw. 1987. Regression analysis of natural selection: Statistical inference and biological interaction. *Evolution* 41:1149–1161. (Chapters 8, 9)

Mitter, C., B. Farrell, and D. J. Futuyma. 1991. Phylogenetic studies of insect-plant interactions: Insights into the genesis of diversity. *Trends in Ecology and Evolution* 6:290–293. (Chapter 5)

Mivart, S. G. 1889. Professor Weisman's essay. *Nature* (London) 41:41. (Chapter 6)

Moeller, R., and M. T. Thogerson 1978. Predation by the downy woodpecker on the goldenrod gall fly larva. *Iowa Bird Life* 48:131–136. (Chapter 2)

Moran, N. 1981. Intraspecific variability in herbivore performance and host quality: A field study of *Uroleucon caligatum* (Homoptera: Aphididae) and its *Solidago* hosts (Asteraceae). *Ecological Entomology* 6:301–306. (Chapter 9)

Moran, N., and W. D. Hamilton. 1981. Low nutritive quality as a defense against herbivore. *Journal of Theoretical Biology* 86:247–254. (Chapter 8)

Mousseau, T. A., and D. A. Roff. 1987. Natural selection and the heritability of fitness components. *Heredity* 59:181–197. (Chapters 6, 9)

Mueller, L. D. 1988. Density-dependent population growth and natural selection in food-limited environments: The *Drosophila* model. *American Naturalist* 132:786–809. (Chapter 9)

Mugnano, J. A., R. E. Lee, Jr., and R. T. Taylor. 1996. Fat body cells and calcium phosphate spherules induce ice nucleation in the freeze-tolerant larvae of the gall fly *Eurosta solidaginis* (Diptera, Tephritidae). *The Journal of Experimental Biology* 199:465–471. (Chapter 2)

Mulligan, G. A., and J. N. Findlay. 1970. Reproductive systems and colonization in Canadian weeds. *Canadian Journal of Botany* 48:859–860. (Chapter 2)

Nei, M. 1978. Analysis of gene diversity in subdivided populations. *Proceedings of the National Academy of Science* 70:3321–3323. (Chapter 7)

Ng, D. 1988. A novel level of interaction in plant-insect systems. *Nature* 334:611–612. (Chapter 11)

Nitao, J. K., and M. R. Berenbaum. 1988. Laboratory rearing of the parsnip webworm *Depressaria pastinacella* (Lepidoptera: Oecophoridae). *Annals of the Entomological Society of America* 81:485–487. (Chapter 4)

Novak, J. A., and B. A. Foote. 1980. Biology and immature stages of fruit flies: The genus *Eurosta* (Diptera: Tephritidae). *Journal of the Kansas Entomological Society* 53:205–216. (Chapter 2)

Nur, N. 1988. The consequences of brood size for breeding blue tits. III. Measuring the cost of reproduction: Survival, future fecundity and differential dispersal. *Evolution* 42:351–366. (Chapter 9)

Nylin, S. 1988. Host plant specialization and seasonality in a polyphagous butterfly, *Polygonia c-album* (Lepidoptera: Nymphalidae). *Oikos* 53:381–386. (Chapter 5)

Nylin, S., and N. Janz. 1993. Oviposition preference and larval performance in *Polygonia c-album* (Lepidoptera: Nymphalidae): The choice between bad and worse. *Ecological Entomology* 18:394–398. (Chapter 5)

Obeso, J. R., and P. J. Grubb. 1994. Interactive effects of extent and timing of defoliation, and nutrient supply on reproduction in a chemically protected annual *Senecio vulgaris*. *Oikos* 71:506–514. (Chapter 4)

Owen, D. F., and R. G. Weigert. 1981. Mutualism between grasses and grazers: An evolutionary hypothesis. *Oikos* 36:367–368. (Chapter 11)

Päclt, J. V. 1972. Zur allgemein-biologischen Deutung der Pflanzengalle. *Beit. Biol. Pflanz.* 48:63–77. (Chapter 6)

Paige, K. N., and T. G. Whitham. 1987. Overcompensation in response to mammalian herbivory: The advantage of being eaten. *American Naturalist* 129:407–416. (Chapter 11)

Palmer, A. R. 1985a. Adaptive value of shell variation in *Thais lamellosa:* Effect of thick shells on vulnerability to and preference by crabs. *The Velliger* 27:349–356. (Chapter 10)

———. 1985b. Quantum changes in gastropod shell morphology need not reflect speciation. *Evolution* 39:699–705. (Chapter 10)

Papaj, D. R., and R. J. Prokopy. 1988. The effect of prior adult experience on components of habitat preference in the apple maggot fly (*Rhagoletis pomonella*). *Oecologia* 76:538–543. (Chapter 7)

Parejko, K. 1992. Embryology of *Chaoborous*-induced spines in *Daphnia pulex*. *Hydrobiologia* 231:77–84. (Chapter 10)

Parson, P. 1987. Evolutionary rates under environmental stress. *Evolutionary Biology* 21:311–347. (Chapter 9)

Payne, J. A., and S. H. Berlocher. 1995 Phenological and electrophoretic evidence for a new blueberry-infesting species in the *Rhagoletis pomonella* sibling species complex. *Entomologia experimentalis et applicata* 75:183–187. (Chapters 7,12)

Pilson, D. 1992. Insect distribution patterns and the evolution of host use. In *Plant Resistance to Herbivores and Pathogens*, ed. R. S. Fritz and E. L. Simms, 120–139. University of Chicago Press, Chicago. (Chapter 9)

Pimentel, D. 1968. Population regulation by genetic feedback. *Science* 159:1432–1437. (Chapter 10)

Pimentel, D., and R. Al-Hafidh. 1965. Ecological control of a parasite population by genetic evolution in the parasite-host system. *Annals of the Entomological Society of America* 58:1–6. (Chapter 10)

Pimentel, D., S. A. Levin, and D. A. Olson. 1978. Coevolution and the stability of exploiter-victim systems. *American Naturalist* 112:119–125. (Chapter 10)

Ping, C. 1915. Some inhabitants of the round gall of goldenrod. *Pomona Journal of Entomology and Zoology* 8:161–179. (Chapters 2, 8)

Pitelka, L. F., and J. W. Ashmun. 1985. Physiology and integration of ramets in clonal plants. In *Population Biology and Evolution of Clonal Organisms*, ed. J.B.C. Jackson, L. W. Buss, and R. E. Cook, 399–435. Yale University Press, New Haven, CT. (Chapter 3)

Plakidas J. D., and A. E. Weis. 1994. Depth associations and utilization patterns in the parasitoid guild of *Asphondylia rudbeckiae conspicua* (Diptera: Cecidomyiidae). *Environmental Entomology* 23:115–121. (Chapters 1, 2)

Pollard, A. J. 1992. The importance of deterrence: Responses of grazing animals to plant variation. In *Plant Resistance to Herbivores and Pathogens: Ecology, Evolution and Genetics*, ed. R. S. Fritz and E. L Simms, 216–238. University of Chicago Press, Chicago. (Chapter 11)

Preszler, R. W., and P. W. Price. 1995. A test of plant-vigor, plant-stress, and plant-genotype effects on leaf-minor oviposition and performance. *Oikos* 74:485–492. (Chapters 5, 12)

Price, G. R. 1970. Selection and covariance. *Nature* 227:520–521. (Chapter 9)

Price, P. W. 1980. *Evolutionary Biology of Parasites*. Princeton University Press, Princeton, NJ. (Chapters 2, 7)

———. 1991. The plant vigor hypothesis and herbivore attack. *Oikos* 62:244–251. (Chapters 5, 11, 12)

———. 1992. Evolution and ecology of gall-inducing sawflies. In *Biology of Insect-Induced Galls*, ed. J. D. Shorthouse and O. Rohfritsch, 208–224. Oxford University Press, Oxford, England. (Chapter 6)

———. 1994. Phylogenetic constraints, adaptive syndromes, and emergent properties: From individuals to population dynamics. *Researches on Population Ecology* 36:3–14. (Chapter 5)

Price, P. W., and K. M. Clancy. 1986. Interactions among three trophic levels: Gall size and parasitoid attack. *Ecology* 67:1593–1600. (Chapter 8)

Price, P. W., and T. Ohgushi. 1995. Preference and performance linkage in a *Phyllocolpa* sawfly on the willow, *Salix miyabeana*, on Hokkaido. *Researches on Population Ecology* 37:23–28. (Chapter 5)

Price, P. W., and M. F. Willson. 1976. Some consequences for a parasitic herbivore, the milkweed longhorn beetle, *Tetraopes tetraophthalmus*, of a host-plant shift from *Asclepias syriaca* to *A. verticilata*. *Oecologia* 25:331–340. (Chapter 7)

Price, P. W., G. L. Waring, and G. W. Fernandes. 1987. Hypotheses on the adaptive nature of galls. *Proceedings of the Entomological Society of Washington* 88:361–363. (Chapters 1, 4, 5, 6)

Price, P. W., C. E. Bouton, P. Gross, B. A. McPheron, J. N. Thompson, and A. E. Weis. 1980. Interactions among three trophic levels: Influences of plants on interactions between herbivores and natural enemies. *Annual Review of Ecology and Systematics* 11:41–65. (Chapters 7, 8, 12)

Price, T., and L. Liou. 1989. Selection on clutch size in birds. *American Naturalist* 134:950–959. (Chapter 9)

Price, T., and D. Schluter. 1991. On the low heritability of life-history traits. *Evolution* 45:853–861. (Chapters 9, 11)

Price, T., M. Kirkpatrick, and S. J. Arnold. 1988. Directional selection and the evolution of breeding date in birds. *Science* 240:798–799. (Chapters 8, 9)

Prokopy, R. J., A. L. Averill, S. S. Cooley, and C. A. Roitberg. 1982. Associate learning in egg laying site selection by apple maggot flies. *Science* 218:76–77. (Chapter 7)

Provine, W. B. 1971. *The Origins of Theoretical Population Genetics*. University of Chicago Press, Chicago. (Chapter 1)

———. 1986. *Sewall Wright and Evolutionary Biology*. University of Chicago Press, Chicago. (Chapter 9)

Purohit, S. D., K. G. Ramawat, and H. C. Arya. 1979. Phenolics, peroxidase and phenolase as related to gall formation in some arid zone plants. *Current Science* 48:714–716. (Chapters 3, 4)

Raman, A. 1993. Chemical ecology of gall insect host plant interactions: Substances that influence the nutrition and resistance of insects and the growth of galls. In *Chemical Ecology of Phytophagous Insects*, ed. R. N. Ananthakrishnan and A. Raman, 227–260. Oxford and IBH Publishing, New Delhi, India. (Chapters 2, 3)

Raman, A., and W. G. Abrahamson. 1995. Morphometric relationships and energy allocation in the apical rosette galls of *Solidago altissima* (Asteraceae) induced by *Rhopalomyia solidaginis* (Diptera: Cecidomyiidae). *Environmental Entomology* 24:635–639. (Chapters 2, 4)

Rausher, M. D. 1978. Search image for leaf shape in a butterfly. *Science* 200:1071–1073. (Chapter 11)

————. 1984. The evolution of habitat preference in subdivided populations. *Evolution* 38:596–688. (Chapter 7)

————. 1985. Variability for host preference in insect populations: Mechanistic and evolutionary models. *Journal of Insect Physiology* 31:873–889. (Chapter 5)

————. 1988a. Is coevolution dead? *Ecology* 69:989–901. (Chapter 7)

————. 1988b. The evolution of habitat preference. III. The evolution of avoidance and adaptation. In *Evolution of Insect Pests: The Patterns of Variations*, ed. K. C. Kim, 259–283. John Wiley and Sons, New York. (Chapter 11)

————. 1992. The measurement of selection on quantitative traits: Biases due to environmental covariances between traits and fitness. *Evolution* 46:616–626. (Chapters 8, 9)

Ray, S. D. 1986. GA, ABA, phenol interaction in the control of growth: Phenolic compounds as effective modulators of GA-ABA interaction in radish seedlings. *Biologia Plantarum* (Praha) 28:361–369. (Chapter 3)

Ray, S. D., K. N. Guruprasad, and M. M. Laloraya. 1980. Antagonistic action of phenolic compounds on abscisic acid-induced inhibition of hypocotyl growth. *Journal of Experimental Botany* 31:1651–1656. (Chapter 3)

Raymond, M., A. Callaghan, P. Fort, and N. Pasteur. 1991. Worldwide migration of amplified insecticide resistance genes in mosquitoes. *Nature* 350:151–153. (Chapter 9)

Reichardt, P. B., F. S. Chapin III, J. P. Bryant, and T. P. Clausen. 1991. Carbon-nutrient balance as a predictor of plant defense in Alaskan balsam poplar: Potential importance of metabolite turnover. *Oecologia* 88:401–406. (Chapter 4)

Reznick, D., and J. A. Endler. 1982. The impact of predation on life history evolution in Trinidadian guppies (*Poecilia reticulata*). *Evolution* 36:160–177. (Chapters 1, 9)

Reznick, D. A., H. Bryga, and J. A. Endler. 1990. Experimentally induced life-history evolution in a natural population. *Nature* 346:357–359. (Chapter 9)

Rhoades, D. F., and R. G. Cates. 1976. Toward a general theory of plant herbivore chemistry. In *Biochemical Interaction between Plants and Insects*, ed. J. W. Wallace and R. L. Mansell, 168–213. Plenum Press, New York. (Chapter 3)

Rice, W. R. 1984. Disruptive selection on habitat preference and the evolution of reproductive isolation: A simulation study. *Evolution* 38:1251–1260. (Chapters 7, 12)

————. 1987. Speciation via habitat specialization. The evolution of reproductive isolation as a correlated character. *Evolutionary Ecology* 1:301–314. (Chapters 7, 12)

Rice, W. R., and E. E. Hostert. 1993. Laboratory experiments on speciation: What have we learned in 40 years? *Evolution* 47:1637–1653. (Chapter 7)

Rice, W. R., and G. W. Salt. 1990. The evolution of reproductive isolation as a correlated character under sympatric conditions: Experimental evidence. *Evolution* 44:1140–1152. (Chapter 7)

REFERENCES

Riley, C. V., and L. O. Howard. 1891. The phylloxera in France and the North American vine. *Insect Life* 4:224–225. (Chapter 3)

Rockwell, R. F., C. S. Findlay, and F. Cooke. 1987. Is there an optimal clutch size in snow geese? *American Naturalist* 130:839–863. (Chapter 9)

Rohfritsch, O. 1977. Ultrastructure of the nutritive tissue of the *Chermes abietis* L. fundatrix on *Picea excelsa* L. *Marcellia* 40:135–149. (Chapter 3)

———. 1981. A "defense" mechanism of *Picea excelsa* L. against the gall former *Chermes abietis* L. (Homoptera, Adelgidae). *Zeitschrift für Angewandte Entomologie* 92:18–26. (Chapters 3, 4)

———. 1992. Patterns in gall development. In *Biology of Insect-Induced Galls,* ed. J. D. Shorthouse and O. Rohfritsch, 60–86. Oxford University Press, New York. (Chapter 3)

Rohfritsch, O., and J. D. Shorthouse. 1982. Insect galls. In *Molecular Biology of Plant Tumors,* ed. G. Kahl and J. Schell, 131–152. Academic Press, New York. (Chapters 3, 6)

Roininen, H., and J. Tahvanainen. 1989. Host selection and larval performance of two willow-feeding sawflies. *Ecology* 70:129–136. (Chapter 5)

Rojas, R. R., R. E. Lee, Jr., T. Luu, and J. G. Baust. 1983. Temperature dependence-independence of antifreeze turnover in *Eurosta solidaginis* (Fitch). *Journal of Insect Physiology* 29:865–869. (Chapter 2)

Romanes, G. 1889. Galls. *Nature* (London) 41:80. (Chapter 6)

Rosenthal, J. P., and P. M. Kotanen. 1994. Terrestrial plant tolerance to herbivory. *Trends in Ecology and Evolution* 9:145–148. (Chapter 11)

Rosenzweig, M. L., and W. M. Shaffer. 1978. Homage to the Red Queen II: Coevolutionary response to enrichment of exploitation ecosystems. *Theoretical Population Biology* 9:158–163. (Chapter 7)

Ross, H. 1966. The use of wild *Solanum* species in German potato breeding of the past and today. *American Potato Journal* 43:63–80. (Chapter 4)

Rossi, A. M., and D. R. Strong. 1991. Effects of host-plant nitrogen on the preference and performance of laboratory populations of *Carneocephala floridana* (Homoptera: Cecadellidae). *Environmental Entomology* 20:1349–1355. (Chapters 4, 5)

Roughgarden, J. 1979. *Theory of Population Genetics and Evolutionary Ecology: An Introduction.* Macmillan, New York. (Chapter 9)

Russell, G. E. 1978. *Plant Breeding for Pest and Disease Resistance.* Butterworths, London, England. (Chapter 4)

Salzman, A. G., and M. A. Parker. 1985. Neighbors ameliorate local salinity stress for a rhizomatus plant in a heterogeneous environment. *Oecologia* 65:273–277. (Chapter 3)

Scheiner, S. M. 1989. Variable selection along an environmental gradient. *Evolution* 43:548–562. (Chapter 8)

Scheiner, S. M., and C. Goodnight. 1984. The comparison of phenotypic plasticity and genetic variation in populations of *Danthonia spicata*. *Evolution* 38:845–855. (Chapter 6)

Scheiner, S. M., and R. F. Lyman. 1989. The genetics of phenotypic plasticity. I. Heritability. *Journal of Evolutionary Biology* 2:94–107. (Chapter 9)

Schlichter, L. 1978. Winter predation by black-capped chickadees and downy woodpeckers on inhabitant of the goldenrod ball gall. *Canadian Field Naturalist* 92:71–74. (Chapters 2, 8, 9)

Schlichting, C. D., and M. Piglicci. 1993. Control of phenotypic plasticity via regulatory genes. *American Naturalist* 142:366–370. (Chapter 9)

Schluter, D. 1988. Estimating the form of natural selection on a quantitative trait. *Evolution* 42:849–861. (Chapter 8)

Schluter, D., and N. M. Smith. 1986. Natural selection on beak and body size in the song sparrow. *Evolution* 40:221–231. (Chapters 8, 9)

Schmalhausen, I. I. 1949. *Factors of Evolution: The Theory of Stabilizing Selection.* Blakiston Company, Philadelphia. (Chapter 9)

Schmid, B., and F. A. Bazzaz. 1987. Clonal integration and population structure in perennials: Effects of severing rhizome connections. *Ecology* 68:2016–2022. (Chapter 3)

Schmid, B., M. Puttick, K. H. Burgess, and F. A. Bazzaz. 1988a. Clonal integration and effects of stimulated herbivory in old-field perennials. *Oecologia* 75:465–471. (Chapters 2, 3)

———. 1988b. Correlations between genet architecture and some life history features in three species of *Solidago*. *Oecologia* 75:459–464. (Chapters 2, 3)

Schoener, T. W. 1970. Non-synchronous spatial overlap of lizards in patchy habitats. *Ecology* 51:408–418. (Chapter 2)

Schultz, J. C., and I. T. Baldwin. 1982. Oak leaf quality declines in response to defoliation by gypsy moth larvae. *Science* 217:149–151. (Chapters 4, 10)

Scriber, J. M. 1984. Host-plant suitability. In *Chemical Ecology of Insects,* ed. W. J. Bell and R. T. Carde, 159–202. Chapman and Hall, London. (Chapter 4)

Scriber, M. S. 1983. Evolution of feeding specialization, physiological efficiency, and host races in selected Papillionidae and Saturnidae. In *Variable Plants and Herbivores in Natural and Managed Systems,* ed. R. F. Denno and M. M. McClure, 373–412. Academic Press, New York. (Chapters 7, 11)

Seibert, T. F. 1993. A nectar-secreting gall wasp and ant mutualism: Selection and counter-selection shaping gall wasp phenology, fecundity and persistence. *Ecological Entomology* 18:247–253. (Chapter 1)

Semple, J. C., and G. S. Ringius. 1983. Goldenrods of Ontario *Solidago* and *Euthamia. University of Waterloo Biology Series,* no. 26 (Chapters 2, 12)

Service, P. 1984. Genotypic interactions in an aphid-host plant relationship: *Uroleucon rudbeckia* and *Rudbeckia laciniata. Oecologia* 61:271–276. (Chapter 9)

Sharma, S., S. S. Sharma, and V. K. Rai. 1986. Reversal by phenolic compounds of abscisic acid-induced inhibition of *in vitro* activity of amylase from seeds of *Triticum aestivum* L. *The New Phytologist* 103:293–297. (Chapter 3)

Shelley, T. E., M. D. Greenfield, and K. R. Downum. 1987. Variation in host plant quality: Influences on the mating system of a desert grasshopper. *Animal Behaviour* 35:1200–1209. (Chapter 5)

429

Shorthouse, J. D., and O. Rohfritsch, eds. 1992. *Biology of Insect-Induced Galls.* Oxford University Press, New York. (Chapter 1)

Simms, E. L. 1990. Examining selection on the multivariate phenotype—plant resistance to herbivores. *Evolution* 44:1177–1188. (Chapter 1)

Simms, E. L. 1992. Costs of plant resistance to herbivory. In *Plant Resistance to Herbivores and Pathogens,* ed. R. S. Fritz and E. L. Simms, 392–424. University of Chicago Press, Chicago. (Chapter 12)

Simms, E. L., and M. D. Rausher. 1989. The evolution of resistance to herbivory in *Ipomoea purpurea.* II. Natural selection by insects on the costs of resistance. *Evolution* 43:573–585. (Chapters 11, 12)

Simms, E. L., and J. Triplett. 1995. Costs and benefits of plant responses to disease: Resistance and tolerance. *Evolution.* (Chapter 11)

Singer, M. C. 1986. The definition and measurement of oviposition preference in plant feeding insects. In *Insect-Plant Relations,* ed. J. Miller and T. A. Miller, 66–94. Springer-Verlag, New York. (Chapter 5)

Singer, M. C., C. D. Thomas, H. L. Billington, and C. Parmesan. 1989. Variation among conspecific insect populations in the mechanistic basis of diet breadth. *Animal Behaviour* 37:751–759. (Chapter 11)

Slade, A. J., and M. J. Hutchings. 1987a. An analysis of the costs and benefits of physiological integration between ramets in the perennial herb *Glechoma hederaceae. Oecologia* 73:425–431. (Chapter 3)

———. 1987b. The effects of nutrient availability on foraging in the clonal herb *Glechoma hederaceae. Journal of Ecology* 75:95–112. (Chapter 3)

———. 1987c. Clonal integration and plasticity in foraging behaviour in *Glechoma hederaceae. Journal of Ecology* 75:1023–1036. (Chapter 3)

Slansky, F. 1986. Nutritional ecology of endoparasitic insects: An overview. *Journal of Insect Physiology* 32:255–261. (Chapter 12)

Slansky, F., and P. Feeny. 1977. Stabilization of the rate of nitrogen accumulation by larvae of the cabbage butterfly on wild and cultivated food plants. *Ecological Monographs* 47:209–228. (Chapter 3)

Smith, A. P., and J. P. Palmer. 1976. Vegetative reproduction and close packing in a successional plant species. *Nature* 261:232–233. (Chapter 2)

Sneath, P.H.A., and R. R. Sokal. 1973. *Numerical Taxonomy.* W. H. Freeman, San Francisco. (Chapter 7)

Stamp, N. E. 1992. Theory of plant-insect herbivore interaction on the inevitable brink of re-synthesis? *Bulletin of the Ecological Society of America* 73:28–34. (Chapter 1)

Stearns, S. C. 1988. The evolutionary significance of phenotypic plasticity. *BioScience* 39:436–446. (Chapters 6, 9)

———. 1992. *The Evolution of Life Histories.* Oxford University Press, New York. (Chapter 1)

Stewart, S. C., and D. J. Schoen. 1987. Pattern of phenotypic variability and fecundity selection in a natural population of *Impatiens pallida. Evolution* 41:1290–1301. (Chapter 8)

Steyskal, G. C., and R. H. Foote. 1977. Revisionary notes on North American Tephritidae (Diptera), with keys and descriptions of new species.

REFERENCES

Proceedings of the Entomological Society of Washington 79:146–155. (Chapter 2)

Stinner, B. R., and W. G. Abrahamson. 1979. Energetics of the *Solidago canadensis*-stem gall insect-parasitoid guild interaction. *Ecology* 60:918–926. (Chapters 2, 3, 8, 9, 12)

Stoltzfus, W. B. 1989. A non-gall forming *Eurosta solidaginis* (Diptera: Tephritidae). *Journal of the Iowa Academy of Science* 96:50–51. (Chapter 2)

Storey, J. M., and K. B. Storey. 1986. Winter survival of the gall fly larva, *Eurosta solidaginis:* Profiles of fuel reserves and cryoprotectants in a natural population. *Journal of Insect Physiology* 32:549–556. (Chapter 2)

Strauss, S. Y. 1990. The role of plant genotype, environment and gender in resistance to a specialist chysomelid herbivore. *Oecologia* 84:111–116. (Chapter 4)

Strong, D. R., S. Larsson, and U. Gullberg. 1993. Heritability of host plant resistance to herbivory changes with gall midge density during outbreak on willow. *Evolution* 47:291–300. (Chapter 11)

Strong, D. R., J. H. Lawton, and T.R.E. Southwood. 1984. Insects on plants, community patterns and mechanisms. Harvard University Press, Cambridge, MA. (Chapter 4)

Sumerford, D. V. 1991. Two studies of tri-trophic level interactions among goldenrods, *Eurosta solidaginis* (Diptera: Tephritidae), and its natural enemies. M. S. thesis, Bucknell University, Lewisburg, PA. (Chapter 2)

Sumerford, D. V., and W. G. Abrahamson. 1995. Georgraphic and host species effects in *Eurosta solidaginis* (Diptera: Tephritidae) mortality. *Environmental Entomology* 24:657–662. (Chapters 3, 7)

Sumerford, D. V., W. G. Abrahamson, and A. E. Weis. In prep. The effects of drought on interactions in the *Solidago altissima–Eurosta solidaginis–*natural enemy complex: Population dynamics, local extinctions, and measures of selection intensity on gall size. (Chapter 8)

Taper, M. L., and T. J. Case. 1987. Interactions between oak tannins and parasite community structure: Unexpected benefits of tannins to cyinipid gall-wasps. *Oecologia* 71:254–261. (Chapter 4)

Tauber, C. A., and M. J. Tauber. 1989. Sympatric speciation in insects: Perception and perspective. In *Speciation and Its Consequences,* ed. D. Otte and J. A. Endler, 307–345. Sinauer Associates, Sunderland, MA. (Chapters 7, 12)

Thompson, J. N. 1986. Constraints on arms races in coevolution. *Trends in Ecology and Evolution* 1:105–107. (Chapter 10)

———. 1988. Evolutionary ecology of the relationship between oviposition preference and performance of offspring in phytophagous insects. *Entomologia experimentalis et applicata* 47:3–14. (Chapters 1, 5, 11)

———. 1989. Concepts of coevolution. *Trends in Ecology and Evolution* 4:179–183. (Chapters 1, 4, 5, 12)

———. 1994. *The Coevolutionary Process.* University of Chicago Press, Chicago. (Chapters 1, 11, 12)

Thompson, J. N., and O. Pellmyr. 1991. Evolution of oviposition behavior and host preference in Lepidoptera. *Annual Review of Entomology* 36: 65–89. (Chapter 5)

Thompson, K., and A.J.A. Stewart. 1981. The measurement and meaning of reproductive effort in plants. *American Naturalist* 177:205–211. (Chapter 3)

Tisdale, R. A., and M. R. Wagner. 1991. Host stress influences oviposition preference and performance of a pine sawfly. *Ecological Entomology* 16: 371–376. (Chapters 4, 5)

Travis, J. 1989. The role of optimizing selection in natural populations. *Annual Review of Ecology and Systematics* 20:279–296. (Chapter 8)

Travisano, M., J. A. Mongold, A. F. Bennett, and R. E. Lenski. 1995. Experimental tests of the roles of adaptation, chance, and history in evolution. *Science* 267:87–90. (Chapters 1, 9)

Troll, R. 1990. Location of tunnels on goldenrod ball galls made by downy woodpeckers, *Picoides pubescens*. *Transactions of the Illinois State Academy of Science* 83:195–196. (Chapter 2)

Trumble, J. T., D. M. Kolodny-Hirsch, and I. P. Ting. 1993. Plant compensation for arthropod herbivory. *Annual Review of Entomology* 38:93–119. (Chapter 11)

Tscharntke, T. 1992. Cascade effects among 4 trophic levels: Bird predation on galls affects density-dependent parasitism. *Ecology* 73:1689–1698. (Chapters 1, 2)

Uhler, L. D. 1951. Biology and ecology of the goldenrod gall fly, *Eurosta solidaginis* (Fitch). *Cornell University Agricultural Station Memoir* 300:1–51. (Chapters 1, 2, 3, 4, 5, 7, 8)

———. 1961. Mortality of the goldenrod gall fly, *Eurosta solidaginis* in the vicinity of Ithaca, NY. *Ecology* 42:215–216. (Chapter 2)

Uhler, L. D., C. R. Crispen, Jr., and D. B. McCormick. 1971. Free amino acid patterns during development of *Eurosta solidaginis* (Fitch). *Comparative Biochemistry and Physiology* 38:87–91. (Chapter 3)

Vail, S. G. 1992. Selection for overcompensatory plant responses to herbivory: A mechanism for the evolution of plant-herbivore mutualism. *American Naturalist* 139:1–8. (Chapter 11)

Valladares, G., and J. H. Lawton. 1991. Host-plant selection in the holly leafminer: Does mother know best? *Journal of Animal Ecology* 60:227–240. (Chapter 5)

van Andel, J., and F. Vera. 1977. Reproductive allocation in *Senecio sylvaticus* and *Chamaenerion angustifolium* in relation to mineral nutrition. *Journal of Ecology* 65:747–758. (Chapter 3)

van Staden, J., and J. E. Davey. 1978. Endogenous cytokinins in the laminae and galls of *Erythrina latissima* leaves. *Botanical Gazette* 139:36–41. (Chapter 3)

van Staden, J., J. E. Davey, and A.R.A. Noel. 1977. Gall formation in *Erythrina altissima*. *Zeitschrift für Pflanzenernährung und Bodenkunde* 84:283–294. (Chapter 3)

van Valen, L. 1973. A new evolutionary law. *Evolutionary Theory* 1:1–30. (Chapter 9)

REFERENCES

Vermeij, G. 1987. *Evolution and Escalation: An Ecological History of Life.* Princeton University Press, Princeton, NJ. (Chapter 10)

Via, S. 1984. The quantitative genetics of polyphagy in an insect herbivore. II. Genetic correlations in larval performance within and among host plants. *Evolution* 38:896–905. (Chapters 1, 7, 9)

———. 1990. Ecological genetics and host adaptation in herbivorous insects: The experimental study of evolution in natural and agricultural systems. *Annual Review of Entomology* 35:421–426. (Chapters 5, 7)

———. 1991. The genetic structure of host plant adaptation in a spatial patchwork: Demographic variability among reciprocally transplanted pea aphid clones. *Evolution* 45:827–852. (Chapters 1, 5, 9)

Via, S., and R. Lande. 1985. Genotype-environment interaction and the evolution of phenotypic plasticity. *Evolution* 39:505–522. (Chapters 9, 11)

Vinson, S. B. 1984. Parasitoid-host relationship. In *Chemical Ecology of Insects,* ed. W. J. Bell and R. Carde, 205–233. Sinauer Associates, Sunderland, MA. (Chapter 8)

Walton, R. 1988. The distribution of risk and density-dependent mortality in the galls of *Eurosta solidaginis,* the goldenrod gall fly. *Ecological Entomology* 13:347–354. (Chapters 2, 9)

Walton, R., A. E. Weis, and J. P. Lichter. 1990. Oviposition behavior and response to plant height by *Eurosta solidaginis* Fitch (Diptera: Tephritidae). *Annals of the Entomological Society of America* 83:509–514. (Chapters 2, 5)

Waring, G. L., and P. W. Price. 1989. Plant water stress and gall formation (Cecidomyiidae: *Asphondylia* spp.) on creosote bush. *Ecology and Entomology* 15:87–95. (Chapter 2)

Waring, G. L., W. G. Abrahamson, and D. J. Howard. 1990. Genetic differentiation among host-associated populations of the gallmaker *Eurosta solidaginis* (Diptera: Tephritidae). *Evolution* 44:1648–1655. (Chapters 2, 7)

Waring, R. H., A.J.S. McDonald, S. Larsson, T. Ericsson, A. Wiren, E. Arwidsson, A. Ericsson, and T. Lohammar. 1985. Differences in chemical composition of plants grown at constant relative growth rates with stable mineral nutrition. *Oecologia* 66:157–160. (Chapter 4)

Wasbauer, M. S. 1972. An annoted host catalog of the fruit flies of America north of Mexico (Diptera: Tephritidae). California Department of Agriculture, *Bureau of Entomology Occasional Paper* 19:1–172. (Chapters 1, 2, 7)

Washburn, J. O. 1984. Mutualism between a cynipid gall wasp and ants. *Ecology* 65:654–656. (Chapter 1)

Washburn, J. O., and H. V. Cornell. 1979. Chalcid parasitoid attack on a gall wasp population *Acraspis hirta* (Hymenoptera: Cynipidae) on *Quercus prinus* (Fagaceae). *Canadian Entomologist* 111:391–400. (Chapters 1, 2, 7)

Waterman, P. G., J. A. M. Ross, and D. B. McKey. 1984. Factors affecting levels of some phenolic compounds, digestibility, and nitrogen content of the mature leaves of *Barteria fistulosa* (Passifloraceae). *Journal of Chemical Ecology* 10:387–401. (Chapter 4)

433

Weis, A. E. 1982a. Resource utilization patterns in a community of gall-attacking parasitoids. *Environmental Entomology* 11:809–815. (Chapters 1, 2, 7)

———. 1982b. Use of symbiotic fungus by the gall maker *Asteromyia carbonifera* to inhibit attack by the parasitoid *Torymus capite*. *Ecology* 63: 1602–1605. (Chapter 7)

———. 1992. Plant variation and the evolution of phenotypic plasticity in herbivore performance. In *Ecology and Evolution of Host Plant Resistance*, ed. R. F. Fritz and E. L. Simms, 140–171. University of Chicago Press, Chicago. (Chapters 6, 9, 10, 11)

———. 1993a. Host gall size predicts host quality for the parasitoid *Eurytoma gigantea* (Hymenoptera: Eurytomidae), but can the parasitoid tell. *Journal of Insect Behavior* 6:591–602. (Chapters 2, 9)

———. 1993b. What can gallmakers tell us about natural selection on the components of plant defense? In *The Ecology and Evolution of Gall-Forming Insects*, ed. P. W. Price, Y. N. Baranchikov, 157–171. USDA Forest Service North Central Experiment Station General Technical Report NC-174. (Chapter 3)

Weis, A. E. 1996. Variable selection on *Eurosta*'s gall size, III: Can an evolutionary response to selection be detected? *Journal of Evolutionary Biology* 9:623–640. (Chapters 9, 10)

Weis, A. E., and W. G. Abrahamson. 1985. Potential selective pressures by parasitoids on the evolution of a plant-herbivore interaction. *Ecology* 66: 1261–1269. (Chapters 2, 3, 5, 6, 7, 8, 12)

———. 1986. Evolution of host-plant manipulation by gall makers: Ecological and genetic factors in the *Solidago-Eurosta* system. *American Naturalist* 127:681–695. (Chapters 2, 4, 6, 7, 8, 9)

Weis, A. E., and M. R. Berenbaum. 1989. Herbivorous insects and green plants. In *Plant-Animal Interactions*, ed. W. G. Abrahamson, 123–162. McGraw Hill, New York. (Chapters 2, 7, 11)

Weis, A. E., and D. R. Campbell. 1992. Plant genotype: A variable factor in insect-plant interactions. In *Effects of Resource Distribution on Animal-Plant Interactions*, ed. M. D. Hunter, T. Ohgushi, and P. W. Price, 75–111. Academic Press, San Diego. (Chapters 4, 11, 12)

Weis, A. E., and W. L. Gorman. 1990. Measuring canalizing selection on reaction norms: An exploration of the *Eurosta-Solidago* system. *Evolution* 44:820–831. (Chapters 4, 6, 9)

Weis, A. E., and A. D. Kapelinski. 1994. Variable selection on *Eurosta*'s gall size. II. A path analysis of the ecological factors behind selection. *Evolution* 48:734–745. (Chapters 7, 8, 9)

Weis, A. E., W. G. Abrahamson, and K. D. McCrea. 1985. Host gall size and oviposition success by the parasitoid *Eurytoma gigantea*. *Ecological Entomology* 10:341–348. (Chapters 2, 3, 4, 7, 8, 9)

Weis, A. E., W. G. Abrahamson, and M. C. Andersen. 1992. Variable selection on *Eurosta*'s gall size. I: The extent and nature of variation in phenotypic selection. *Evolution* 46:1674–1697. (Chapters 2, 3, 4, 5, 7, 8, 9)

Weis, A. E., H. G. Hollenbach, and W. G. Abrahamson. 1987. Genetic and maternal effects on seedling characters of *Solidago altissima* (Compositae). *American Journal of Botany* 74:1476–1486. (Chapter 6)

Weis, A. E., K. D. McCrea, and W. G. Abrahamson. 1989. Can there be an escalating arms race without coevolution? Implications from a host-parasitoid simulation. *Evolutionary Ecology* 3:361–370. (Chapter 10)

Weis, A. E., P. W. Price, and M. Lynch. 1983. Selective pressures on clutch size of the gall-maker *Asteromyia carbonifera*. *Ecology* 64:688–695 (Chapter 11)

Weis, A. E., R. L. Walton, and C. L. Crego. 1988. Reactive plant tissue sites and the population biology of gall makers. *Annual Review of Entomology* 33:467–486. (Chapters 5, 6, 12)

Weis, A. E., C. L. Wolfe, and W. L. Gorman. 1989. Genotypic variation and integration in histological features of the goldenrod ball gall. *American Journal of Botany* 76:1541–1550. (Chapters 2, 7)

Welter, S. C. 1989. Arthropod impact on plant gas exchange. In *Insect-Plant Interactions*, ed. E. A. Bernays, 135–150. CRC Press, Boca Raton, FL. (Chapter 11)

Werner, P. A., and W. J. Platt. 1976. Ecological relationships of co-occurring goldenrods (*Solidago:* Compositae). *American Naturalist* 110:959–971. (Chapter 2)

Werner, P. A., I. K. Bradbury, and R. S. Gross. 1980. The biology of Canadian weeds. 45. *Solidago canadensis* L. *Canadian Journal of Plant Science* 60:1393–1409. (Chapter 2)

Westphal, E., R. Bronner, and M. LeRet. 1981. Changes in leaves of susceptible and resistant *Solanum dulcamara* infested by the gall mite *Eriophyes cladophthirus* (Acarina, Eriophyoidea). *Canadian Journal of Botany* 59:875–882. (Chapters 3, 4)

Westphal, E., M. J. Perrot-Minnot, S. Kreiter, and J. Gutierrez. 1992. Hypersensitive reaction of *Solanum dulcamara* to the gall mite *Aceria cladophthirus* causes an increased susceptibility to *tetranychus urticae*. *Experimental and Applied Acarology* 15:15–26. (Chapters 3, 4)

Whigham, D. F. 1984. Biomass and nutrient allocation of *Tipularia discolor* (Orchidaceae). *Oikos* 42:303–313. (Chapter 3)

Whitaker, R. H., and P. P. Feeny. 1971. Allelochemicals: Chemical interactions between species. *Science* 171:757–770. (Chapter 10)

Whitham, T. 1978. Habitat selection by *Pemphigus* aphids in response to resource limitation and competition. *Ecology* 59:1164–1176. (Chapter 5)

———. 1980. The theory of habitat selection: Examined and extended using *Pemiphigus* aphids. *American Naturalist* 115:449–466. (Chapters 5, 11)

———. 1986. Costs and benefits of territoriality: Behavioral and reproductive release by competing aphids. *Ecology* 67:139–147. (Chapter 5)

———. 1992. Ecology of *Pemphigus* gall aphids. In *Biology of Insect-Induced Galls*, ed. J. D. Shorthouse and O. Rohfritsch, 225–237. Oxford University Press, New York. (Chapters 5, 12)

435

Whitham, T. G., J. Maschinski, K. C. Larson, and K. N. Paige. 1991. Plant responses to herbivory: The continuum from negative to positive and underlying physiological mechanisms. In *Plant-Animal Interactions: Evolutionary Ecology in Tropical and Temperate Regions*, ed. P. W. Price, T. M. Lewinsohn, G. W. Fernandes, and W. W. Benson, 227–256. John Wiley and Sons, New York. (Chapter 3)

Whitman, D. W. 1988. Allelochemical interactions among plants, herbivores and their predators. In *Novel Aspects of Insect-Plant Interactions*, ed. P. Barbosa and D. K. Letourneau, 11–64. John Wiley and Sons, New York. (Chapter 12)

Wiegert, R. G., and F. C. Evans. 1967. Investigations of secondary productivity in grasslands. In *Secondary Productivity of Terrestrial Ecosystems*, ed. K. Petrusewicz, 447–457. Panstwowe Wydawnictwo Naukowe, Warszawa-Krakow, Poland. (Chapter 3)

Williams, G. C. 1992. *Natural Selection: Domains, Levels and Challenges.* Oxford University Press, New York. (Chapters 1, 9, 10)

Williams, H. J., G. W. Elzen, and S. B. Vinson. 1988. Parasitoid-host-plant interactions, emphasizing cotton (Gossypium). In *Novel Aspects of Insect-Plant Interactions*, ed. P. Barbosa and D. K. Letourneau, 171–200. John Wiley and Sons, New York. (Chapter 12)

Wilson, D. S., and M. Turelli. 1986. Stable underdominance and the evolutionary invasion of empty niches. *American Naturalist* 127:835–850. (Chapter 7)

Windle, P. N., and E. H. Franz. 1979. The effects of insect parasitism on plant competition: Breen bugs and barley. *Ecology* 60:521–529. (Chapter 12)

Wodehouse, R. P. 1945. *Hayfever Plants.* Chronica Botanica Co., Waltham, MA. (Chapter 2)

Wolterek, R. 1909. Weitere experimentelle Untersuchungen über Artveränderung, speziell über das Wesen quantitativer Artunterschiede bei Daphniden. *Verhandlungen der deutschen zoologischen Gesellschaft* 1090: 110–172. (Chapter 10)

Wood, R. J., and J. A. Bishop. 1981. Insecticide resistance: Populations and evolution. In *Genetic Consequences of Man-Made Change*, ed. J. A. Bishop and L. M. Cook, 97–128. Academic Press, New York. (Chapter 9)

Wood, T. K. 1980. Divergence in the *Enchenopa binotata* Say complex (Homoptera: Membracidae) effected by host plant adaptation. *Evolution* 34: 147–160. (Chapter 7)

———. 1986. Host plant shifts and speciation in the *Enchenopa binotata* Say complex. In *Proceedings of the Second International Workshop on Leafhoppers and Planthoppers of Economic Importance*, ed. M. R. Wilson and L. R. Nault, 361–368. CIE, London. (Chapter 7)

Wood, T. K., and S. I. Guttman. 1983. *Enchenopa binotata* complex: Sympatric speciation? *Science* 220:310–312. (Chapter 7)

Wood, T. K., and M. C. Keese. 1990. Host-plant induced assortative mating in *Enchenopa* treehoppers. *Evolution* 44:619–628. (Chapter 7)

Wood, T. K., K. L. Olmstead, and S. I. Guttman. 1990. Insect phenology mediated by host-plant water relations. *Evolution* 44:629–636. (Chapter 7)

Workman, P. L., and J. D. Niswander. 1970. Population studies on southwestern Indian tribes. II. Local genetic differentiation in the Papago. *American Journal of Human Genetics* 22:24–49. (Chapter 7)

Wright, R. J. 1984. Evaluation of crop rotation for control of Colorado potato beetles (Coleoptera: Chrysomelidae) in commercial potato fields on Long Island. *Journal of Economic Entomology* 77:1254–1259. (Chapter 4)

Wright, S. 1968. *Evolution and the Genetics of Populations: A Treatise.* Vol. 1: *Genetic and Biometric Foundations.* University of Chicago Press, Chicago. (Chapter 9)

———. 1978. *The Evolution and Genetics of Populations: A Treatise.* Vol. 4: *Variability within and among Natural Populations.* University of Chicago Press, Chicago. (Chapter 9)

Zangerl, A. 1990. Furanocoumarin induction in wild parsnip: Evidence for an induced defense against herbivores. *Ecology* 71:1926–1932. (Chapters 4, 10)

Zangerl, A. R., and F. A. Bazzaz. 1992. Theory and pattern in plant defense allocation. In *Plant Resistance to Herbivores and Pathogens,* ed. R. S. Fritz and E. L. Simms, 363–391. University of Chicago Press, Chicago. (Chapter 12)

Zucker, W. V. 1982. How aphids choose leaves: The roles of phenolics in host selection by a galling aphid. *Ecology* 63:972–981. (Chapter 3)

Author Index

439

Subject Index

MONOGRAPHS IN POPULATION BIOLOGY

(continued)

Warren G. Abrahamson is David Burpee Professor of Plant Genetics at Bucknell University. Arthur E. Weis is Associate Professor of Ecology and Evolutionary Biology at the University of California, Irvine.